Strain Hardening Cementitious Composites

RILEM BOOKSERIES

Volume 39

RILEM, The International Union of Laboratories and Experts in Construction Materials, Systems and Structures, founded in 1947, is a non-governmental scientific association whose goal is to contribute to progress in the construction sciences, techniques and industries, essentially by means of the communication it fosters between research and practice. RILEM's focus is on construction materials and their use in building and civil engineering structures, covering all phases of the building process from manufacture to use and recycling of materials. More information on RILEM and its previous publications can be found on www.RILEM.net.

Indexed in SCOPUS, Google Scholar and SpringerLink.

More information about this series at https://link.springer.com/bookseries/8781

Minoru Kunieda · Toshiyuki Kanakubo ·
Tetsushi Kanda · Koichi Kobayashi
Editors

Strain Hardening Cementitious Composites

SHCC5

Springer

Editors
Minoru Kunieda
Department of Civil Engineering
Gifu University
Gifu, Japan

Tetsushi Kanda
Kajima Institute of Technology
Chofu, Tokyo, Japan

Toshiyuki Kanakubo
Division of Engineering Mechanics
and Energy
University of Tsukuba
Tsukuba, Ibaraki, Japan

Koichi Kobayashi
Department of Civil Engineering
Gifu University
Gifu, Japan

ISSN 2211-0844 ISSN 2211-0852 (electronic)
RILEM Bookseries
ISBN 978-3-031-15804-9 ISBN 978-3-031-15805-6 (eBook)
https://doi.org/10.1007/978-3-031-15805-6

This Springer imprint is published by the registered company Springer Nature Switzerland AG
The registered company address is: Gewerbestrasse 11, 6330 Cham, Switzerland

Preface

This book presents the proceedings of the 5th International RILEM Workshop on Strain Hardening Cementitious Composites (SHCC5), which was to be held at the Gifu University, Japan, from September 11 to 13, 2022. The SHCC5 Workshop would have been a follow-up to the four previous successful international events in Stellenbosch (South Africa, 2009), Rio de Janeiro (Brazil, 2011), Dordrecht (The Netherlands, 2014), and Dresden (Germany, 2017), all of them focusing on strain hardening cementitious composites and other types of advanced fiber-reinforced concrete materials such as high-performance fiber-reinforced cementitious composites (HPFRCC). The use of such types of fiber-reinforced concrete can help to accelerate the innovation of concrete and concrete structures including the strengthening or repair of existing buildings and structures. In particular, SHCC have the potential to contribute to a resilient and sustainable society.

Although the meeting was unfortunately canceled owing to the COVID-19 pandemic, the editors believe that the contributions of SHCC, which include the latest findings and research from the authors, should be published in a timely manner. This volume contains 33 contributions from eight countries to SHCC5, with those papers covering the latest findings.

I would like to thank RILEM and Japan Concrete Institute (JCI) for their informational support on this activity. I would like to also thank Gifu University and the Maeda Engineering Foundation for their financial support to publish the proceedings of SHCC5. Furthermore, I extend my personal thanks to Prof. Toshiyuki Kanakubo, Dr. Tetsushi Kanda, and Prof. Koichi Kobayashi, who contributed in co-editing this book. I would like to extend my thanks to the local organizing committee members, in particular to Prof. Tomoya Nishiwaki, Dr. Atsuhisa Ogawa, Dr. Hiroki Ogura, and Prof. Naoshi Ueda for their commitment.

Finally, I hope the SHCC conference series will be continued in the future and that the members of this community will keep in touch with one another.

Minoru Kunieda

Contents

RILEM Publications

The following list is presenting the global offer of RILEM Publications, sorted by series. Each publication is available in printed version and/or in online version.

RILEM Proceedings (PRO)

PRO 1: Durability of High Performance Concrete (ISBN: 2-912143-03-9; e-ISBN: 2-351580-12-5; e-ISBN: 2351580125); *Ed. H. Sommer*

PRO 2: Chloride Penetration into Concrete (ISBN: 2-912143-00-04; e-ISBN: 2912143454); *Eds. L.-O. Nilsson and J.-P. Ollivier*

PRO 3: Evaluation and Strengthening of Existing Masonry Structures (ISBN: 2-912143-02-0; e-ISBN: 2351580141); *Eds. L. Binda and C. Modena*

PRO 4: Concrete: From Material to Structure (ISBN: 2-912143-04-7; e-ISBN: 2351580206); *Eds. J.-P. Bournazel and Y. Malier*

PRO 5: The Role of Admixtures in High Performance Concrete (ISBN: 2-912143-05-5; e-ISBN: 2351580214); *Eds. J. G. Cabrera and R. Rivera-Villarreal*

PRO 6: High Performance Fiber Reinforced Cement Composites - HPFRCC 3 (ISBN: 2-912143-06-3; e-ISBN: 2351580222); *Eds. H. W. Reinhardt and A. E. Naaman*

PRO 7: 1st International RILEM Symposium on Self-Compacting Concrete (ISBN: 2-912143-09-8; e-ISBN: 2912143721); *Eds. Å. Skarendahl and Ö. Petersson*

PRO 8: International RILEM Symposium on Timber Engineering (ISBN: 2-912143-10-1; e-ISBN: 2351580230); *Ed. L. Boström*

PRO 9: 2nd International RILEM Symposium on Adhesion between Polymers and Concrete ISAP '99 (ISBN: 2-912143-11-X; e-ISBN: 2351580249); *Eds. Y. Ohama and M. Puterman*

PRO 10: 3rd International RILEM Symposium on Durability of Building and Construction Sealants (ISBN: 2-912143-13-6; e-ISBN: 2351580257); *Eds. A. T. Wolf*

PRO 11: 4th International RILEM Conference on Reflective Cracking in Pavements (ISBN: 2-912143-14-4; e-ISBN: 2351580265); *Eds. A. O. Abd El Halim, D. A. Taylor and El H. H. Mohamed*

PRO 12: International RILEM Workshop on Historic Mortars: Characteristics and Tests (ISBN: 2-912143-15-2; e-ISBN: 2351580273); *Eds. P. Bartos, C. Groot and J. J. Hughes*

PRO 13: 2nd International RILEM Symposium on Hydration and Setting (ISBN: 2-912143-16-0; e-ISBN: 2351580281); *Ed. A. Nonat*

PRO 14: Integrated Life-Cycle Design of Materials and Structures - ILCDES 2000 (ISBN: 951-758-408-3; e-ISBN: 235158029X); (ISSN: 0356-9403); *Ed. S. Sarja*

PRO 15: Fifth RILEM Symposium on Fibre-Reinforced Concretes (FRC) - BEFIB'2000 (ISBN: 2-912143-18-7; e-ISBN: 291214373X); *Eds. P. Rossi and G. Chanvillard*

PRO 16: Life Prediction and Management of Concrete Structures (ISBN: 2-912143-19-5; e-ISBN: 2351580303); *Ed. D. Naus*

PRO 17: Shrinkage of Concrete – Shrinkage 2000 (ISBN: 2-912143-20-9; e-ISBN: 2351580311); *Eds. V. Baroghel-Bouny and P.-C. Aïtcin*

PRO 18: Measurement and Interpretation of the On-Site Corrosion Rate (ISBN: 2-912143-21-7; e-ISBN: 235158032X); *Eds. C. Andrade, C. Alonso, J. Fullea, J. Polimon and J. Rodriguez*

PRO 19: Testing and Modelling the Chloride Ingress into Concrete (ISBN: 2-912143-22-5; e-ISBN: 2351580338); *Eds. C. Andrade and J. Kropp*

PRO 20: 1st International RILEM Workshop on Microbial Impacts on Building Materials (CD 02) (e-ISBN 978-2-35158-013-4); *Ed. M. Ribas Silva*

PRO 21: International RILEM Symposium on Connections between Steel and Concrete (ISBN: 2-912143-25-X; e-ISBN: 2351580346); *Ed. R. Eligehausen*

PRO 22: International RILEM Symposium on Joints in Timber Structures (ISBN: 2-912143-28-4; e-ISBN: 2351580354); *Eds. S. Aicher and H.-W. Reinhardt*

PRO 23: International RILEM Conference on Early Age Cracking in Cementitious Systems (ISBN: 2-912143-29-2; e-ISBN: 2351580362); *Eds. K. Kovler and A. Bentur*

PRO 24: 2nd International RILEM Workshop on Frost Resistance of Concrete (ISBN: 2-912143-30-6; e-ISBN: 2351580370); *Eds. M. J. Setzer, R. Auberg and H.-J. Keck*

PRO 25: International RILEM Workshop on Frost Damage in Concrete (ISBN: 2-912143-31-4; e-ISBN: 2351580389); *Eds. D. J. Janssen, M. J. Setzer and M. B. Snyder*

PRO 26: International RILEM Workshop on On-Site Control and Evaluation of Masonry Structures (ISBN: 2-912143-34-9; e-ISBN: 2351580141); *Eds. L. Binda and R. C. de Vekey*

PRO 27: International RILEM Symposium on Building Joint Sealants (CD03; e-ISBN: 235158015X); *Ed. A. T. Wolf*

PRO 28: 6th International RILEM Symposium on Performance Testing and Evaluation of Bituminous Materials - PTEBM'03 (ISBN: 2-912143-35-7; e-ISBN: 978-2-912143-77-8); *Ed. M. N. Partl*

PRO 29: 2nd International RILEM Workshop on Life Prediction and Ageing Management of Concrete Structures (ISBN: 2-912143-36-5; e-ISBN: 2912143780); *Ed. D. J. Naus*

PRO 30: 4th International RILEM Workshop on High Performance Fiber Reinforced Cement Composites - HPFRCC 4 (ISBN: 2-912143-37-3; e-ISBN: 2912143799); *Eds. A. E. Naaman and H. W. Reinhardt*

PRO 31: International RILEM Workshop on Test and Design Methods for Steel Fibre Reinforced Concrete: Background and Experiences (ISBN: 2-912143-38-1; e-ISBN: 2351580168); *Eds. B. Schnütgen and L. Vandewalle*

PRO 32: International Conference on Advances in Concrete and Structures 2 vol. (ISBN (set): 2-912143-41-1; e-ISBN: 2351580176); *Eds. Ying-shu Yuan, Surendra P. Shah and Heng-lin Lü*

PRO 33: 3rd International Symposium on Self-Compacting Concrete (ISBN: 2-912143-42-X; e-ISBN: 2912143713); *Eds. Ó. Wallevik and I. Níelsson*

PRO 34: International RILEM Conference on Microbial Impact on Building Materials (ISBN: 2-912143-43-8; e-ISBN: 2351580184); *Ed. M. Ribas Silva*

PRO 35: International RILEM TC 186-ISA on Internal Sulfate Attack and Delayed Ettringite Formation (ISBN: 2-912143-44-6; e-ISBN: 2912143802); *Eds. K. Scrivener and J. Skalny*

PRO 36: International RILEM Symposium on Concrete Science and Engineering – A Tribute to Arnon Bentur (ISBN: 2-912143-46-2; e-ISBN: 2912143586); *Eds. K. Kovler, J. Marchand, S. Mindess and J. Weiss*

PRO 37: 5th International RILEM Conference on Cracking in Pavements – Mitigation, Risk Assessment and Prevention (ISBN: 2-912143-47-0; e-ISBN: 2912143764); *Eds. C. Petit, I. Al-Qadi and A. Millien*

PRO 38: 3rd International RILEM Workshop on Testing and Modelling the Chloride Ingress into Concrete (ISBN: 2-912143-48-9; e-ISBN: 2912143578); *Eds. C. Andrade and J. Kropp*

PRO 39: 6th International RILEM Symposium on Fibre-Reinforced Concretes - BEFIB 2004 (ISBN: 2-912143-51-9; e-ISBN: 2912143748); *Eds. M. Di Prisco, R. Felicetti and G. A. Plizzari*

PRO 40: International RILEM Conference on the Use of Recycled Materials in Buildings and Structures (ISBN: 2-912143-52-7; e-ISBN: 2912143756); *Eds. E. Vázquez, Ch. F. Hendriks and G. M. T. Janssen*

PRO 41: RILEM International Symposium on Environment-Conscious Materials and Systems for Sustainable Development (ISBN: 2-912143-55-1; e-ISBN: 2912143640); *Eds. N. Kashino and Y. Ohama*

PRO 42: SCC'2005 - China: 1st International Symposium on Design, Performance and Use of Self-Consolidating Concrete (ISBN: 2-912143-61-6; e-ISBN: 2912143624); *Eds. Zhiwu Yu, Caijun Shi, Kamal Henri Khayat and Youjun Xie*

PRO 43: International RILEM Workshop on Bonded Concrete Overlays (e-ISBN: 2-912143-83-7); *Eds. J. L. Granju and J. Silfwerbrand*

PRO 44: 2nd International RILEM Workshop on Microbial Impacts on Building Materials (CD11) (e-ISBN: 2-912143-84-5); *Ed. M. Ribas Silva*

PRO 45: 2nd International Symposium on Nanotechnology in Construction, Bilbao (ISBN: 2-912143-87-X; e-ISBN: 2912143888); *Eds. Peter J. M. Bartos, Yolanda de Miguel and Antonio Porro*

PRO 46: ConcreteLife'06 - International RILEM-JCI Seminar on Concrete Durability and Service Life Planning: Curing, Crack Control, Performance in Harsh Environments (ISBN: 2-912143-89-6; e-ISBN: 291214390X); *Ed. K. Kovler*

PRO 47: International RILEM Workshop on Performance Based Evaluation and Indicators for Concrete Durability (ISBN: 978-2-912143-95-2; e-ISBN: 9782912143969); *Eds. V. Baroghel-Bouny, C. Andrade, R. Torrent and K. Scrivener*

PRO 48: 1st International RILEM Symposium on Advances in Concrete through Science and Engineering (e-ISBN: 2-912143-92-6); *Eds. J. Weiss, K. Kovler, J. Marchand, and S. Mindess*

PRO 49: International RILEM Workshop on High Performance Fiber Reinforced Cementitious Composites in Structural Applications (ISBN: 2-912143-93-4; e-ISBN: 2912143942); *Eds. G. Fischer and V. C. Li*

PRO 50: 1st International RILEM Symposium on Textile Reinforced Concrete (ISBN: 2-912143-97-7; e-ISBN: 2351580087); *Eds. Josef Hegger, Wolfgang Brameshuber and Norbert Will*

PRO 51: 2nd International Symposium on Advances in Concrete through Science and Engineering (ISBN: 2-35158-003-6; e-ISBN: 2-35158-002-8); *Eds. J. Marchand, B. Bissonnette, R. Gagné, M. Jolin and F. Paradis*

PRO 52: Volume Changes of Hardening Concrete: Testing and Mitigation (ISBN: 2-35158-004-4; e-ISBN: 2-35158-005-2); *Eds. O. M. Jensen, P. Lura and K. Kovler*

PRO 53: High Performance Fiber Reinforced Cement Composites - HPFRCC5 (ISBN: 978-2-35158-046-2; e-ISBN: 978-2-35158-089-9); *Eds. H. W. Reinhardt and A. E. Naaman*

PRO 54: 5th International RILEM Symposium on Self-Compacting Concrete (ISBN: 978-2-35158-047-9; e-ISBN: 978-2-35158-088-2); *Eds. G. De Schutter and V. Boel*

PRO 55: International RILEM Symposium Photocatalysis, Environment and Construction Materials (ISBN: 978-2-35158-056-1; e-ISBN: 978-2-35158-057-8); *Eds. P. Baglioni and L. Cassar*

PRO 56: International RILEM Workshop on Integral Service Life Modelling of Concrete Structures (ISBN 978-2-35158-058-5; e-ISBN: 978-2-35158-090-5); *Eds. R. M. Ferreira, J. Gulikers and C. Andrade*

PRO 57: RILEM Workshop on Performance of cement-based materials in aggressive aqueous environments (e-ISBN: 978-2-35158-059-2); *Ed. N. De Belie*

PRO 58: International RILEM Symposium on Concrete Modelling - CONMOD'08 (ISBN: 978-2-35158-060-8; e-ISBN: 978-2-35158-076-9); *Eds. E. Schlangen and G. De Schutter*

PRO 59: International RILEM Conference on On Site Assessment of Concrete, Masonry and Timber Structures - SACoMaTiS 2008 (ISBN set: 978-2-35158-061-5; e-ISBN: 978-2-35158-075-2); *Eds. L. Binda, M. di Prisco and R. Felicetti*

PRO 60: Seventh RILEM International Symposium on Fibre Reinforced Concrete: Design and Applications - BEFIB 2008 (ISBN: 978-2-35158-064-6; e-ISBN: 978-2-35158-086-8); *Ed. R. Gettu*

PRO 61: 1st International Conference on Microstructure Related Durability of Cementitious Composites 2 vol., (ISBN: 978-2-35158-065-3; e-ISBN: 978-2-35158-084-4); *Eds. W. Sun, K. van Breugel, C. Miao, G. Ye and H. Chen*

PRO 62: NSF/ RILEM Workshop: In-situ Evaluation of Historic Wood and Masonry Structures (e-ISBN: 978-2-35158-068-4); *Eds. B. Kasal, R. Anthony and M. Drdácký*

PRO 63: Concrete in Aggressive Aqueous Environments: Performance, Testing and Modelling, 2 vol., (ISBN: 978-2-35158-071-4; e-ISBN: 978-2-35158-082-0); *Eds. M. G. Alexander and A. Bertron*

PRO 64: Long Term Performance of Cementitious Barriers and Reinforced Concrete in Nuclear Power Plants and Waste Management - NUCPERF 2009 (ISBN: 978-2-35158-072-1; e-ISBN: 978-2-35158-087-5); *Eds. V. L'Hostis, R. Gens, C. Gallé*

PRO 65: Design Performance and Use of Self-consolidating Concrete - SCC'2009 (ISBN: 978-2-35158-073-8; e-ISBN: 978-2-35158-093-6); *Eds. C. Shi, Z. Yu, K. H. Khayat and P. Yan*

PRO 66: 2nd International RILEM Workshop on Concrete Durability and Service Life Planning - ConcreteLife'09 (ISBN: 978-2-35158-074-5; ISBN: 978-2-35158-074-5); *Ed. K. Kovler*

PRO 67: Repairs Mortars for Historic Masonry (e-ISBN: 978-2-35158-083-7); *Ed. C. Groot*

PRO 68: Proceedings of the 3rd International RILEM Symposium on 'Rheology of Cement Suspensions such as Fresh Concrete (ISBN 978-2-35158-091-2; e-ISBN: 978-2-35158-092-9); *Eds. O. H. Wallevik, S. Kubens and S. Oesterheld*

PRO 69: 3rd International PhD Student Workshop on 'Modelling the Durability of Reinforced Concrete (ISBN: 978-2-35158-095-0); *Eds. R. M. Ferreira, J. Gulikers and C. Andrade*

PRO 70: 2nd International Conference on 'Service Life Design for Infrastructure' (ISBN set: 978-2-35158-096-7, e-ISBN: 978-2-35158-097-4); *Ed. K. van Breugel, G. Ye and Y. Yuan*

PRO 71: Advances in Civil Engineering Materials - The 50-year Teaching Anniversary of Prof. Sun Wei' (ISBN: 978-2-35158-098-1; e-ISBN: 978-2-35158-099-8); *Eds. C. Miao, G. Ye, and H. Chen*

PRO 72: First International Conference on 'Advances in Chemically-Activated Materials – CAM'2010' (2010), 264 pp, ISBN: 978-2-35158-101-8; e-ISBN: 978-2-35158-115-5, *Eds. Caijun Shi and Xiaodong Shen*

PRO 73: 2nd International Conference on 'Waste Engineering and Management - ICWEM 2010' (2010), 894 pp, ISBN: 978-2-35158-102-5; e-ISBN: 978-2-35158-103-2, *Eds. J. Zh. Xiao, Y. Zhang, M. S. Cheung and R. Chu*

PRO 74: International RILEM Conference on 'Use of Superabsorsorbent Polymers and Other New Addditives in Concrete' (2010) 374 pp., ISBN: 978-2-35158-104-9; e-ISBN: 978-2-35158-105-6; *Eds. O. M. Jensen, M. T. Hasholt, and S. Laustsen*

PRO 75: International Conference on 'Material Science - 2nd ICTRC - Textile Reinforced Concrete - Theme 1' (2010) 436 pp., ISBN: 978-2-35158-106-3; e-ISBN: 978-2-35158-107-0; *Ed. W. Brameshuber*

PRO 76: International Conference on 'Material Science - HetMat - Modelling of Heterogeneous Materials - Theme 2' (2010) 255 pp., ISBN: 978-2-35158-108-7; e-ISBN: 978-2-35158-109-4; *Ed. W. Brameshuber*

PRO 77: International Conference on 'Material Science - AdIPoC - Additions Improving Properties of Concrete - Theme 3' (2010) 459 pp., ISBN: 978-2-35158-110-0; e-ISBN: 978-2-35158-111-7; *Ed. W. Brameshuber*

PRO 78: 2nd Historic Mortars Conference and RILEM TC 203-RHM Final Workshop – HMC2010 (2010) 1416 pp., e-ISBN: 978-2-35158-112-4; *Eds. J. Válek, C. Groot, and J. J. Hughes*

PRO 79: International RILEM Conference on Advances in Construction Materials Through Science and Engineering (2011) 213 pp., ISBN: 978-2-35158-116-2, e-ISBN: 978-2-35158-117-9; *Eds. Christopher Leung and K.T. Wan*

PRO 80: 2nd International RILEM Conference on Concrete Spalling due to Fire Exposure (2011) 453 pp., ISBN: 978-2-35158-118-6, e-ISBN: 978-2-35158-119-3; *Eds. E. A. B. Koenders and F. Dehn*

PRO 81: 2nd International RILEM Conference on Strain Hardening Cementitious Composites (SHCC2-Rio) (2011) 451 pp., ISBN: 978-2-35158-120-9, e-ISBN: 978-2-35158-121-6; *Eds. R.D. Toledo Filho, F. A. Silva, E. A. B. Koenders and E. M. R. Fairbairn*

PRO 82: 2nd International RILEM Conference on Progress of Recycling in the Built Environment (2011) 507 pp., e-ISBN: 978-2-35158-122-3; *Eds. V. M. John, E. Vazquez, S. C. Angulo and C. Ulsen*

PRO 83: 2nd International Conference on Microstructural-related Durability of Cementitious Composites (2012) 250 pp., ISBN: 978-2-35158-129-2; e-ISBN: 978-2-35158-123-0; *Eds. G. Ye, K. van Breugel, W. Sun and C. Miao*

PRO 84: CONSEC13 - Seventh International Conference on Concrete under Severe Conditions – Environment and Loading (2013) 1930 pp., ISBN: 978-2-35158-124-7; e-ISBN: 978-2- 35158-134-6; *Eds. Z. J. Li, W. Sun, C. W. Miao, K. Sakai, O. E. Gjorv & N. Banthia*

PRO 85: RILEM-JCI International Workshop on Crack Control of Mass Concrete and Related issues concerning Early-Age of Concrete Structures – ConCrack 3 – Control of Cracking in Concrete Structures 3 (2012) 237 pp., ISBN: 978-2-35158-125-4; e-ISBN: 978-2-35158-126-1; *Eds. F. Toutlemonde and J.-M. Torrenti*

PRO 86: International Symposium on Life Cycle Assessment and Construction (2012) 414 pp., ISBN: 978-2-35158-127-8, e-ISBN: 978-2-35158-128-5; *Eds. A. Ventura and C. de la Roche*

PRO 87: UHPFRC 2013 – RILEM-fib-AFGC International Symposium on Ultra-High Performance Fibre-Reinforced Concrete (2013), ISBN: 978-2-35158-130-8, e-ISBN: 978-2-35158-131-5; *Eds. F. Toutlemonde*

PRO 88: 8th RILEM International Symposium on Fibre Reinforced Concrete (2012) 344 pp., ISBN: 978-2-35158-132-2, e-ISBN: 978-2-35158-133-9; *Eds. Joaquim A. O. Barros*

PRO 89: RILEM International workshop on performance-based specification and control of concrete durability (2014) 678 pp, ISBN: 978-2-35158-135-3, e-ISBN: 978-2-35158-136-0; *Eds. D. Bjegović, H. Beushausen and M. Serdar*

PRO 90: 7th RILEM International Conference on Self-Compacting Concrete and of the 1st RILEM International Conference on Rheology and Processing of Construction Materials (2013) 396 pp, ISBN: 978-2-35158-137-7, e-ISBN: 978-2-35158-138-4; *Eds. Nicolas Roussel and Hela Bessaies-Bey*

PRO 91: CONMOD 2014 - RILEM International Symposium on Concrete Modelling (2014), ISBN: 978-2-35158-139-1; e-ISBN: 978-2-35158-140-7; *Eds. Kefei Li, Peiyu Yan and Rongwei Yang*

PRO 92: CAM 2014 - 2nd International Conference on advances in chemically-activated materials (2014) 392 pp., ISBN: 978-2-35158-141-4; e-ISBN: 978-2-35158-142-1; *Eds. Caijun Shi and Xiadong Shen*

PRO 93: SCC 2014 - 3rd International Symposium on Design, Performance and Use of Self-Consolidating Concrete (2014) 438 pp., ISBN: 978-2-35158-143-8; e-ISBN: 978-2-35158-144-5; *Eds. Caijun Shi, Zhihua Ou, Kamal H. Khayat*

PRO 94 (online version): HPFRCC-7 - 7th RILEM conference on High performance fiber reinforced cement composites (2015), e-ISBN: 978-2-35158-146-9; *Eds. H. W. Reinhardt, G. J. Parra-Montesinos, H. Garrecht*

PRO 95: International RILEM Conference on Application of superabsorbent polymers and other new admixtures in concrete construction (2014), ISBN: 978-2-35158-147-6; e-ISBN: 978-2-35158-148-3; *Eds. Viktor Mechtcherine, Christof Schroefl*

PRO 96 (online version): XIII DBMC: XIII International Conference on Durability of Building Materials and Components(2015), e-ISBN: 978-2-35158-149-0; *Eds. M. Quattrone, V. M. John*

PRO 97: SHCC3 – 3rd International RILEM Conference on Strain Hardening Cementitious Composites (2014), ISBN: 978-2-35158-150-6; e-ISBN: 978-2-35158-151-3; *Eds. E. Schlangen, M. G. Sierra Beltran, M. Lukovic, G. Ye*

PRO 98: FERRO-11 – 11th International Symposium on Ferrocement and 3rd ICTRC - International Conference on Textile Reinforced Concrete (2015), ISBN: 978-2-35158-152-0; e-ISBN: 978-2-35158-153-7; *Ed. W. Brameshuber*

PRO 99 (online version): ICBBM 2015 - 1st International Conference on Bio-Based Building Materials (2015), e-ISBN: 978-2-35158-154-4; *Eds. S. Amziane, M. Sonebi*

PRO 100: SCC16 - RILEM Self-Consolidating Concrete Conference (2016), ISBN: 978-2-35158-156-8; e-ISBN: 978-2-35158-157-5; *Ed. Kamal H. Kayat*

PRO 101 (online version): III Progress of Recycling in the Built Environment (2015), e-ISBN: 978-2-35158-158-2; *Eds. I. Martins, C. Ulsen and S. C. Angulo*

PRO 102 (online version): RILEM Conference on Microorganisms-Cementitious Materials Interactions (2016), e-ISBN: 978-2-35158-160-5; *Eds. Alexandra Bertron, Henk Jonkers, Virginie Wiktor*

PRO 103 (online version): ACESC'16 - Advances in Civil Engineering and Sustainable Construction (2016), e-ISBN: 978-2-35158-161-2; *Eds. T.Ch. Madhavi, G. Prabhakar, Santhosh Ram and P. M. Rameshwaran*

PRO 104 (online version): SSCS'2015 - Numerical Modeling - Strategies for Sustainable Concrete Structures (2015), e-ISBN: 978-2-35158-162-9

PRO 105: 1st International Conference on UHPC Materials and Structures (2016), ISBN: 978-2-35158-164-3, e-ISBN: 978-2-35158-165-0

PRO 106: AFGC-ACI-fib-RILEM International Conference on Ultra-High-Performance Fibre-Reinforced Concrete – UHPFRC 2017 (2017), ISBN: 978-2-35158-166-7, e-ISBN: 978-2-35158-167-4; *Eds. François Toutlemonde & Jacques Resplendino*

PRO 107 (online version): XIV DBMC – 14th International Conference on Durability of Building Materials and Components (2017), e-ISBN: 978-2-35158-159-9; *Eds. Geert De Schutter, Nele De Belie, Arnold Janssens, Nathan Van Den Bossche*

PRO 108: MSSCE 2016 - Innovation of Teaching in Materials and Structures (2016), ISBN: 978-2-35158-178-0, e-ISBN: 978-2-35158-179-7; *Ed. Per Goltermann*

PRO 109 (2 volumes): MSSCE 2016 - Service Life of Cement-Based Materials and Structures (2016), ISBN Vol. 1: 978-2-35158-170-4, Vol. 2: 978-2-35158-171-4, Set Vol. 1&2: 978-2-35158-172-8, e-ISBN : 978-2-35158-173-5; *Eds. Miguel Azenha, Ivan Gabrijel, Dirk Schlicke, Terje Kanstad and Ole Mejlhede Jensen*

PRO 110: MSSCE 2016 - Historical Masonry (2016), ISBN: 978-2-35158-178-0, e-ISBN: 978-2-35158-179-7; *Eds. Inge Rörig-Dalgaard and Ioannis Ioannou*

PRO 111: MSSCE 2016 - Electrochemistry in Civil Engineering (2016), ISBN: 978-2-35158-176-6, e-ISBN: 978-2-35158-177-3; *Ed. Lisbeth M. Ottosen*

PRO 112: MSSCE 2016 - Moisture in Materials and Structures (2016), ISBN: 978-2-35158-178-0, e-ISBN: 978-2-35158-179-7; *Eds. Kurt Kielsgaard Hansen, Carsten Rode and Lars-Olof Nilsson*

PRO 113: MSSCE 2016 - Concrete with Supplementary Cementitious Materials (2016), ISBN: 978-2-35158-178-0, e-ISBN: 978-2-35158-179-7; *Eds. Ole Mejlhede Jensen, Konstantin Kovler and Nele De Belie*

PRO 114: MSSCE 2016 - Frost Action in Concrete (2016), ISBN: 978-2-35158-182-7, e-ISBN: 978-2-35158-183-4; *Eds. Marianne Tange Hasholt, Katja Fridh and R. Doug Hooton*

PRO 115: MSSCE 2016 - Fresh Concrete (2016), ISBN: 978-2-35158-184-1, e-ISBN: 978-2-35158-185-8; *Eds. Lars N. Thrane, Claus Pade, Oldrich Svec and Nicolas Roussel*

PRO 116: BEFIB 2016 – 9th RILEM International Symposium on Fiber Reinforced Concrete (2016), ISBN: 978-2-35158-187-2, e-ISBN: 978-2-35158-186-5; *Eds. N. Banthia, M. di Prisco and S. Soleimani-Dashtaki*

PRO 117: 3rd International RILEM Conference on Microstructure Related Durability of Cementitious Composites (2016), ISBN: 978-2-35158-188-9, e-ISBN: 978-2-35158-189-6; *Eds. Changwen Miao, Wei Sun, Jiaping Liu, Huisu Chen, Guang Ye and Klaas van Breugel*

PRO 118 (4 volumes): International Conference on Advances in Construction Materials and Systems (2017), ISBN Set: 978-2-35158-190-2, Vol. 1: 978-2-35158-193-3, Vol. 2: 978-2-35158-194-0, Vol. 3: ISBN:978-2-35158-195-7, Vol. 4: ISBN:978-2-35158-196-4, e-ISBN: 978-2-35158-191-9; *Eds. Manu Santhanam, Ravindra Gettu, Radhakrishna G. Pillai and Sunitha K. Nayar*

PRO 119 (online version): ICBBM 2017 - Second International RILEM Conference on Bio-based Building Materials, (2017), e-ISBN: 978-2-35158-192-6; *Ed. Sofiane Amziane*

PRO 120 (2 volumes): EAC-02 - 2nd International RILEM/COST Conference on Early Age Cracking and Serviceability in Cement-based Materials and Structures, (2017), Vol. 1: 978-2-35158-199-5, Vol. 2: 978-2-35158-200-8, Set: 978-2-35158-197-1, e-ISBN: 978-2-35158-198-8; *Eds. Stéphanie Staquet and Dimitrios Aggelis*

PRO 121 (2 volumes): SynerCrete18: Interdisciplinary Approaches for Cement-based Materials and Structural Concrete: Synergizing Expertise and Bridging Scales of Space and Time, (2018), Set: 978-2-35158-202-2, Vol.1: 978-2-35158-211-4, Vol.2: 978-2-35158-212-1, e-ISBN: 978-2-35158-203-9; *Eds. Miguel Azenha, Dirk Schlicke, Farid Benboudjema, Agnieszka Knoppik*

PRO 122: SCC'2018 China - Fourth International Symposium on Design, Performance and Use of Self-Consolidating Concrete, (2018), ISBN: 978-2-35158-204-6, e-ISBN: 978-2-35158-205-3; *Eds. C. Shi, Z. Zhang, K. H. Khayat*

PRO 123: Final Conference of RILEM TC 253-MCI: Microorganisms-Cementitious Materials Interactions (2018), Set: 978-2-35158-207-7, Vol.1: 978-2-35158-209-1, Vol.2: 978-2-35158-210-7, e-ISBN: 978-2-35158-206-0; *Ed. Alexandra Bertron*

PRO 124 (online version): Fourth International Conference Progress of Recycling in the Built Environment (2018), e-ISBN: 978-2-35158-208-4; *Eds. Isabel M. Martins, Carina Ulsen, Yury Villagran*

PRO 125 (online version): SLD4 - 4th International Conference on Service Life Design for Infrastructures (2018), e-ISBN: 978-2-35158-213-8; *Eds. Guang Ye, Yong Yuan, Claudia Romero Rodriguez, Hongzhi Zhang, Branko Savija*

PRO 126: Workshop on Concrete Modelling and Material Behaviour in honor of Professor Klaas van Breugel (2018), ISBN: 978-2-35158-214-5, e-ISBN: 978-2-35158-215-2; *Ed. Guang Ye*

PRO 127 (online version): CONMOD2018 - Symposium on Concrete Modelling (2018), e-ISBN: 978-2-35158-216-9; *Eds. Erik Schlangen, Geert de Schutter, Branko Savija, Hongzhi Zhang, Claudia Romero Rodriguez*

PRO 128: SMSS2019 - International Conference on Sustainable Materials, Systems and Structures (2019), ISBN: 978-2-35158-217-6, e-ISBN: 978-2-35158-218-3

PRO 129: 2nd International Conference on UHPC Materials and Structures (UHPC2018-China), ISBN: 978-2-35158-219-0, e-ISBN: 978-2-35158-220-6;

PRO 130: 5th Historic Mortars Conference (2019), ISBN: 978-2-35158-221-3, e-ISBN: 978-2-35158-222-0; *Eds. José Ignacio Álvarez, José María Fernández, Íñigo Navarro, Adrián Durán, Rafael Sirera*

PRO 131 (online version): 3rd International Conference on Bio-Based Building Materials (ICBBM2019), e-ISBN: 978-2-35158-229-9; *Eds. Mohammed Sonebi, Sofiane Amziane, Jonathan Page*

PRO 132: IRWRMC'18 - International RILEM Workshop on Rheological Measurements of Cement-based Materials (2018), ISBN: 978-2-35158-230-5, e-ISBN: 978-2-35158-231-2; *Eds. Chafika Djelal and Yannick Vanhove*

PRO 133 (online version): CO2STO2019 - International Workshop CO2 Storage in Concrete (2019), e-ISBN: 978-2-35158-232-9; *Eds. Assia Djerbi, Othman Omikrine-Metalssi and Teddy Fen-Chong*

PRO 134: 3rd ACF/HNU International Conference on UHPC Materials and Structures - UHPC'2020, ISBN: 978-2-35158-233-6, e-ISBN: 978-2-35158-234-3; *Eds. Caijun Shi and Jiaping Liu*

PRO 135: Fourth International Conference on Chemically Activated Materials (CAM2021), ISBN: 978-2-35158-235-0, e-ISBN: 978-2-35158-236-7; *Eds. Caijun Shi and Xiang Hu*

RILEM Reports (REP)

Report 19: Considerations for Use in Managing the Aging of Nuclear Power Plant Concrete Structures (ISBN: 2-912143-07-1); *Ed. D. J. Naus*

Report 20: Engineering and Transport Properties of the Interfacial Transition Zone in Cementitious Composites (ISBN: 2-912143-08-X); *Eds. M. G. Alexander, G. Arliguie, G. Ballivy, A. Bentur and J. Marchand*

Report 21: Durability of Building Sealants (ISBN: 2-912143-12-8); *Ed. A. T. Wolf*

Report 22: Sustainable Raw Materials - Construction and Demolition Waste (ISBN: 2-912143-17-9); *Eds. C. F. Hendriks and H. S. Pietersen*

Report 23: Self-Compacting Concrete state-of-the-art report (ISBN: 2-912143-23-3); *Eds. Å. Skarendahl and Ö. Petersson*

Report 24: Workability and Rheology of Fresh Concrete: Compendium of Tests (ISBN: 2-912143-32-2); *Eds. P. J. M. Bartos, M. Sonebi and A. K. Tamimi*

Report 25: Early Age Cracking in Cementitious Systems (ISBN: 2-912143-33-0); *Ed. A. Bentur*

Report 26: Towards Sustainable Roofing (Joint Committee CIB/RILEM) (CD 07) (e-ISBN 978-2-912143-65-5); *Eds. Thomas W. Hutchinson and Keith Roberts*

Report 27: Condition Assessment of Roofs (Joint Committee CIB/RILEM) (CD 08) (e-ISBN 978-2-912143-66-2); *Ed. CIB W 83/RILEM TC166-RMS*

Report 28: Final report of RILEM TC 167-COM 'Characterisation of Old Mortars with Respect to Their Repair (ISBN: 978-2-912143-56-3); *Eds. C. Groot, G. Ashall and J. Hughes*

Report 29: Pavement Performance Prediction and Evaluation (PPPE): Interlaboratory Tests (e-ISBN: 2-912143-68-3); *Eds. M. Partl and H. Piber*

Report 30: Final Report of RILEM TC 198-URM 'Use of Recycled Materials' (ISBN: 2-912143-82-9; e-ISBN: 2-912143-69-1); *Eds. Ch. F. Hendriks, G. M. T. Janssen and E. Vázquez*

Report 31: Final Report of RILEM TC 185-ATC 'Advanced testing of cement-based materials during setting and hardening' (ISBN: 2-912143-81-0; e-ISBN: 2-912143-70-5); *Eds. H. W. Reinhardt and C. U. Grosse*

Report 32: Probabilistic Assessment of Existing Structures. A JCSS publication (ISBN 2-912143-24-1); *Ed. D. Diamantidis*

Report 33: State-of-the-Art Report of RILEM Technical Committee TC 184-IFE 'Industrial Floors' (ISBN 2-35158-006-0); *Ed. P. Seidler*

Report 34: Report of RILEM Technical Committee TC 147-FMB 'Fracture mechanics applications to anchorage and bond' Tension of Reinforced Concrete Prisms – Round Robin Analysis and Tests on Bond (e-ISBN 2-912143-91-8); *Eds. L. Elfgren and K. Noghabai*

Report 35: Final Report of RILEM Technical Committee TC 188-CSC 'Casting of Self Compacting Concrete' (ISBN 2-35158-001-X; e-ISBN: 2-912143-98-5); *Eds. Å. Skarendahl and P. Billberg*

Report 36: State-of-the-Art Report of RILEM Technical Committee TC 201-TRC 'Textile Reinforced Concrete' (ISBN 2-912143-99-3); *Ed. W. Brameshuber*

Report 37: State-of-the-Art Report of RILEM Technical Committee TC 192-ECM 'Environment-conscious construction materials and systems' (ISBN: 978-2-35158-053-0); *Eds. N. Kashino, D. Van Gemert and K. Imamoto*

Report 38: State-of-the-Art Report of RILEM Technical Committee TC 205-DSC 'Durability of Self-Compacting Concrete' (ISBN: 978-2-35158-048-6); *Eds. G. De Schutter and K. Audenaert*

Report 39: Final Report of RILEM Technical Committee TC 187-SOC 'Experimental determination of the stress-crack opening curve for concrete in tension' (ISBN 978-2-35158-049-3); *Ed. J. Planas*

Report 40: State-of-the-Art Report of RILEM Technical Committee TC 189-NEC 'Non-Destructive Evaluation of the Penetrability and Thickness of the Concrete Cover' (ISBN 978-2-35158-054-7); *Eds. R. Torrent and L. Fernández Luco*

Report 41: State-of-the-Art Report of RILEM Technical Committee TC 196-ICC 'Internal Curing of Concrete' (ISBN 978-2-35158-009-7); *Eds. K. Kovler and O. M. Jensen*

Report 42: 'Acoustic Emission and Related Non-destructive Evaluation Techniques for Crack Detection and Damage Evaluation in Concrete' - Final Report of RILEM Technical Committee 212-ACD (e-ISBN: 978-2-35158-100-1); *Ed. M. Ohtsu*

Report 45: Repair Mortars for Historic Masonry - State-of-the-Art Report of RILEM Technical Committee TC 203-RHM (e-ISBN: 978-2-35158-163-6); *Eds. Paul Maurenbrecher and Caspar Groot*

Report 46: Surface delamination of concrete industrial floors and other durability related aspects guide - Report of RILEM Technical Committee TC 268-SIF (e-ISBN: 978-2-35158-201-5); *Ed. Valerie Pollet*

New Materials and Process Technology

Mechanical Performance of Strain Hardening Limestone Calcined Clay Cementitious Composites (SHLC4) Subject to Wet-Dry Cycles

Ameer Hamza Ahmed, Marco Liebscher(✉), and Viktor Mechtcherine

Institute of Construction Materials, Technische Universität Dresden, 01062 Dresden, Germany
marco.liebscher@tu-dresden.de

Abstract. The article at hand focuses on the influence of wet-dry cycles on the mechanical properties and crack formation in strain-hardening cementitious composites (SHCC) made of limestone calcined clay cement (LC3) and high-performance polyethylene (PE) fibers. Specimens for uniaxial tension tests were produced with 2 vol.% fiber content and preloaded until 1% strain at the age of 28 days. Subsequently, the specimens were weekly exposed to three days of wetting and four days of drying cycles for 12 consequent weeks, followed by further tension tests, this time until ultimate failure. In addition, a series of preloaded specimens were kept under a controlled environment (20 °C and 65% RH) in a climate chamber and characterized as well. The mechanical properties of the two curing conditions were compared with each other, 28 days reference specimens, and the respected preloaded samples. The analysis of the mechanical properties showed a pronounced recovery in Young's modulus, first crack stress, and tensile strength due to the wet-dry exposure. Furthermore, a detailed crack analysis via DIC analysis and optical microscopy revealed the crack closure phenomenon to some extent due to wet-dry cycles. The results indicate that strain-hardening limestone calcined clay cementitious composites (SHLC4) exhibit considerable mechanical performance and self-healing capacity.

Keywords: SHCC · LC3 · SHLC4 · Wet-dry cycles · Self-healing

1 Introduction

Strain hardening cementitious composites (SHCC) is a special class of quasi-ductile materials [1]. These micro-mechanically designed composites utilize short polymer fibers as micro-reinforcement to bridge multiple microcracks [2]. Unlike ordinary concrete (OC), which comprises aggregates (usually 70% by volume fraction) and cement paste (30% by volume fraction), SHCC consists of only a fine-grained matrix and fiber. Hence on an equal volume basis, the SHCC contains more cement clinker than OC. This additional clinker content is one of the reasons for the higher cost and larger CO_2 footprint of SHCC [1]. To mitigate these issues, in earlier studies, the cement clinker in the composition of SHCC has been partially replaced by supplementary cementitious materials (SCMs) such as fly ash and silica fume. However, the long-term availability

M. Kunieda et al. (Eds.): SHCC 2022, RILEM Bookseries 39, pp. 3–12, 2023.
https://doi.org/10.1007/978-3-031-15805-6_1

of these SCMs is questionable. In contrast, limestone and calcined clay are geographically omnipresent, abundant SCMs, enabling high substitution levels of cement clinker without compromising the mechanical performance [3–5]. The partial substitution of cement clinker with limestone calcined clay cement (LC3) leads to a greener ductile composite termed in this paper as strain-hardening limestone calcined clay cementitious composites (SHLC4).

SHLC4 opens a new research field with respect to durability and sustainability. One way to enhance durability relies on self-healing (autogenous healing) of damaged (cracked) structural elements [6]. Especially the narrow crack widths, an inherent feature of SHCC, allow it to self-heal under proper environmental conditions [7]. Water is a critical requirement for self-healing to occur. Studies have shown that cracks in SHCC can partially or completely seal themselves depending on the initial crack width and exposure conditions [6–12]. This limits the ingress of harmful agents inside the material and therefore enhances its durability.

The mechanisms responsible for autogenous healing include: (i) the crystallization of calcium carbonate (white residue in the crack flanks) resulting from the reaction between free calcium ions and dissolved carbon dioxide (also known as carbonation); (ii) hydration of unreacted cementitious materials (in relatively young concrete); (iii) blocking of cracks due to impurities in water and spalling of loose concrete particles; and (iv) swelling of C-S-H in the crack vicinity [13].

The objective of this research is to evaluate the self-healing capability of SHLC4. For this purpose, uncracked and cracked (preloaded) samples are exposed to wet-dry cycles for up to three months. The recovery in mechanical properties and self-healing of cracks are investigated and reported.

2 Experimental Program

2.1 Materials and Composition

The SHLC4 composition under investigation comprises high early strength cement, CEM I 52.5 R-SR3 (na), from Holcim Germany. Ultra-high molecular weight polyethylene (UHMWPE, short for "PE") fibers, produced by DSM, the Netherlands, in 2% volume fraction are used as micro-reinforcement. Calcined clay, limestone powder, and gypsum are added as SCMs in 30%, 15%, and 5% by weight of cement, respectively. This enables the replacement of 50% of the cement clinker in the mix proportion. The quartz sand-to-binder ratio is 0.45. PCE-based superplasticizer and viscosity modifying agent are used to adjust the workability and facilitate fiber dispersion in the fresh mix. Table 1 shows the composition of the SHLC4. The properties of the fibers are summarized in Table 2.

2.2 Specimen Preparation and Testing Configuration

An established SHCC mixing protocol was applied, details can be found elsewhere [15]. After mixing, the fresh SHLC4 was cast into molds for producing dumbbell-shaped specimens; see Fig. 1a. The following day, specimens were de-molded, sealed in plastic bags, and stored in a climatic chamber at 65% RH and 20 °C for 28 days. Before testing, a speckle pattern was applied to the reference specimens for crack width measurements using digital image correlation (DIC). At the age of 28 days, reference samples were tested under a displacement-controlled uniaxial tension regime in Instron 8802 testing machine; see Fig. 1b. The displacement rate was 0.05 mm/s. Two cameras by GOM with a frequency of 2 Hz were used to capture the images for DIC analysis.

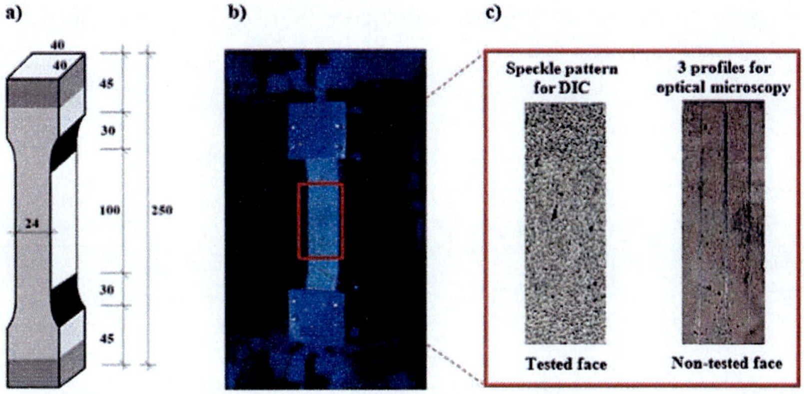

Fig. 1. Geometry of dumbbell-shaped specimen [16]; b) specimen clamped in Instron 8802 for quasi-static uniaxial tension test, and c) speckle pattern for DIC analysis of crack widths in the loaded state (left) and three control lines on a preloaded specimen for measuring residual crack widths using an optical microscope (right).

Table 1. Mixture composition of SHLC4 under investigation (in kg/m^3).

Material	Designation	Quantity
Cement	CEM I 52.5R-SR3/NA	599
Limestone	Soxadol 90 LE	190
Calcined clay	Liapor	379
Gypsum	Fluka-Honeywell	30
Quartz Sand	BCS 413	536
Water	–	359
Superplasticizer	MG ACE 460	11
Viscosity modifying agent	MasterMatrix UW 420	2

Table 2. Geometric and mechanical properties of PE fiber according to the manufacturer [14].

Length	Diameter	Tensile strength	Elongation at break	Density
12 mm	18 μm	3400 MPa	3.5%	0.97 g/cm3

After 28 days of normal curing, two series of specimens were strained up to 1% in tension; they are referred to as Preloaded (P). From each group of preloaded specimens, some specimens were kept in lab conditions (referred to as Dry curing (DC)), and others underwent weekly wet-dry cycles (referred to as Wet-dry (WD)). The wet-dry cycles comprised 3 days of submergence in the water bath and 4 days of drying in lab conditions. The total duration of DC and WD exposures was 3 months. Subsequently, the specimens were reloaded until failure (referred to as Reloaded (R)). Figure 2 shows the flow chart of the experimental program.

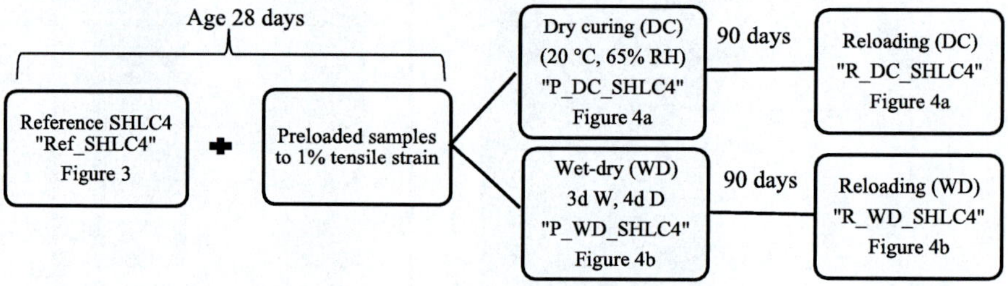

Fig. 2. Flow chart for the experimental program.

3 Results and Discussion

3.1 Uniaxial Tension Test of Reference SHLC4

First, uniaxial tension tests on SHLC4 were performed at an age of 28 days. Figure 3 shows the stress-strain curves along with crack development observed using DIC. The average values of the corresponding numerical data are summarized in Table 3.

The standard deviations in the averaged mechanical characteristics indicate relatively small scattering in the recorded stress-strain responses. The average first crack stress and tensile strength were 3.6 MPa and 6.4 MPa, respectively. The average tangent Young's modulus derived from stress-strain curves was 25 GPa, and the average ultimate strain was approximately 4%. Such high strain capacity resulted in high work-to-fracture energy of 205 kJ/m^3, suggesting that SHLC4 has high damage tolerance and might therefore perform well specifically under high strain rates, such as under impact or blast [17].

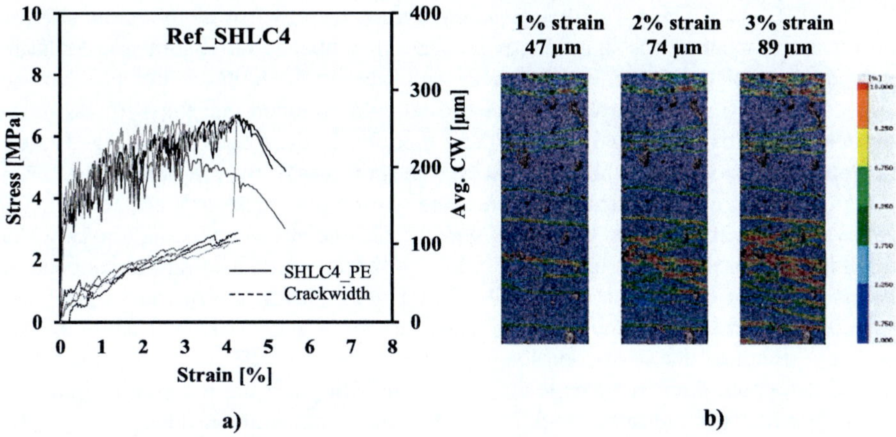

Fig. 3. The stress-strain curves and average crack width curves for Ref_SHLC4 plotted on the primary and secondary axis, respectively, and b) representative DIC images of cracks in the gauge length of the specimen at different strain levels; the numbers on the top indicate the strain level and corresponding average crack opening.

Crack width analysis of the composite under investigation depicted minimum, average, and maximum crack widths of 56.6 µm, 105 µm, and 177.2 µm at 4% strain in the loaded state, respectively. Given such high deformation level, these values are significantly lower as compared to many other fiber-reinforced composites. Additionally, the composite showed a very pronounced multiple cracking with an average of 37 cracks appearing within the gauge length resulting in a total of 370 cracks per meter. This excessive deformation and saturated crack pattern might be due to the higher overall porosity of LC3-based SHCC which facilitates the initiation of microcracks [18]. The initial flaw size distribution influences the stress level to initiate the matrix crack.

The relatively low average crack width values, even at higher strain levels, see Fig. 3b, suggest a strong mechanical bond strength at the fiber/matrix interface which results in higher pull-out energy and crack bridging stresses in the cracked flanks. This is in agreement with the previous studies conducted by Wang et al. (2021) [15].

3.2 Autogenous Healing in SHLC4

Figure 4 shows the stress-strain curves of the preloaded specimens at the age of 28 days and those of the corresponding reloaded specimens after 90 days of exposure to dry curing (DC) and wet-dry curing (WD). Table 3 highlights the key mechanical parameters derived from these diagrams.

The stiffness of SHLC4 in the reloading regimes as provided by the initial slope of the stress-strain curve was significantly reduced in comparison to that measured during preloading or for the reference specimens. Considering the high level of pre-damage, such a decrease in stiffness was expected. It should be noted that the stiffness of the virgin specimen is dominated by the Young modulus of the matrix (θ_1 in Fig. 5) since the volume fraction of fiber is low. However, the stiffness of the pre-damaged samples in the reloading regime indicates the presence of multiple cracks in the specimen. The applied loading first leads to reopening/widening of the already existing cracks which didn't heal either due to too large crack widths or because self-healing did not occur in the given exposure condition such as DC. In this initial stage, the stiffness is very low since it is governed by a few bridging fibers that are already stretched while the rest are being stretched due to applied loading. This stiffness is referred to as θ_2 and θ_5 in Fig. 5. Hereupon, there is a steepening in the ascending curve indicating the stiffness of bridging fibers in the same crack flanks containing no healing products (θ_3 and θ_6). Finally, the stiffness decreases if some healing products are present at different cracking sites (θ_4), as can be seen in the WD curve below and the first crack strength is reached and the strain hardening phase initiates. Nonetheless, the stiffness of the healing product is considerably lower compared to that of the matrix; see Fig. 5.

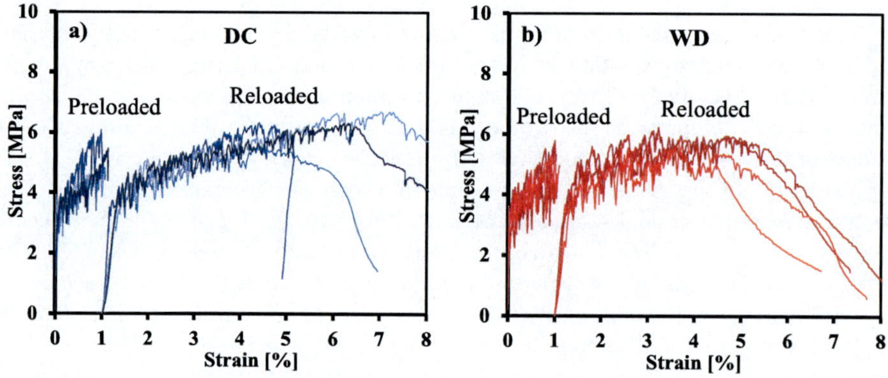

Fig. 4. Stress-strain curves and respective crack analysis of preloaded specimens exposed to a) wet-dry curing (WD) and b) dry curing (DC).

The healing degree of cracks is affected by the crack widths. The residual crack widths for the preloaded DC and WD cured samples, as measured by optical microscopy, are given in Table 3 and visually plotted and presented in Fig. 6. The preloaded DC and WD samples depicted average residual crack widths of 25.1 μm and 31.7 μm. Each crack passing the three control lines drawn in the gauge length of the samples, cf. Figure 1c, was considered and its crack width was averaged. It was found that the cracks opened non-uniformly and the crack widths varied for a single crack. Moreover, comparing the crack widths obtained from DIC analysis for reference specimens at 1% strain, see Fig. 3b, the residual crack widths for preloaded DC and WD samples decreased by 46% and 35%, respectively. This can be traced back to the fact that the residual crack widths

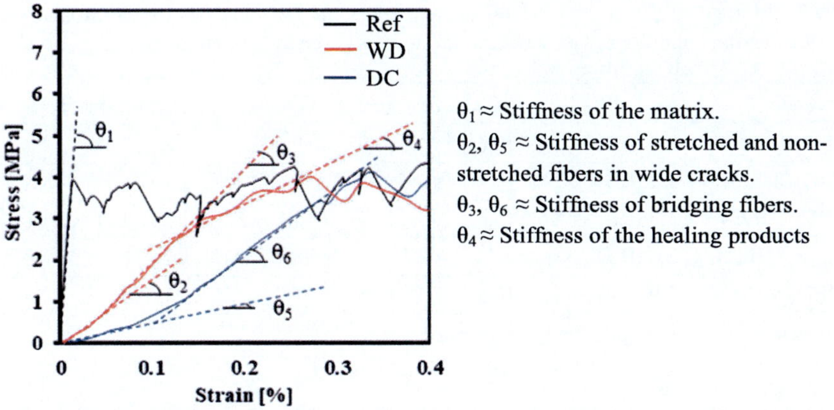

Fig. 5. The initial stiffness of Ref_SHLC4, reloaded WD samples, and reloaded DC samples.

were measured after the removal of load that led to the closing of the cracks. Since the strains in the composite are the function of each crack opening, it is noteworthy to mention that the actual preloaded tension strain differed from that measured using LVDTs. The actual preloading strain was 0.54% for DC samples, i.e., 46% less, and 0.65% for WD samples, i.e., 35% less.

The crack healing phenomenon was observed in all the samples that underwent WD cycles; an example of such a crack is shown in Fig. 6. As mentioned above, the crack opens non-uniformly, i.e., the crack width for a single crack varies significantly. The healing products were formed within the openings of each crack which were roughly less than 30 μm. Also, many of the crack segments with a crack opening below 15–20 μm showed pronounced self-healing.

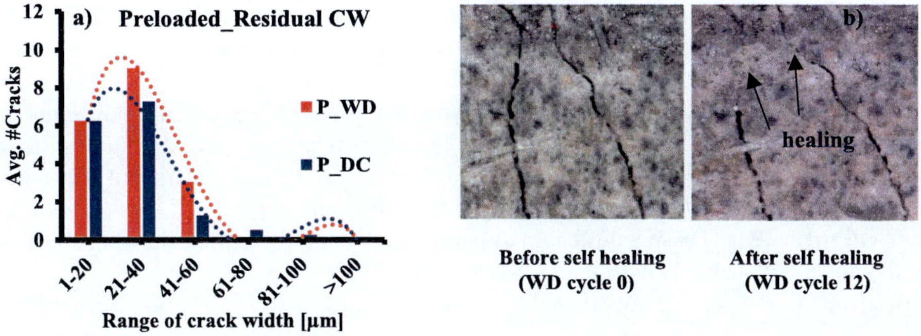

Fig. 6. Residual crack width in preloaded samples for DC and WD curing and b) crack healing due to exposure to WD cycles.

Furthermore, the DC samples retained their first crack strength while the WD samples showed even slightly higher first crack strength than the preloaded and reference

Table 3. Average values of the mechanical parameters and DIC analysis for the SHCL4 subject to various curing conditions. The standard deviation is given in the parenthesis.

Specimen	σ_{fc} [MPa]	σ_t [MPa]	ε_{ult} [%]	CW [μm]	Crack spacing [mm]
Ref_SHLC4	3.6 (0.3)	6.4 (0.6)	4.0 (0.6)	105.0 (6.2)	2.7 (0.4)
P_DC_SHLC4	3.3 (0.5)	–	–	25.1 (1.4)*	5.8 (1.5)*
R_DC_SHLC4	3.3 (0.4)	6.2 (0.6)	5.0 (1.4)	104.3 (2.7)	2.2 (0.7)
P_WD_SHLC4	2.9 (0.7)	–	–	30.7 (3.1)*	5.4 (1.1)*
R_WD_SHLC4	3.9 (0.2)	5.9 (0.3)	2.8 (0.6)	91.7 (12.1)	3.4 (0.6)

* values measured from optical microscope in unloaded state.

specimens. In this case, reloaded samples, the first crack refers to the formation of additional crack(s), besides the already existing cracks, in the undamaged matrix parts. Since the first crack stress is a function of matrix strength exclusively, the continued hydration triggered by the WD cycles leads to an increase in uncracked matrix strength, hence the formation of additional cracks was at a higher stress level in WD reloaded samples. While in DC cured samples, the quick loss of bound water due to available cracks ceased the further hydration and resulted in preserving the original matrix strength.

The reloaded DC and WD samples retained 97% and 92% of their tensile strength, respectively, in comparison to the 28d reference specimen. The strain capacity exceeded the reference specimens by 25% for the reloaded DC samples, while the reloaded WD experienced a 30% decrease in the strain capacity compared to the reference. This decrease can be attributed to the continued hydration due to WD cycles, resulting in a stronger fiber-matrix bond, particularly important in pre-damaged fiber areas. Also, the formation of healing products at the interface can result in stronger fiber/matrix interlocking. A higher pull-out force might result in a predominant fiber rupture causing fracture localization in the SHLC4 at a lower ultimate strain. This smaller strain can be directly related to the smaller crack widths of 91.7 μm as average depicted by DIC analysis at the ultimate strength, in comparison to the reference case with an average crack width of 105 μm. In contrast, for reloaded DC samples the synergetic action between a weaker fiber/matrix interface due to pre-damaging resulted in more pronounced fiber debonding and partial fiber pull-out which led to higher ultimate strain. Future studies should investigate the effect of the curing conditions at various scales to quantify the bond strength and crack bridging strength using single fiber pull-out test and tension tests on notched specimens. These would help to better understand the self-healing capabilities of the novel SHCC made with the LC3 binder (Fig. 7).

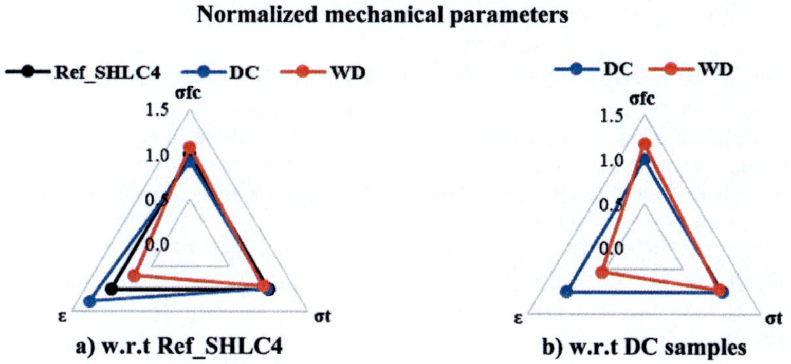

Fig. 7. Normalized mechanical parameters a) for different curing conditions with respect to reference specimens, and b) with respect to dry-cured samples.

4 Conclusion

The effects of dry curing and wet-dry cycle curing were investigated with respect to the uniaxial tensile behavior of SHCC made with LC3 binder and PE fibers. The results suggest that even at a high level of pre-damaging, i.e., 1% tensile strain, the specimens retained their original tensile strength. After 90 days of exposure, the tensile strength of pre-damaged specimens was similar to that obtained for the reference specimens at an age of 28 days.

Moreover, self-healing only took place in the presence of water. Very pronounced crack healing was observed in the section of the crack openings below 15 to 20 μm, however, the healing products were formed in crack sections below 30 μm. A slight recovery in the stiffness of pre-damaged WD samples was noticed and it can be traced back to the additionally formed healing products in the crack flanks and fiber/matrix interface. Matrix cracking strength increased significantly when exposed to WD cycles, indicating continued hydration of the unhydrated cement and supplementary cementitious materials inside the composite. The recovery in mechanical parameters and self-healing phenomena occur only when specimens were exposed to wet-dry cycles. In the case of dry curing, all the pre-damaged specimens retained their original and inherent mechanical response.

Acknowledgment. The authors greatly acknowledge the financial support by the German Research Foundation (DFG) for funding the project 455631638. Moreover, this research was co-financed by tax funds on the basis of the budget adopted by the Saxon State Parliament.

References

1. Li, V. C.: On Engineered cementitious composites (ECC) a eeview of the material and its applications. J. Adv. Concr. Technol. **1** (2003)
2. Li, V.C.: From micromechanics to structural engineering - the design of cementitious composites for civil engineering applications. Struct. Eng. Eng. **10**, 1–34 (1994)

3. Scrivener, K., Martirena, F., Bishnoi, S., Maity, S.: Calcined clay limestone cements (LC3). Cem. Concr. Res. **114**, 49–56 (2018)
4. Zunino, F., Scrivener, K.: The reaction between metakaolin and limestone and its effect in porosity refinement and mechanical properties. Cem. Concr. Res. **140** (2021)
5. Sharma, M., Bishnoi, S., Martirena, F., Scrivener, K.: Limestone calcined clay cement and concrete: A state-of-the-art review. Cem. Concr. Res. **149** (2021)
6. Nguyễn, H.H., et al.: Autogenous healing of high strength engineered cementitious composites (ECC) using calcium-containing binders. Constr. Build. Mater. **265** (2020)
7. Li, V.C., Yang, E.: Self Healing in Concrete Materials, pp. 161–193 (2007)
8. Zhang, P., et al.: Self-healing behaviour of multiple microcracks of strain hardening cementitious composites (SHCC). Constr. Build. Mater. **169**, 705–715 (2018)
9. Ma, H., Herbert, E., Ohno, M., Li, V.C.: Scale-linking model of self-healing and stiffness recovery in Engineered Cementitious Composites (ECC). Cem. Concr. Compos. **95**, 1–9 (2019)
10. Sisomphon, K., Copuroglu, O., Koenders, E.A.B.: Effect of exposure conditions on self healing behavior of strain hardening cementitious composites incorporating various cementitious materials. Constr. Build. Mater. **42**, 217–224 (2013)
11. Yang, Y., Lepech, M.D., Yang, E.H., Li, V.C.: Autogenous healing of engineered cementitious composites under wet-dry cycles. Cem. Concr. Res. **39**, 382–390 (2009)
12. Zhu, H., Zhang, D., Wang, T., Wu, H., Li, V. C.: Mechanical and self-healing behavior of low carbon engineered cementitious composites reinforced with PP-fibers. Constr. Build. Mater. **259** (2020)
13. Wu, M., Johannesson, B., Geiker, M.: A review : Self-healing in cementitious materials and engineered cementitious composite as a self-healing material. Constr. Build. Mater. **28**, 571–583 (2012)
14. Eurofibers.: Fact Sheet, Ultra High Molecular Weight Polyethylene Fiber From DSM Dyneema. https://issuu.com/eurofibers/docs/name8f0d44 (2010)
15. Wang, L., et al.: On the use of limestone calcined clay cement (LC3) in high-strength strain-hardening cement-based composites (HS-SHCC). Cem. Concr. Res. (2021)
16. Curosu, I.: Influence of fiber type and matrix composition on the tensile behavior of strain-hardening cement-based composites (shcc) under impact loading zum einfluss der faserart und matrixzusammensetzung auf das zugverhalten von (2017)
17. Curosu, I., Mechtcherine, V., Forni, D., Cadoni, E.: Performance of various strain-hardening cement-based composites (SHCC) subject to uniaxial impact tensile loading. Cem. Concr. Res. **102**, 16–28 (2017)
18. Zhang, D., Jaworska, B., Zhu, H., Dahlquist, K., Li, V.C.: Engineered Cementitious Composites (ECC) with limestone calcined clay cement (LC3). Cem. Concr. Compos. **114**, 103766 (2020)

The Use of Ultra-high Volume of Lime Stone Calcine Clay to Produce Basalt Fiber Reinforced Strain Hardening Cementitious Composites

Avik Kumar Das[1,2](✉) and Christopher K Y Leung[2]

[1] Shenzhen International Graduate School, Tsinghua University, Shenzhen, China
akdas@connect.ust.hk

[2] Hong Kong University of Science and Technology, Hong Kong, China

Abstract. One of the remarkable attributes of strain hardening cementitious composites (SHCCs) is the hardening region of the material. Because of this even if the material is suddenly overloaded, the structure may still be functional. One of the major component which drives the hardening behavior of the material is the fiber. The problems with polymer fibers which is the commonly used for making SHCCs is that they have a low melting point. Unlike polymer fibers, basalt fibers have high melting point thus, more resistant to fire. In this work, we investigated usage of basalt fibers in making SHCCs. More specifically, SHCCs developed through a combination of basalt fiber with ultra high volume LCC cement blend (UHVLC3-BF-SHCC). We investigated the effect of composition of ultra high volume of SCM in the binder on the mechanical performance UHVLC3-BF-SHCC. We observe that UHVLC3-BF-SHCC has ~ 1% tensile strain while cracks widths are controlled to very value of 20 microns and below. By improving mechanical properties, UHVLC3-BF-SHCC may be an attractive green SHCC for applications subjected to high temperatures.

Keywords: Basalt Fiber · Ultra high volume LC3 · Strain Hardening Cementitious Composites · Sustainable Concrete

1 Introduction

Strain hardening cementitious composites (SHCCs)/ Engineered Cementitious Composites(ECCs) are a special type of fiber reinforced composites which shows few percent of strain hardening through opening of multiple cracks, which are intrinsically controlled to small value. [1]. This achieves many beneficial properties such as sudden overloading does not lead to failure[2], lowers the transport of deleterious material [3–5] and promotes autogenous self-healing [6–8]. The utilization of SHCC had been seriously contemplated to enable durable civil infrastructure to exploit these beneficial properties [9, 10]. Research has shown that by integrating SHCCs judiciously it is possible to get a significant reduction of life cycle emissions of greenhouse gasses. This aligns itself within the rising concern of sustainable constructions [11–13]. Many of these remarkable properties can be accreted to the presence of the fibers. Polymer fibers such as

M. Kunieda et al. (Eds.): SHCC 2022, RILEM Bookseries 39, pp. 13–22, 2023.
https://doi.org/10.1007/978-3-031-15805-6_2

polypropylene (PP), polyvinyl alcohol (PVA), poly ethylene (PE) are common types of fiber used in SHCCs. Compared to polymer fibers basalt fiber has substantially higher melting point thus, more resistant to fire.

The developmental history, science and various applications of SHCCs in references [1, 14]. Typically the matrix of SHCCs does not incorporate coarse aggregates [14], thus requires relatively large amount of fine aggregates leading to higher carbon footprint. A complementary approach towards greener construction material is by supplementing the key ingredients such as cement, with supplementary cementitious materials (SCMs) are reduces overall greenhouse gas emissions and makes construction sustainable. It has been said that among other supplementary materials such as fly ash, the availability of limestone and calcine clay is more equitably distributed and abundant to match the volume of the concrete consumption of the world. In the literature using of high volume of SCMs (50 percent or higher) has been utilized with polymer fibers to create green SHCCs reviewed in ref. [15]. In this work, we investigated the possibility of basalt fiber SHCC with ultra-high volume LCC (80 percent or higher) cement blend to develop a novel SHCC called UHVLC3-BF-SHCC.

2 Materials and Testing

2.1 Constituents

The mortar used in this study are composed of cement, silica fumes, limestone and calcine clay blend (LCC blend), silica sand as filler and basalt fiber. The cement is Portland cement (OPC) grade 52.5, LC2 blend was prepared by 2 parts of calcine clay with 1 part of limestone. Particle size of this well graded silica sand is 80–120 μm. In this work, our focus is on the possibility of using ultra high volume of supplementary cementitious materials is investigated. Therefore, in all cases the amount of SCMs are 80 percent or higher. The water/binder ratio for this study is fixed at 0.4. The SP used here is ADVA 189 and amount is added to ensure enough workability for casting. In all cases, the fiber volume is 2.25%. Previous research has shown that fillers play an important role in determining the mechanical performance of the matrix. In order to elucidate effect of mechanical performance by changes in relative concentration of SCMs in such a high volume only a nominal quantity of the silica sand is utilized, and this quantity was kept constant in all cases. The matrix design is reported in Table 1.

2.2 Specimen Preparation

In the first step the binder ingredients weighted and put into a 4-L Hobart mixer for 2min dry mixing in a lower speed. This is done for binders are homogenously mixed. Then moisture is added to the dry mix in form of superplasticizer (SP) and water. We have created a homogenous mix containing part of the super-plasticizer (SP) with tap water. This mix was then poured into mixer for 4-6min wet mixing in a higher speed. BF was then added slowly to the mixed until the fibers are well dispersed. After dispersion was complete residual fluid is added to ensure good workability and homogeneity for casting the mix. The mix was then cast in molds prepared. In casting, mixture was added part

Table 1. Matrix Design of UHVLC3-BF SHCC

Mix ID	Binder				Silica Sand	Water
	OPC	LCC	Fly ash	Silica fume		
L80C10SF10	0.1	0.8	–	0.1		0.4
L85C5SF10	0.05	0.85	–	0.1	–	0.4
L85C7SF8	0.07	0.85	–	0.08	–	0.4
L85C10SF5	0.1	0.85	–	0.05	–	0.4
F80C10SF10	0.1	–	0.8	0.1	–	0.4

by part from the top side these molds. We then use fingers to to fill mold to ensure good dispersion of fibers. Then specimens were covered with plastic sheets and steel plates to make surface of samples is flat. After 24 h, specimens become set enough to demold. These specimens were than demolded and kept for curing. In our previous experience it was found that curing in the hot water tank at 60 °C reaches 28 days equivalent curing at a much lower time. The dimension of the specimen should also be mentioned. For tensile test the dimensions were 350×25×10 mm while that for cube test was 40×40×40 mm respectively.

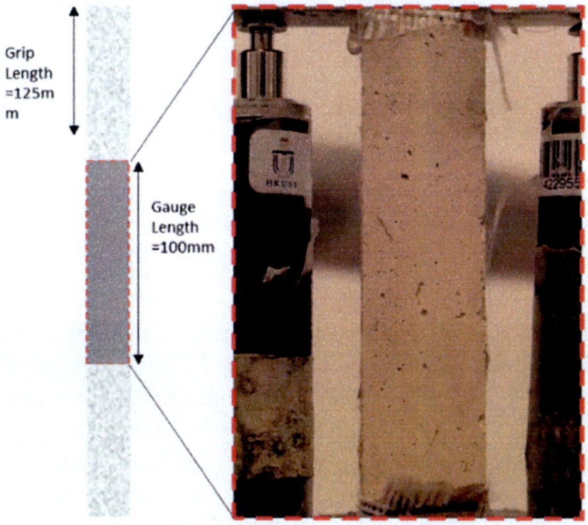

Fig. 1. Test setup for tensile testing

2.3 Testing Details

Tensile Test: Sample preparation and testing technique is based on our experience of testing such samples. To ensure sufficient length at the two ends for proper gripping of the specimen during testing, the middle gauge length is 100mm as shown in Fig. 1. During testing, two external linear variable displacement transducer (LVDT) was attached on the sides of the middle part of the tensile specimen to measure the elongation (Fig. 1), for calculating the tensile strain. The tensile test was performed in a Lloyd-Ametek EZ50 (50kN) testing machine, with a loading rate of 0.2 mm/min. The grips of the machine restrained the rotation and transverse displacements of the specimen both ends can be considered fixed. Test was stopped when loading reach to at least 0.8 post-peak.

Compression Test: The compressive strength test follows the code (ASTM, 2013). Three cubic samples after curing were measured 40 mm × 40 mm × 40 mm dimension. The load rate setting in testing machine is 0.5 kN/s (Fig. 2).

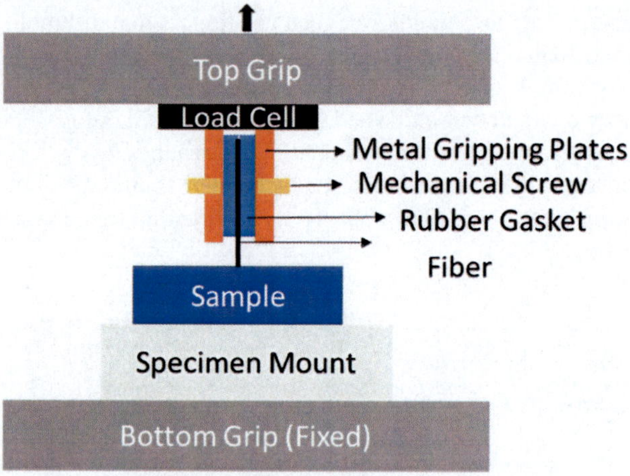

Fig. 2. Schematic of fiber pull out test

Fiber Pull Out Test: For the single fiber pullout tests, a single basalt fiber was embedded into the matrix with an embedded length of 2 mm. The specimen dimensions and the test setup of the fiber pullout test are shown in Fig. 3. This specimen after casting went through the same curing scheme as that of other samples. Before testing, the bottom of the specimen was glued with AA glue to a heavy metal plate. The free end i.e. the fiber end was tightly hold within a mechanical clamp. To facilitate gentle clamping and avoid sliding of the thin fiber rubber pads were attached to the clamps. A load cell was used to monitor the pullout load and the pullout distance of fibers, respectively. During the test a displacement rate of 0.1mm/min was applied.

3 Results and Discussions

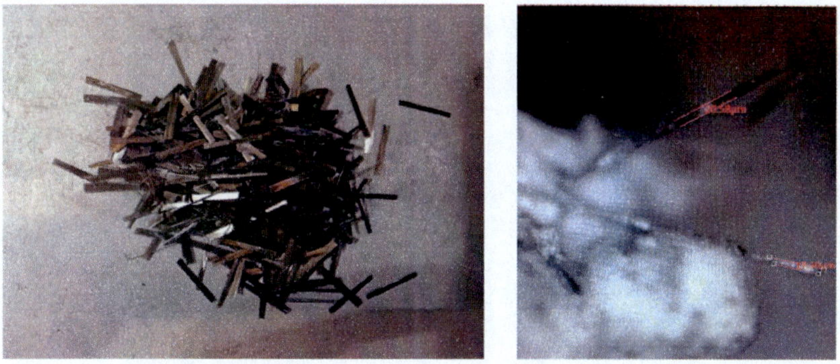

(a) (b)

Fig. 3. a) Bundles of basalt fiber b) Fiber diameter from optical microscopy

3.1 Fiber Properties

Basalt fiber used in this test are bundled by polymers as shown in Fig. 3. Therefore, a novel sequence of mixing technique was used as described in Sect. 2 for proper fiber dispersion. We investigated the results of the this dispersion through optical microscopy. With respect to the results in Fig. 3 shows that this technique is successful in separating the bundles of fiber as well as dispersing it well within the matrix. With this technique, geometrical properties were measured (Fig. 3). The nominal fiber properties are reported in Table 2.

Table 2. Nominal fiber properties

Length (mm)	Diameter (μm)	Elastic modulus (GPa)	Fiber strength (MPa)	Fiber density (g/cm^3)
9	16	91	4800	2.65

3.2 Compressive Strength

The compressive strength is reported in the following chart Fig. 4. It is expectedly observed that in all cases the strength of the matrix is fairly low. This stems from the fact that the matrix lacks filler i.e. sand. Generally, speaking the results show that in general increasing cement content increases compressive strength.

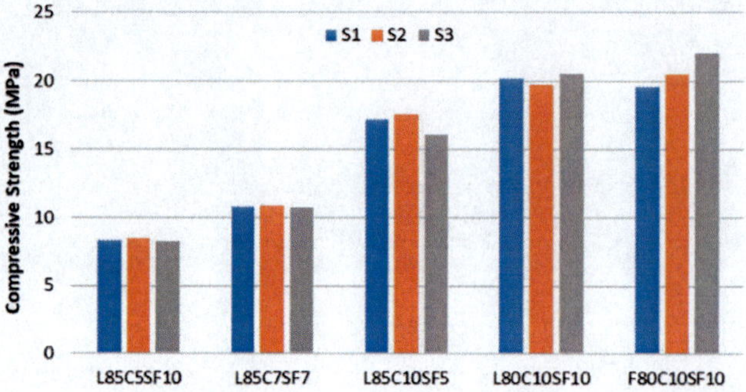

Fig. 4. Compressive test result

3.3 Pullout Results

The samples in the pullout consists are of single fibers. This fiber are embedded perpendicular to the matrix. The fibers are then pulled out using the test process described in the Sect. 2C. The fiber embedding depth was approximately 2.5mm. It should be noted that pullout test are performed on LC80C10SF10 matrix. The results are reported in the Fig. 6. By matching this results with that in the literature. It seems the fiber-matrix interface seems to showcase constant friction or slip softening but barely any slip-hrdening was observed.

3.4 Tensile Test Result

Ultimate strain capacity and ultimate tensile stress capacity is calculated from the tensile test graph is shown in Fig. 5A and Fig. 5B respectively. This strength also shows a decreasing trend as cement content in UHVLC3 matrix decreases. Conversely, for the strain capacity the behavior is opposite, which is expected. That being said in both of.

these cases, the trends are not very significant. This could be because, tensile properties in these SHCCs are generally governed by fiber-matrix interface because the matrix in all these cases are weak (see the compressive strength), and because majority of the matrix is LC blend therefore, small changes in proportion of other binding component may not affect the overall behavior significantly.

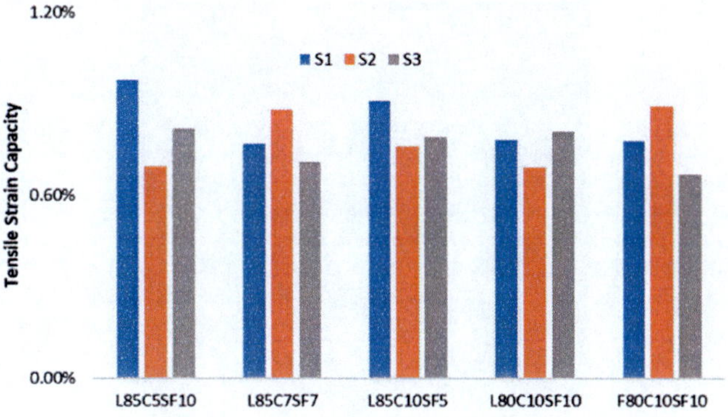

(a)

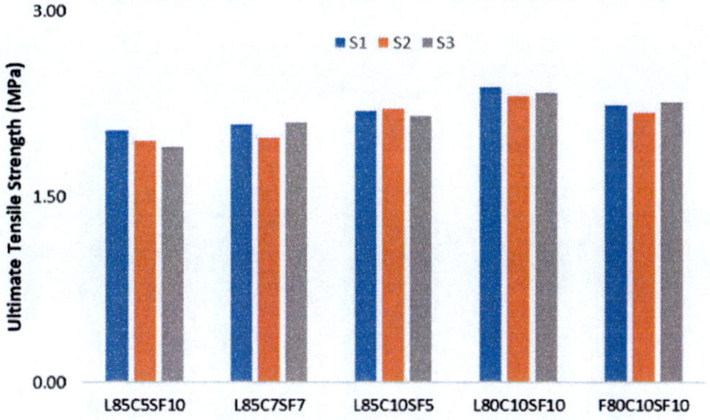

(b)

Fig. 5. A) Strain capacity B) Tensile strength of different mixes

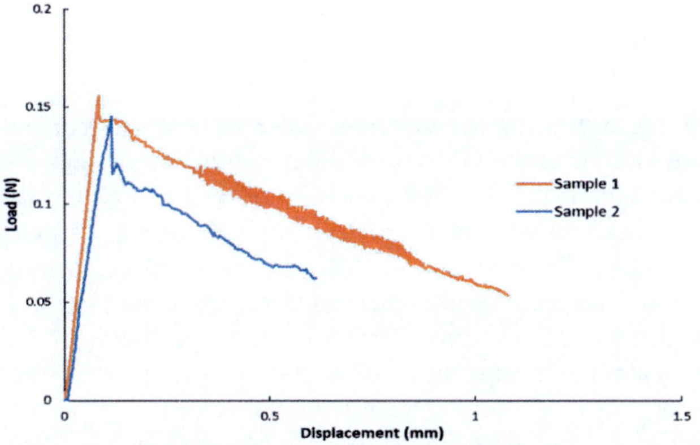

Fig. 6. Single fiber pullout results

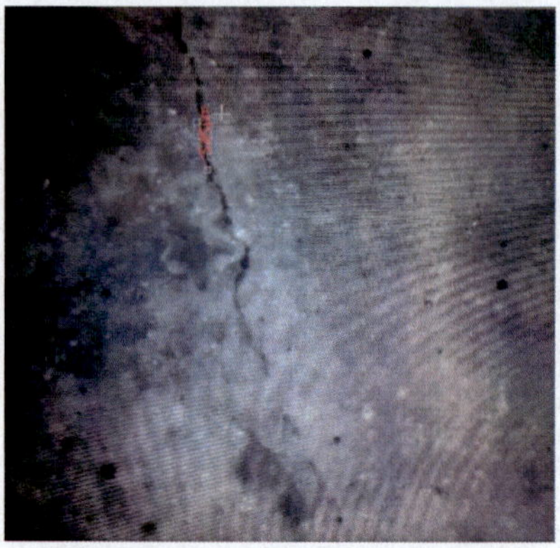

Fig. 7. Surface crack measurement through optical microscopy

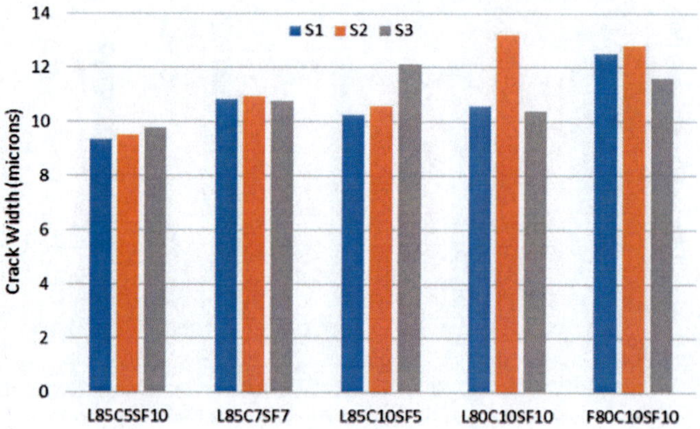

Fig. 8. Average Crack Widths of the samples

3.5 Durability Assessment

Generally speaking, the transfer of deleterious materials within the concrete is associated with the deterioration of concrete. Experimental results thus associate durability with crack widths in the sample [14, 16]. Because the micro cracks are extremely thin in these specimens therefore, we investigate the crack widths with the help of optical microscope. After the tension test is finished the samples were unloaded and then investigated for optical microscopy. One of the results from this test is reported in Fig. 7, in this example the measured width is just 5.6 microns. Figure 8 reports the finding of the average width of the samples. In almost all cases the crack widths are ~ 10 microns. It should be noted that the widths are measured after the sample was unloaded therefore, crack width under load will be higher. A non-significant trend was observed that with decreasing cement content the crack width decreases this could be correlated with the matrix strength.

4 Conclusions

Strain hardening cementitious composites (SHCCs) are being used in many structural applications to create more durable infrastructures. In this work we study the possibility of using basalt fiber (BF) and ultra-high volume limestone calcine clay blend for creating SHCCs (UHVLC3-BF-SHCC). A novel mixing sequence is introduced to separate and disperse the polymer coated BF fibers. We studied mechanical and durability properties. The results are as follow. Expectedly with decreasing content of cement the compressive strength decreases; however, after a critical value a significant drop was observed. No such thing was observed in tensile test result, in all cases the ultimate strength was approx. 2–2.5 MPa and strain capacity ~ 0.7–1%. This is because in weak UHV-LC3 small changes in binding component does not significantly affect these results. The fiber pull-out test also confirms that for these fibers no slip hardening behavior was observed. The durability of SHCCs is associated with the characteristics of surface cracks. It was found that in all cases the average crack widths are ~ 10 microns. A non-significant decrease in average widths with decreasing compressive strength was observed. Further experiments will be undertaken to better understand the implications of the BF in LC3 matrix. In the future, further tailoring of the mechanical properties may be required to make it a good alternative for structural applications.

References

1. Li, V.C.: Engineered Cementitious Composites (ECC): Bendable Concrete for Sustainable and Resilient Infrastructure, 1st ed. (2019) https://doi.org/10.1007/978-3-662-58438-5
2. Das, A.K., Leung, C.K.Y.: A fundamental method for prediction of failure of strain hardening cementitious composites without prior information. Cement Concr. Compos. **114**, 103745 (2020). https://doi.org/10.1016/j.cemconcomp.2020.103745
3. Wang, K., Jansen, D.C., Shah, S.P., Karr, A.F.: Permeability study of cracked concrete. Cem. Concr. Res. **27**(3), 381–393 (1997). https://doi.org/10.1016/S0008-8846(97)00031-8
4. Djerbi, A., Bonnet, S., Khelidj, A., Baroghel-Bouny, V.: Influence of traversing crack on chloride diffusion into concrete. Cement Concr. Res. **38**(6) (2008) https://doi.org/10.1016/j.cemconres.2007.10.007
5. Lepech, M.D., Li, V.C.: Water permeability of engineered cementitious composites. Cement Concr. Compos. **31**(10), 744–753 (2009). https://doi.org/10.1016/j.cemconcomp.2009.07.002
6. Herbert, E.: Development and Application of Self-healing Engineered Cementitious Composites (ECC) for Durable and Sustainable Infrastructure PhD (2016)
7. Li, V., Herbert, E.: Robust self-healing concrete for sustainable infrastructure. J. Adv. Conc. Technol. **10**(6) (2012) https://doi.org/10.3151/jact.10.207.
8. Das, A.K., Leung, C.K.Y.: A strategy for in situ determination of self-healing state for strain hardening cementitious composites. Cement Concr. Compos. **112**, 103641 (2020). https://doi.org/10.1016/j.cemconcomp.2020.103641
9. Lepech, M.D., Li, V.C.: Design and field demonstration of ECC link slabs for jointless bridge decks (2005) https://www.michigan.gov/documents/mdot/MDOT_Research_Report_RC1471_200102_7.pdf
10. Rokugo, K: Applications of SHCC in Japan - Tools and Tips for Promoting its Use (2017)

11. Das, A.K., Mishra, D.K., Jing, Y., Leung, C.K.Y.: Smart self-healing and self-sensing cementitious composites—recent developments, challenges, and prospects. Adv. Civil Eng. Mater. **8**(3), 20190023 (2019). https://doi.org/10.1520/ACEM20190023
12. Keoleian, G.A.: Life cycle modeling of concrete bridge design: comparison of Engineered cementitious composite Link Slabs and Conventional Steel Expansion Joints. J. Infrastr. Syst. **11**(1) (2005) https://doi.org/10.1061/(ASCE)1076-0342(2005)11:1(51)
13. Lepech, M.D., Li, V.C.: Long term durability performance of engineered cementitious composites /LangzeitbestÃ¤ndigkeit systematisch entwickelter zusammengesetzter Zement gebundener Werkstoffe. Restoration of Buildings and Monuments **12** (2014) https://doi.org/10.1515/rbm-2006-6038
14. van Zijl, Gideon, P.A.G., Slowik, V.: A Framework for Durability Design with Strain-Hardening Cement-Based Composites (SHCC): State-of-the-Art Report of the RILEM Technical Committee 240-FDS, vol. 22 (2017) https://doi.org/10.1007/978-94-024-1013-6
15. Shoji, D., He, Z., Zhang, D., Li, V.C.: The greening of engineered cementitious composites (ECC): A review. Constr. Build. Mater. **327**, 126701 (2022). https://doi.org/10.1016/j.conbuildmat.2022.126701
16. Das, A.K., Christopher, K.Y.: Application of deep convolutional neural networks for automated and rapid identification and computation of crack statistics of thin cracks in strain hardening cementitious composites (SHCCs). Cement Concr. Compos. **122**, 104159 (2021). https://doi.org/10.1016/j.cemconcomp.2021.104159

Utilization of Artificial Geopolymer Aggregates in High-Strength Engineered Cementitious Composites (HS-ECC)

Ling-Yu Xu, Bo-Tao Huang(✉), and Jian-Guo Dai(✉)

The Hong Kong Polytechnic University, Kowloon, Hong Kong, China
botaohuang@zju.edu.cn, cejgdai@polyu.edu.hk

Abstract. In this study, high-strength high-ductility Engineered/Strain-Hardening Cementitious Composites (ECC/SHCC) were developed with the combined use of ultra-high-strength cementitious matrix, artificial geopolymer aggregates (GPA), and ultra-high-molecular-weight (UHMW) polyethylene (PE) fibers. Apart from short-term characteristics, the long-term mechanical properties of GPA-ECC were evaluated by an accelerated aging test. It was found that GPA could behave as "additional flaws" in the high-strength matrix, leading to a better strain-hardening ability of ECC. Compared with fine silica sand ECC (FSS-ECC) whose strength indices increased but both tensile ductility and crack resistance decreased after accelerated aging, GPA-ECC showed improved long-term performances in all aspects. Furthermore, the multiple cracks were found to propagate through GPA in GPA-ECC, and the microhardness analysis revealed that the hardness growth of GPA was slower than that of cementitious matrix during the accelerated ageing test, ensuring the role of GPA as "additional flaws" in improving the long-term performance of the ECC material.

Keywords: Strain-Hardening Cementitious Composites (SHCC) · Engineered Cementitious Composites (ECC) · Geopolymer aggregates (GPA) · Ductility · Flaw effect

1 Introduction

As a special category of cement-based materials, Engineered Cementitious Composites (ECC) [1, 2], which are alternatively termed as "Strain-Hardening Cementitious Composites" (SHCC) [3, 4] or "Ultra-High Toughness Cementitious Composites" [5, 6], can possess outstanding tensile ductility, multiple cracking, and strain-hardening behavior on the basis of micromechanical design principles [1, 7]. In recent years, following the research trends of high-performance and green concrete developments, substantial efforts have been focused on further improving the mechanical properties, durability, and sustainability of ECC as well [18]. By combining the ultra-high-performance concrete (UHPC) technology with the micromechanics theory of ECC, high-strength ECC (HS-ECC) with a compressive strength of 120–210 MPa and a tensile ductility over 3% have been achieved [8, 9]. However, exploitation and production of fine silica sand

M. Kunieda et al. (Eds.): SHCC 2022, RILEM Bookseries 39, pp. 23–33, 2023.
https://doi.org/10.1007/978-3-031-15805-6_3

(FSS), which is usually utilized in HS-ECC, not only causes irreversible damage to the environment, but also increases the cost of HS-ECC materials.

Geopolymer aggregates (GPA) is a kind of the recently developed waste-based artificial aggregates, which can provide an attractive one-stone two-bird solution to reduce the natural rock excavation and relieve the burden of waste landfills on the environment [10]. As solid wastes (e.g., fly ash, red mud, palm oil fuel ash, and rice husk ash) can be taken as raw materials and only ambient temperature is required for the GPA production, such type of aggregates can possess the advantages of cost efficiency, high sustainability and low self-weight, although they inevitably have relatively lower strength and rigidity compared to natural aggregates [11]. Therefore, considering the safety issues, some limitations may be imposed on the practical applications of GPA in concrete structures. However, taking advantage of the comparatively low strength and modulus of GPA, both the cracking strength and fracture toughness of HS-ECC matrix can be lowered, which facilitates saturated multiple cracking phenomena and enhances the tensile strain capacity of ECC.

In this study, GPA were innovatively utilized as the replacement of FSS to develop HS-ECC. As the knowledge of the long-term performance of ECC is critical for the practical application [12], both the short- and long-term mechanical performances of GPA-ECC were investigated under normal curing and accelerated aging conditions, respectively. Furthermore, X-ray computed tomography (X-CT) and microhardness tests were conducted to assess the role of GPA in the HS-ECC matrix. The results of this study demonstrated the feasibility of using GPA to produce high-strength high-ductility ECC with the stabilized long-term mechanical performance.

2 Experimental Program

2.1 Raw Materials

Class F fly ash (FA) and ground granulated blast-furnace slag (GGBS) were used as the precursors to produce GPA while Type I 52.5 N Portland cement and silica fume (SF) were utilized as the binder materials for ECC production. Anhydrous sodium metasilicate (Na_2SiO_3-Anhydrous) particles in industrial grade, whose silica modulus [M = Mol (SiO_2)/Mol (Na_2O)] was 0.94, were taken as the alkaline activator for GPA formation. Ultra-high-molecular-weight PE fibers with the strength, modulus, and density of 3000 MPa, 100 GPa, and 0.97 g/cm^3 were used as the reinforcement in HS-ECC matrix. The diameter and length of the PE fibers were 24 μm and 18 mm, respectively. In addition, Polycarboxylate ether type super-plasticizer (SP) with the solid content of 22% was used to increase the flowability of HS-ECC during the mixing procedure.

2.2 GPA Production

For the GPA mix proportion, the precursors were composed of 80% FA and 20% GGBS, with the activator to precursor ratio of 0.12 and the water to precursor ratio of 0.35 by weight. During the production procedure, FA, GGBS, and solid activator were firstly dry-mixed together for 5 min before the water addition. Then, the slurry was stirred for another 4 min, and the fresh pastes were poured into cubic molds (100 mm × 100 mm × 100 mm) and vibrated for 30 s. After 24 h, the specimens were crushed into fragments smaller than 4.75 mm and kept in sealed plastic bags for six-month storage. The particle sizes of GPA for ECC production are shown in Table 1 and the photograph of GPA is presented in Fig. 1. The specific gravity, water absorption and water content of the GPA were tested as 2.05, 24.6% and 19.2%, respectively.

Table 1. Particle size distribution of GPA for ECC casting.

Particle size (mm)	2.36–4.75	1.18–2.36	0.6–1.18	0.30–0.60	0.15–0.30	0–0.15
GPA	30%	28%	13%	10%	7%	12%

Fig. 1. Photograph of geopolymer fine aggregates.

2.3 ECC Production

In addition to the ECC incorporating GPA (GPA-ECC), FSS with the specific gravity of 2.67, water absorption of 0.8%, and average particle size below 300 μm was also adopted to produce fine silica sand ECC (FSS-ECC) as a comparison. The mix proportions of both FSS- and GPA-ECC are listed in Table 2. It is noted that the volume of GPA and FSS were kept the same in the two mixes. Before mixing, both GPA and FSS were pre-wetted to the saturated surface dry (SSD) condition. Cement, SF and fine aggregates were firstly dry-mixed for 5 min, and then water and SP were added with the stirring continued for another 10 min. After the fresh mixture was formed, PE fibers were added and the mixing was continued for 5 min. Afterwards, the mixture was poured into six 50 mm × 50 mm × 50 mm cubic and six dumbbell molds. After 24 h, the specimens were demolded with the demolding density measured (Table 2), and then stored in water (23 °C). After 28 d, half of the specimens (for each mix) were mechanically tested, while the other half were transferred into 60 °C hot water for another four-week curing, which is approximately equivalent to 10.9 years of natural weathering [12]. After the accelerated aging, the specimens were also tested by mechanical methods.

Table 2. Mix proportions of FSS- and GPA-ECC

Raw Materials	FSS-ECC	GPA-ECC
Cement	0.800	0.800
Silica Fume	0.200	0.200
SSD FSS (Dry FSS)	0.250 (0.248)	0
SSD GPA (Dry GPA)	0	0.238 (0.191)
Water	0.171	0.171
Superplasticizer (in solid)	0.013	0.013
PE Fiber (Vol %)	2.0	2.0
Demolding Density (g/cm^3)	2.22	2.12

2.4 Testing Methods

Direct tensile tests and compressive tests were conducted to evaluate the mechanical performances of FSS- and GPA-ECC, with the loading rate of 0.5 mm/min and 1.0 MPa/s, respectively. In order to capture the maximal crack width at the ultimate stage, one digital camera was used in the direct tensile test to photograph the dumbbell specimens at an interval of 3 s. The samples were obtained from the middle portion of short-term GPA-ECC dumbbell specimens for X-CT analysis. Finally, the samples cut from both FSS- and GPA-ECC were polished for microhardness tests.

3 Mechanical Properties

3.1 Tensile Performance

The short- and long-term tensile stress–strain curves of FSS- and GPA-ECC are shown in Fig. 2, and the tensile properties are summarized in Fig. 3. Obviously, GPA-ECC showed a better strain-hardening ability than FSS-ECC. As presented in Fig. 3a, the tensile strength of GPA-ECC was lower than that of FSS-ECC in both short- and long-term conditions probably because the adopted large GPA size (2.36–4.75 mm) negatively influenced the uniformity of fiber distribution in the dumbbell specimens. Besides, after accelerated aging, the cracking and tensile strengths of both FSS- and GPA-ECC increased, due to the more complete hydration of the cementitious matrix and the resultant higher matrix toughness as well as fiber/matrix bond strength. On the other hand, the short-term tensile strain capacity of GPA-ECC (7.6%) was evidently higher than that of FSS-ECC (4.2%), as shown in Fig. 3b. The reason for the above phenomenon was that the utilization of GPA as the replacement of FSS could function as "additional flaws" and reduce of matrix toughness of HS-ECC, which was beneficial for enhancing the multiple cracking behavior of ECC according to the micromechanics theory. It is noted that the matrix toughness means the fracture toughness of GPA-matrix blended composite. In addition, the tensile strain capacity of FSS-ECC dropped to 3.1%, while that of GPA-ECC increased to 8.3% after accelerated aging, which demonstrated a retained role of GPA as "additional flaws" in HS-ECC and an excellent long-term performance of GPA-ECC. In Fig. 3c, the strain energy density values were calculated based on the area under the ascending tensile stress–strain curves. In the short-term condition, GPA-ECC showed a comparatively larger strain energy density than FSS-ECC. However, after accelerated aging, the strain energy density value of FSS-ECC slightly dropped, but that of GPA-ECC significantly increased, which also demonstrated the excellent long-term performance of GPA-ECC.

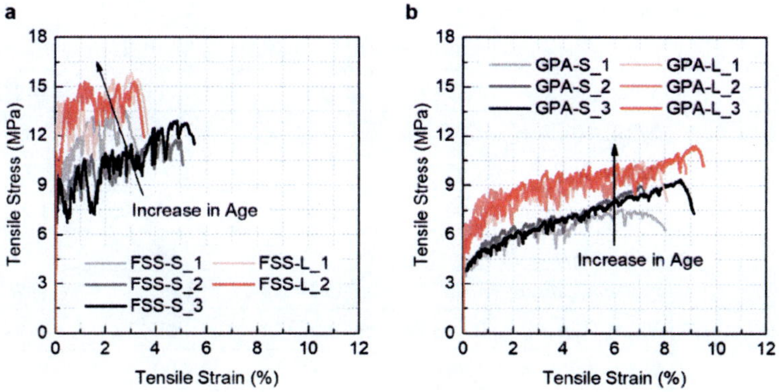

Fig. 2 Tensile stress–strain curves (a) FSS-ECC, and (b) GPA-ECC

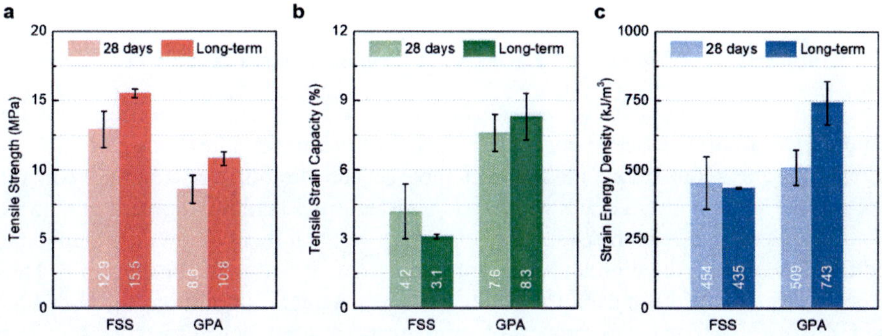

Fig. 3 Comparison of FSS- and GPA-ECC properties (a) tensile strength, (b) tensile strain capacity, and (c) strain energy density

3.2 Compressive Strength

The short- and long-term compressive strengths of FSS- and GPA-ECC are listed in Table 3. Both the 28-d compressive strengths of FSS- and GPA-ECC are higher than 130 MPa, although the strength value of GPA-ECC was lower than that of FSS-ECC due to the weakness nature of GPA. On the other hand, the strength gain of FSS-ECC was 21.7% due to the further hydration of the cementitious matrix after accelerated aging, which was larger than that of GPA-ECC (11.3%). The reason for the above phenomenon may be that the lower long-term strength gain of GPA compared to the cementitious matrix still could not narrow the strength difference between the two materials, which made GPA retain their function as flaws and resulted in less compressive strength improvement in GPA-ECC.

Table 3. Summary of short- and long-term properties of FSS- and GPA-ECC

Mechanical properties		FSS-ECC		GPA-ECC	
		Short-Term	Long-Term	Short-Term	Long-Term
Compressive strength	Value (MPa)	152.4	185.5	134.0	149.2
	Normalization	1.00	1.22	1.00	1.11
Tensile strength	Value (MPa)	12.9	15.5	8.6	10.8
	Normalization	1.00	1.20	1.00	1.26
Tensile ductility	Value (%)	4.20	3.09	7.64	8.31
	Normalization	1.00	0.74	1.00	1.09
Crack resistance, $1/w_{max}$	Value (1/μm)	0.00399	0.00344	0.00426	0.00459
	Normalization	1.00	0.86	1.00	1.08

Note: Normalization means the values normalized by their respective short-term properties

3.3 Comparison of Short- and Long-Term Overall Performance

After the mechanical performance tests, the short- and long-term properties, including compressive strength, tensile strength, tensile ductility and crack resistance of FSS- and GPA-ECC were summarized and listed in Table 3. The durability of structural concrete is related to the maximum crack width, and a larger crack width will result in a poorer durability. Here, the crack resistance is defined by the reciprocal of the maximum crack width obtained from the digital photograph captured at the ultimate tensile strain. Thus, the smaller crack width means the better crack resistance. In addition, for both FSS- and GPA-ECC, the long-term properties were normalized by their respective short-term properties, and the results were presented in a radar graph (Fig. 4). Obviously, although FSS-ECC showed increased long-term strength indices, its long-term tensile ductility and crack resistance decreased. In comparison, improvements were observed in all the four properties of GPA-ECC after accelerated aging, which proved their excellent long-term characteristics.

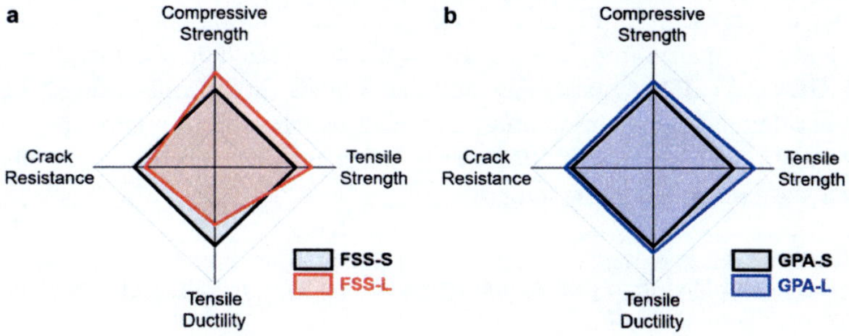

Fig. 4 Radar graphs of short- and long-term performances (a) FSS-ECC, and (b) GPA-ECC

4 Ductility Enhancement Mechanism of GPA in ECC

On the basis of the findings in mechanical tests, GPA can function as "additional flaws" in GPA-ECC due to their weaker strength compared to the cementitious matrix. Here, in order to further demonstrate the ductility enhancement mechanism of GPA in HS-ECC, X-CT analysis was conducted after short-term curing, and the respective two-dimensional X-CT image of GPA-ECC is shown in Fig. 5. As presented in the figure, the initial flaws (voids, in black), GPA (in dark grey) and cementitious matrix (in light grey) can be easily distinguished. In addition, from the cracking pattern observed, the multiple cracks mainly passed through the initial flaws and GPA, which demonstrated the role of GPA as "additional flaws" in helping induce more saturated cracking. Furthermore, almost no cracks propagated through the GPA/matrix interface, which indicated an excellent bond performance achieved between GPA and the cementitious matrix.

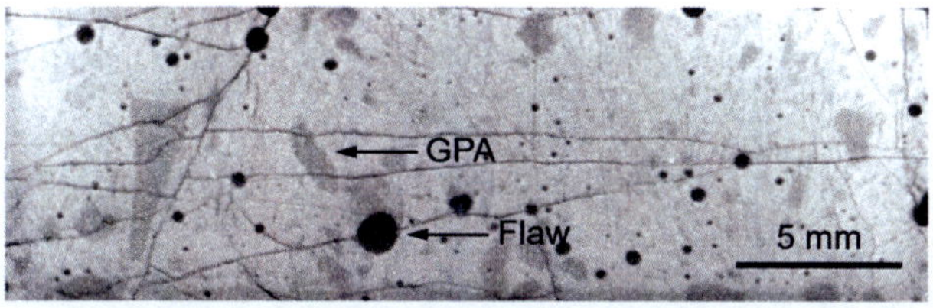

Fig. 5 Two-dimensional X-CT image of GPA-ECC after short-term curing

In order to further demonstrate the flaw effect of GPA, the short- and long-term hardness values of GPA, GPA/matrix interface and the cementitious matrix obtained from microhardness tests were analyzed by the Weibull distribution function. It is noted that from the optical microscope images, a boundary between GPA and matrix can be clearly distinguished, which can be defined as the interface between GPA and the cementitious matrix. All the fitting results exhibited good correlation coefficients over 0.98, as seen in Table 4 and Fig. 6. For the short-term case, the average microhardness value of GPA (54.0 HV) was lower than those of both GPA/matrix interface (78.1 HV) and cementitious matrix (85.6 HV). After the accelerated aging, the average microhardness values of GPA, GPA/matrix interface, and cementitious matrix all increased, with the distribution curves shifting right as presented in Fig. 6. Considering that the microhardness gain of GPA (20.0%) and GPA/matrix interface (19.2%) was lower than that of cementitious matrix (57.6%), GPA can still be regarded as "additional flaws" in the high-strength ECC even after the accelerated aging. In addition, although the short-term microhardness values of the cementitious matrix and GPA/matrix interface were close to each other, their difference was broadened in the long run.

Table 4. Short- and long-term distribution functions and average values of the microhardness values.

Age	Testing Area	Fitted Cumulative Distribution Function	Correlation Coefficient	Average Value (HV)
Short-term	GPA	$F(x) =$ 1-exp(-(x/58.0)^6.4)	0.990	54.0
	GPA/Matrix Interface	$F(x) =$ 1-exp(-(x/82.2)^9.7)	0.995	78.1
	Matrix	$F(x) =$ 1-exp(-(x/91.3)^7.4)	0.983	85.6
Long-term	GPA	$F(x) =$ 1-exp(-(x/69.2)^7.2)	0.983	64.8
	GPA/Matrix Interface	$F(x) =$ 1-exp(-(x/98.3)^9.1)	0.979	93.1
	Matrix	$F(x) =$ 1-exp(-(x/145.3)^6.1)	0.994	134.9

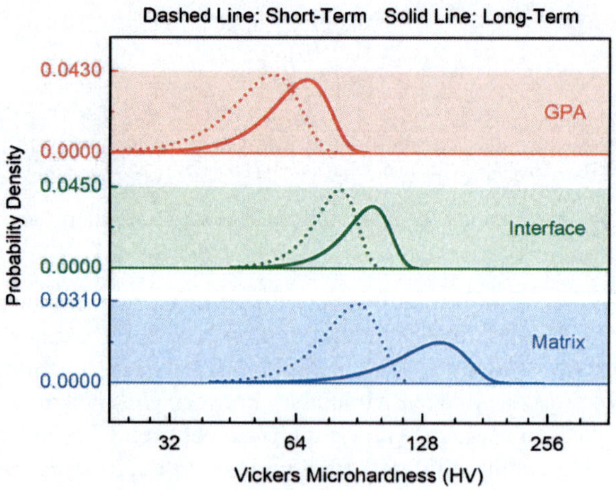

Fig. 6 Comparisons of GPA, GPA/matrix interface and matrix hardness values

Figure 7 presents the schematic graph of the ductility enhancement mechanism of GPA-ECC. For HS-ESS with FSS as aggregates, although initial flaws also exist, the matrix toughness of FSS-ECC is still comparatively high, and the existing flaws cannot promote enough multiple crack initiations in FSS-ECC. When GPA are used as the replacement of FSS, these aggregates cannot only function as fillers with a good bond with the cementitious matrix, but also work as "additional flaws" in the high-strength matrix. With the assistance of low-strength GPA, the fracture toughness of HS-ECC is effectively lowered and more inert initial flaws are activated with more cracks generated. As a result, higher tensile ductility as well as better saturated cracking condition can be achieved in HS-ECC with GPA as fine aggregates. After accelerated aging, although matrix fracture toughness increased, the fiber-bridging stress also increased. For GPA-ECC, because GPA still functioned as additional flaws after accelerated aging, the increase in matrix fracture toughness of GPA-ECC was smaller than that of FSS-ECC. Therefore, due to the improved fiber-bridging stress and a comparatively low fracture toughness of GPA-ECC, it is possible for GPA-ECC to keep the tensile ductility after accelerated aging.

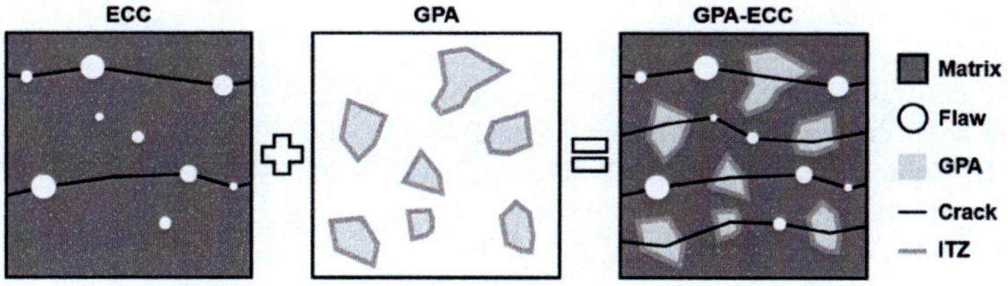

Fig. 7 Ductility enhancement mechanism of GPA-ECC

5 Conclusions

In this study, artificial geopolymer aggregates (GPA) were utilized to develop high-strength high-ductility Engineered Cementitious Composites (ECC). Both the short- and long-term properties of fine silica sand ECC (FSS-ECC) and GPA-ECC were studied by mechanical tests, X-CT analysis, and microhardness tests. From the obtained results, some conclusions are drawn as follows.

(1) The tensile strength of GPA-ECC was lower than that of FSS-ECC in both short- and long-term conditions. In addition, both the tensile strengths of FSS- and GPA-ECC increased after accelerated aging. However, the long-term tensile ductility and strain energy density of FSS-ECC decreased while those of GPA-ECC increased compared to the short-term conditions.
(2) The compressive strength of GPA-ECC was inevitably lower than that of FSS-ECC for both short- and long-term conditions. After accelerated aging, the compressive strengths of both FSS- and GPA-ECC increased due to the further hydration reaction of the cementitious matrix.

(3) After accelerated aging, FSS-ECC showed enhanced strength indices, but the tensile ductility and crack resistance decreased. In comparison, all the long-term mechanical properties of GPA-ECC were improved compared to the short-term ones, indicating the excellent performance of the GPA-ECC after accelerated aging.
(4) From the X-CT results, the cracks were found to propagate through GPA, indicating the role of GPA as "additional flaws" in both short- and long-term conditions, leading to a more saturated multiple cracking in ECC.

More detailed information about GPA-ECC can be found in the authors' recent journal publications (see [13–17]). In addition, as a newly-developed ECC material, the additional investigations on GPA-ECC are being carried out by the authors and will be reported in the future.

References

1. Li, V.C.: Engineered Cementitious Composites (ECC) - Bendable Concrete for Sustainable and Resilient Infrastructure. Verlag GmbH Germany: Springer, Berlin (2019). https://doi.org/10.1007/978-3-662-58438-5
2. Huang, B.T., Wu, J.Q., Yu, J., Dai, J.G., Leung, C.K., Li, V.C.: Seawater sea-sand engineered/strain-hardening cementitious composites (ECC/SHCC): assessment and modeling of crack characteristics. Cement Concr. Res. **140**, 106292 (2021)
3. Qian, S., Zhou, J., De Rooij, M.R., Schlangen, E., Ye, G., Van Breugel, K.: Self-healing behavior of strain hardening cementitious composites incorporating local waste materials. Cement Concr. Compos. **31**(9), 613–621 (2009)
4. Curosu, I., Liebscher, M., Mechtcherine, V., Bellmann, C., Michel, S.: Tensile behavior of high-strength strain-hardening cement-based composites (HS-SHCC) made with high-performance polyethylene, aramid and PBO fibers. Cement Concr. Res. **98**, 71–81 (2017)
5. Huang, B.T., Li, Q.H., Xu, S.L., Zhou, B.: Strengthening of reinforced concrete structure using sprayable fiber-reinforced cementitious composites with high ductility. Compos. Struct. **220**, 940–952 (2019)
6. Huang, B.T., Li, Q.H., Xu, S.L.: Fatigue deformation model of plain and fiber-reinforced concrete based on Weibull function. J. Struct. Eng. **145**(1), 04018234 (2019)
7. Li, V.C., Leung, C.K.: Steady-state and multiple cracking of short random fiber composites. J. Eng. Mech. **118**(11), 2246–2264 (1992)
8. Ranade, R., Li, V.C., Stults, M.D., Heard, W.F., Rushing, T.S.: Composite properties of high-strength, high-ductility concrete. ACI Mater. J. **110**(4), 413–422 (2013)
9. Huang, B.T., Zhu, J.X., Weng, K.F., Li, V.C., Dai, J.G.: Ultra-high-strength engineered/strain-hardening cementitious composites (ECC/SHCC): Material design and effect of fiber hybridization. Cement Concr. Compos. **129**, 104464 (2022)
10. Qian, L.P., Xu, L.Y., Alrefaei, Y., Wang, T., Ishida, T., Dai, J.G.: Artificial alkali-activated aggregates developed from wastes and by-products: a state-of-the-art review. Resour. Conserv. Recycl. **177**, 105971 (2022)
11. Xu, L.Y., Qian, L.P., Huang, B.T., Dai, J.G.: Development of artificial one-part geopolymer lightweight aggregates by crushing technique. J. Clean. Prod. **315**, 128200 (2021)
12. Li, V.C., Horikoshi, T., Ogawa, A., Torigoe, S., Saito, T.: Micromechanics-based durability study of polyvinyl alcohol-engineered cementitious composite. ACI Mater. J. **101**(3), 242–248 (2004)

13. Xu, L.Y., Huang, B.T., Dai, J.G.: Development of engineered cementitious composites (ECC) using artificial fine aggregates. Constr. Build. Mater. **305**, 124742 (2021)
14. Xu, L.Y., Huang, B.T., Li, V.C., Dai, J.G.: High-strength high-ductility engineered/Strain-hardening cementitious composites (ECC/SHCC) incorporating geopolymer fine aggregates. Cement Concr. Compos. **125**, 104296 (2022)
15. Xu, L.Y., Huang, B.T., Lao, J.C., Dai, J.G.: Tailoring strain-hardening behavior of high-strength engineered Cementitious Composites (ECC) using hybrid silica sand and artificial geopolymer aggregates. Mater. Des. **220**, 110876 (2022)
16. Xu, L.Y., Huang, B.T., Lan-Ping, Q., Dai, J.G.: Enhancing long-term tensile performance of Engineered Cementitious Composites (ECC) using sustainable artificial geopolymer aggregates. Cement Concr. Compos. **133**, 104676 (2022)
17. Xu, L.Y., Huang, B.T., Lao, J.C., Yao, J., Li, V.C., Dai, J.G.: Tensile over-saturated cracking of ultra-high-strength engineered cementitious composites (UHS-ECC) with artificial geopolymer aggregates. Cement Concr. Compos. **136**, 104896 (2023). https://doi.org/10.1016/j.cemconcomp.2022.104896
18. Lao, J.C., Huang, B.T., Fang, Y., Xu, L.Y., Dai, J.G., Shah, S.P.: Strain-hardening alkali-activated fly ash/slag composites with ultra-high compressive strength and ultra-high tensile ductility. Cement Concr. Res. **165**, 107075 (2023). https://doi.org/10.1016/j.cemconres.2022.107075

Engineered Geopolymer Composites (EGC) with Ultra-high Strength and Ductility

Jian-Cong Lao[1], Bo-Tao Huang[1](✉), Ling-Yu Xu[1], Jian-Guo Dai[1](✉), and Surendra P. Shah[2,3]

[1] The Hong Kong Polytechnic University, Kowloon, Hong Kong, China
botaohuang@zju.edu.cn, cejgdai@polyu.edu.hk
[2] University of Texas at Arlington, Arlington, TX 76019, USA
[3] Northwestern University, Evanston, IL 60201, USA

Abstract. Engineered Geopolymer Composites (EGC), also known as Strain-Hardening Geopolymer Composites (SHGC), are considered more environmentally friendly than their cement-based counterpart. This study for the first time presents EGC with an ultra-high compressive strength (i.e., over 150 MPa) and an ultra-high tensile ductility (i.e., over 9%) simultaneously. The blended use of fly ash (FA), ground granulated blast slag (GGBS), silica fume, alkali activator, and ultra-high-molecular-weight polyethylene fibers led to the successful development of "Ultra-high-strength & ductility EGC (UHSD-EGC)". The UHSD-EGC were characterized with excellent multiple cracking and strain-hardening features. In addition, it was found that microstructures of FA-rich geopolymer matrix were looser than those with lower FA/GGBS ratios. The findings arising from this study provided a sound basis for developing EGC materials with ultra-high mechanical properties for sustainable and resilient infrastructure.

Keywords: Strain-Hardening Geopolymer Composite (SHGC) · Engineered Geopolymer Composites (EGC) · Compressive strength · Tensile ductility · Multiple cracking

1 Introduction

As a special type of high-performance fiber-reinforced cementitious composites, Engineered Cementitious Composites (ECC) possess a unique feature of strain-hardening capacity and high tensile ductility [1–4]. ECC is also known as Strain-Hardening Cementitious Composites (SHCC) [5–8] or Ultra-High Toughness Cementitious Composites (UHTCC) [9–11]. In the recent decade, high- and ultra-high-strength ECC have been successfully developed to enhance the compressive performance of ECC materials [12, 13]. Generally, high-strength ECC inevitably involve the heavier use of cement binders than ordinary ECC in order to achieve higher compressive strength, which is however not beneficial for reducing the carbon footprint and improving the material greenness. Although the utilization of sustainable aggregates (e.g., artificial aggregates [14–17] and sea-sand [18–20]) can help improve the greenness of high-strength ECC, the heavy consumption

M. Kunieda et al. (Eds.): SHCC 2022, RILEM Bookseries 39, pp. 34–42, 2023.
https://doi.org/10.1007/978-3-031-15805-6_4

of Portland cement still cannot be avoided. Therefore, it is essential to develop more sustainable binders to fully or partially replace the traditional Portland cement in ECC production. Taking advantage of the geopolymer technology, a counterpart of the conventional ECC, i.e., Engineered Geopolymer Composites (EGC) or Strain-Hardening Geopolymer Composites (SHGC), can be produced with the non-clinker feature and better sustainability [21, 22].

Similar to ECC, EGC can also show remarkable strain-hardening and multiple-cracking characteristics [23, 24]. Substantial efforts have been focused on achieving higher mechanical properties, durability, and greenness of EGC. Coal fly ash (FA) is the most widely used precursor for the geopolymer production because of its aluminosilicate nature and wider availability. Up to now, the highest tensile ductility of EGC (13.7%) was achieved using FA as the precursor, but the compressive strength was extremely low, which would hinder the application of EGC in construction industry [25]. Although highly reactive materials such as granulated blast-furnace slag (GGBS) have the potential to improve the compressive strength of the alkali-activated cementitious system [26, 27], they may however not enhance the strain-hardening ability of EGC due to the improved fracture toughness of the geopolymer matrix, as demonstrated in previous studies with the GGBS replacement ratio up to 30% [28]. Currently, there are still no publications in the development of EGC with both high compressive strength and high tensile ductility although enhancing the overall mechanical properties of EGC is beneficial for more resilient and sustainable infrastructure.

In this study, ultra-high strength & ductility EGC (UHSD-EGC) were successfully developed by the proper mix design with the blended use of FA, GGBS, and SF as precursors, sodium metasilicate and waterglass as alkali activators, and ultra-high-molecular-weight polyethylene (PE) fibers as the reinforcement. The mechanical properties of UHSD-EGC including compressive strength, tensile strength, tensile ductility, and strain energy density were obtained through compressive and direct tensile tests. Finally, the microstructures of UHSD-EGC matrices were investigated by Backscattered electron (BSE) tests.

2 Materials and Test Program

2.1 Raw Materials

FA, GGBS, and silica fume (SF) were used as the precursors for UHSD-EGC production in this study. Anhydrous sodium metasilicate (Na_2SiO_3-Anhydrous) particles with 50.75% Na_2O, 46.52% SiO_2, and 2.73% impurities, and waterglass composed of 8.7% Na_2O, 27.7% SiO_2, and 56.8% H_2O were used as the alkaline activators. Following the experience of ultra-high-strength ECC development [13], fine silica sand with the average size smaller than 300 μm was used as the fine aggregates, and ultra-high-molecular-weight PE fibers with a diameter of 24 μm and a length of 18 mm were used as the reinforcement in UHSD-EGC. In addition, borax ($Na_2B_4O_7 \cdot 10H_2O$) of analytical reagent grade with the purification of 99.5% was used as the retarder during UHSD-EGC mixing.

2.2 Preparation of UHSD-EGC

Three UHSD-EGC mixes with different FA-to-GGBS ratios were prepared in the study, and the mix proportions are shown in Table 1. Here, the mix ID "FaSb" was used to represent the UHSD-EGC with the FA/GGBS ratio of a/b. Before the specimen casting, Na_2SiO_3-Anhydrous, waterglass, borax, and extra water were mixed and stirred together until a uniform alkaline activator solution was formed. Next, FA, GGBS, silica fume, and fine silica sand were dry-mixed together for the first 5 min, and then the prepared alkaline activator solution was poured into the mixer and stirred for another 10 min until a uniform fresh geopolymer matrix was observed. After that, PE fibers were added and the mixing was continued for the final 5 min. Finally, the uniform fresh UHSD-EGC were cast into specific molds with a plastic sheet covered to prevent water evaporation. After 24 h, the specimens were demolded, sealed with plastic sheets and heat-cured in an 80 °C oven for 72 h. After the heat curing, the specimens were taken out for mechanical tests.

Table 1. Mix proportions of UHSD-EGC (weight ratio).

Raw materials	F8S2	F5S5	F2S8
FA	0.760	0.475	0.190
GGBS	0.190	0.475	0.760
Silica fume	0.050		
Fine silica sand	0.300		
Na_2SiO_3-Anhydrous	0.095		
Waterglass	0.141		
Borax	0.038		
Extra water	0.172		
PE fibers	2.0% (Vol.)		

2.3 Testing Methods

For each UHSD-EGC mix, three 50 mm × 50 mm × 50 mm cubes were cast for the compressive test, and the loading rate was set as 1.0 MPa/s. In addition, three dumbbell specimens (Fig. 1) [29, 30] were tested under direct tensile tests with a loading rate of 0.5 mm/min, and the middle portion (80-mm length) was set as the measuring range in the study. Furthermore, fresh UHSD-EGC pastes (without fine silica sand and PE fibers) were produced and cured following the procedure described in the previous section, and these samples were tested by BSE analysis for a better characterization of reaction degrees and components in the pastes.

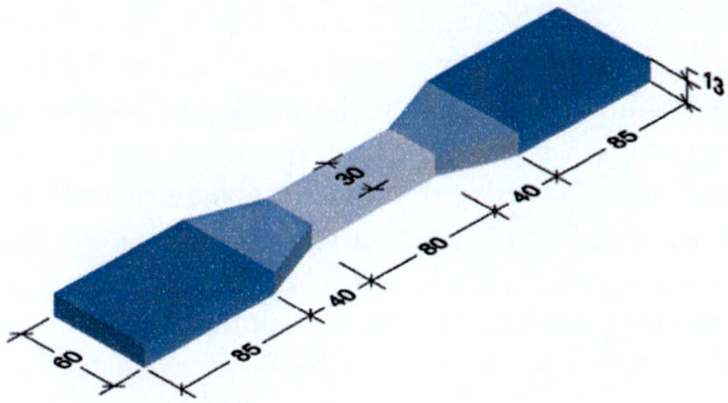

Fig. 1. The geometry of dumbbell specimen for direct tensile tests.

3 Results and Discussions

3.1 Compressive Strength

Figure 2A shows the compressive strengths of UHSD-EGC with different mix proportions. Obviously, the compressive strength of UHSD-EGC increased with the decrease of FA/GGBS ratio due to the lower reactivity of FA compared to GGBS, and the highest compressive strength was over 150.0 MPa (i.e., Mix F2S8). Therefore, it is feasible to adjust the compressive strength of UHSD-EGC by changing the FA/GGBS ratio.

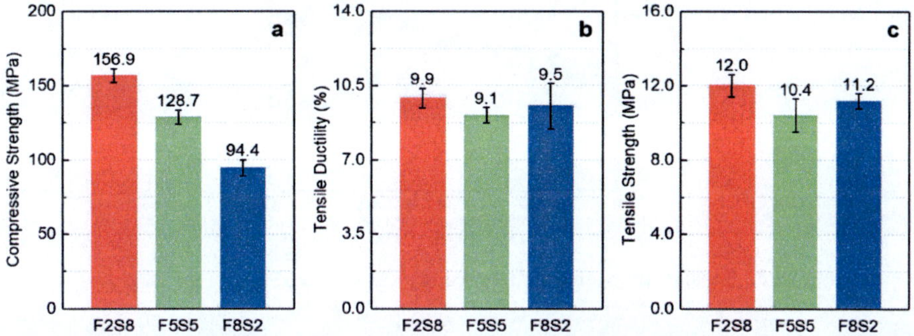

Fig. 2. Mechanical properties of UHSD-EGC (a) compressive strength, (b) tensile ductility, and (c) tensile strength.

3.2 Tensile Performance

The stress-strain curves of F2S8, F5S5, and F8S2 under direct tension are presented in Fig. 3a, Fig. 3b, and Fig. 3c, respectively. The summarized tensile ductility and tensile strength are shown in Fig. 2b and Fig. 2c, respectively. Similar to the ECC materials [31], distinguished strain-hardening behaviors of UHSD-EGC were observed in the study. From Fig. 2b, it was found that ultra-high tensile ductility (over 9.0%) was achieved for all the mixes, and the FA/GGBS ratio had a marginal influence on the deformability of UHSD-EGC. However, as observed from the tensile stress-strain curves (Fig. 3a, Fig. 3b, and Fig. 3c), larger stress drops were found in Mix F2S8, which indicated the formations of cracks with larger widths, while the tensile stress-strain curves of F8S2 showed smaller stress drops. Since the tensile deformation of EGC materials is attributed to both crack number and crack width, it is inferred that the use of higher FA/GGBS ratio could result in more saturated cracks with narrowed widths. In terms of tensile strength, all the mixes showed a high value (over 10.0 MPa), indicating a good bond achieved between PE fibers and the geopolymer matrix in all the mixes.

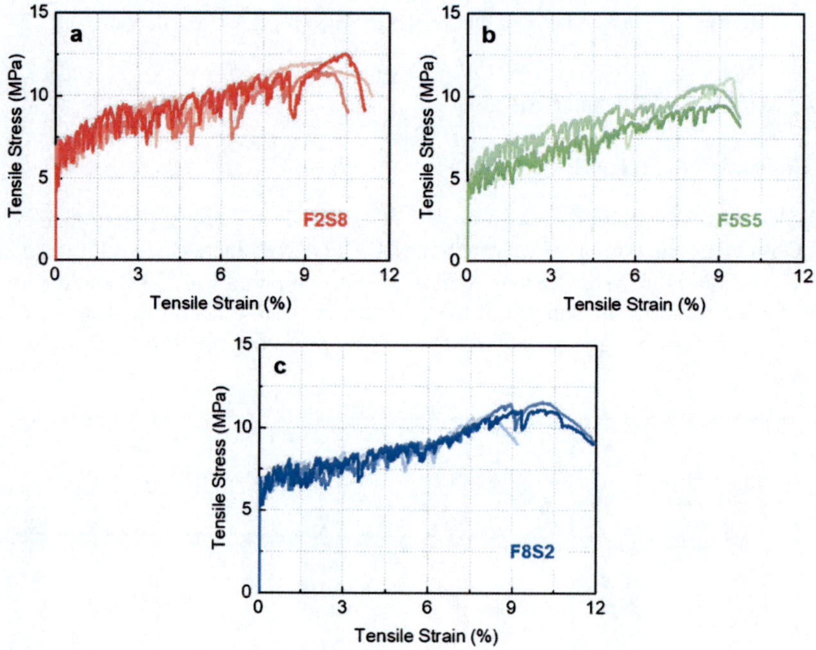

Fig. 3. Tensile stress–strain curves of (a) F2S8, (b) F5S5, and (c) F8S2.

It is worth mentioning that F8S2 has lower compressive strength but comparable tensile strength and tensile ductility with F2S8 and F5S5. This tendency is different from ordinary ECC materials using Portland cement. For strain-hardening geopolymer composites, the tensile ductility is related to many micromechanical parameters, including matrix parameters of geopolymer (e.g., tensile strength, fracture toughness, elastic modulus, and spalling parameter), fiber/geopolymer interface parameters (e.g., bond strength, snubbing coefficient, and Cook-Gordon effect parameter), and fiber parameters (e.g., length, diameter, modulus, and strength). To better understand the aforementioned tensile behavior of UHSD-EGC, the above parameters should be measured and micromechanical analysis are needed in the future study.

3.3 Strain Energy Density Under Tension

The strain energy density of UHSD-EGC was calculated on the basis of the area under the ascending tensile stress-strain curves, as seen in Fig. 4. Among the three mix proportions, Mix F2S8 showed a highest strain energy density value over 900 kJ/m^3. This trend was in accordance with those of tensile ductility and strength values shown in Fig. 2b and Fig. 2c. On the other hand, Mix F5S5 had the lowest strain energy density value. Although the difference of strain energy density among the three mixes was not so prominent, the mechanism how the FA/GGBS ratio affects the tensile strength/ductility as well as the energy absorption capacity still needs further investigation. Overall, all the produced UHSD-EGC mixes showed excellent strain energy density values over 700 kJ/m^3.

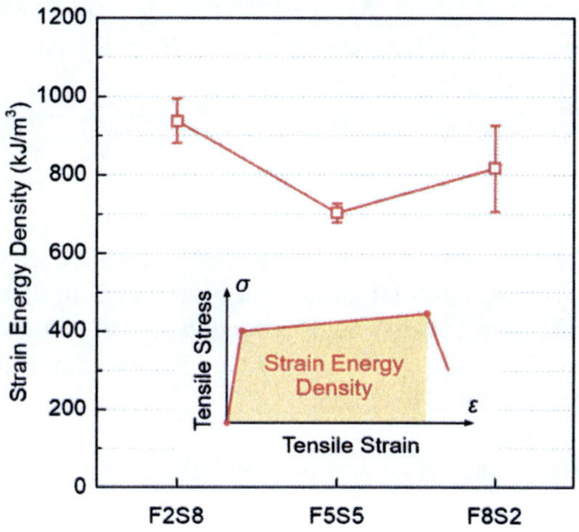

Fig. 4. Strain energy density of UHSD-EGC under tension.

3.4 BSE Results of Matrix

Figure 5 shows the microstructures of the UHSD-EGC pastes from BSE observations. In this mode, many unreacted raw materials were found due to the low water-to-binder ratio (0.27), the different components could be easily distinguished: the round particles were unreacted FA, the gray angular particles were unreacted GGBS, the black pores were the initial flaws, and the remaining regions were the reaction products. As the unreacted particles were densely distributed in the pastes, they could function as rigid fillers and densified the geopolymer matrices. Among the three mixes, Mix F8S2 showed a looser paste microstructure with more heterogeneous flaws compared to Mix F2S8, as the N-A-S-H gel formed by FA-based system is less space-filling than the C-(A)-S-H products generated from the activation of GGBS. This phenomenon was also found in Mix F5S5, where denser pastes were observed in locations with more unreacted GGBS particles, while the regions near unreacted FA particles seemed to be looser. Therefore, apart from the fact that FA showed a lower reactivity than GGBS, the looser paste structure formed in FA-rich system can also be a possible reason for the comparatively lower strength of UHSD-EGC with higher FA/GGBS ratio.

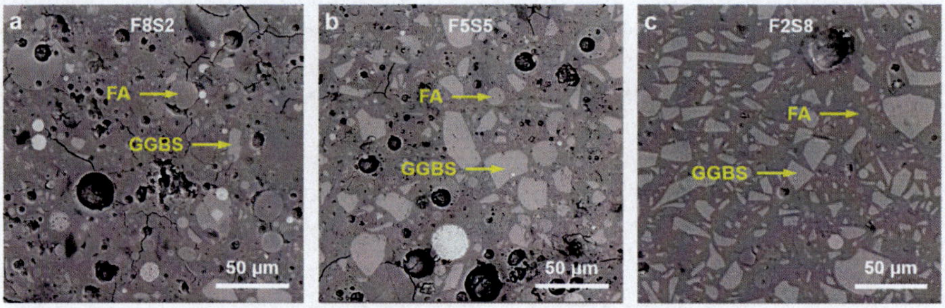

Fig. 5. Microstructures of the UHSD-EGC pastes from BSE observation (a) F8S2, (b) F5S5 and (c) F2S8.

4 Conclusions

IN this study, ultra-high-strength & ductility EGC (UHSD-EGC) were successfully developed with fly ash (FA), ground granulated blast slag (GGBS), silica fume, alkali activator, and ultra-high-molecular-weight polyethylene (PE) fibers. Through the findings from compressive tests, direct tensile tests, and Backscattered electron (BSE) tests, the following conclusions could be drawn:

(1) Higher compressive strength could be achieved in the UHSD-EGC system with lower FA/GGBS ratios due to the higher reactivity of GGBS compared to FA, and the highest compressive strength reached over 150 MPa in this study.
(2) All the UHSD-EGC specimens showed an excellent tensile ductility over 9.0% and a good tensile strength over 10.0 MPa. For UHSD-EGC with higher FA/GGBS ratio, the stress drops in the strain-hardening curves were smaller, indicating a more saturated cracking phenomenon with smaller crack width in the FA-rich UHSD-EGC system.
(3) Mix F2S8 showed the highest strain energy density while Mix F5S5 had the lowest value. Overall, all the produced UHSD-EGC mixes showed excellent strain energy density values over 700 kJ/m^3.
(4) The microstructures of the FA-rich geopolymer paste were found to be much looser than that the ones with lower FA/GGBS ratio, which indicated a less spacing filling effect of the N-A-S-H gel formed by FA-based system than the C-(A)-S-H products generated from the activation of GGBS. The above phenomenon could also be a possible reason for the lower compressive strength of UHSD-EGC with higher FA/GGBS ratio.

As EGC with both ultra-high strength (over 150.0 MPa) and ultra-high ductility (9.0%) were firstly developed in the study, more extensive investigations are being carried out by the authors' research group and the results will be reported in the future. It is worth mentioned that the authors also developed strain-hardening ultra-high-performance geopolymer concrete (UHPGC) reinforced by steel fibers. The compressive strength of the strain-hardening UHPGC reached 163–222 MPa, the tensile ductility varied within the range of 0.35%–0.55%, and the residual crack width after the tensile test was approximately 10–20 μm only. More details can be found in the latest publication [32, 33].

References

1. Li, V.C.: Engineered Cementitious Composites (ECC) - Bendable Concrete for Sustainable and Resilient Infrastructure. Springer, Heidelberg (2019). https://doi.org/10.1007/978-3-662-58438-5_8
2. Leung, C.K.Y.: Design criteria for pseudoductile fiber-reinforced composites. J. Eng. Mech. **122**(1), 10–18 (1996)
3. Huang, B.T., Wu, J.Q., Yu, J., Dai, J.G., Leung, C.K., Li, V.C.: Seawater sea-sand engineered/strain-hardening cementitious composites (ECC/SHCC): assessment and modeling of crack characteristics. Cem. Concr. Res. **140**, 106292 (2021)
4. Zhu, J.X., Xu, L.Y., Huang, B.T., Weng, K.F., Dai, J.G.: Recent developments in Engineered/Strain-Hardening Cementitious Composites (ECC/SHCC) with high and ultra-high strength. Constr. Build. Mater. **342**, 127956 (2022)
5. van Zijl, G.P., et al.: Durability of strain-hardening cement-based composites (SHCC). Mater. Struct. **45**(10), 1447–1463 (2012)
6. Qian, S., Zhou, J., De Rooij, M.R., Schlangen, E., Ye, G., Van Breugel, K.: Self-healing behavior of strain hardening cementitious composites incorporating local waste materials. Cement Concr. Compos. **31**(9), 613–621 (2009)
7. Mechtcherine, V., Millon, O., Butler, M., Thoma, K.: Mechanical behaviour of strain hardening cement-based composites under impact loading. Cement Concr. Compos. **33**(1), 1–11 (2011)
8. Li, Q., Yin, X., Huang, B., Zhang, Y., Xu, S.: Strengthening of the concrete face slabs of dams using sprayable strain-hardening fiber-reinforced cementitious composites. Front. Struct. Civ. Eng. **16**, 145–160 (2022). https://doi.org/10.1007/s11709-022-0806-4
9. Li, Q.H., et al.: Shear interfacial fracture of strain-hardening fiber-reinforced cementitious composites and concrete: a novel approach. Eng. Fract. Mech. **253**, 107849 (2021)
10. Huang, B.T., Li, Q.H., Xu, S.L., Liu, W., Wang, H.T.: Fatigue deformation behavior and fiber failure mechanism of ultra-high toughness cementitious composites in compression. Mater. Des. **157**, 457–468 (2018)
11. Huang, B.T., Li, Q.H., Xu, S.L.: Fatigue deformation model of plain and fiber-reinforced concrete based on Weibull function. J. Struct. Eng. **145**(1), 04018234 (2019)
12. Xu, L.Y., Huang, B.T., Lao, J.C., Yao, J., Li, V.C., Dai, J.G.: Tensile over-saturated cracking of ultra-high-strength engineered cementitious composites (UHS-ECC) with artificial geopolymer aggregates. Cement Concr. Compos. **136**, 104896 (2023). https://doi.org/10.1016/j.cemconcomp.2022.104896
13. Huang, B.T., Zhu, J.X., Weng, K.F., Li, V.C., Dai, J.G.: Ultra-high-strength engineered/strain-hardening cementitious composites (ECC/SHCC): material design and effect of fiber hybridization. Cement Concr. Compos. **129**, 104464 (2022)
14. Xu, L.Y., Huang, B.T., Dai, J.G.: Development of engineered cementitious composites (ECC) using artificial fine aggregates. Constr. Build. Mater. **305**, 124742 (2021)
15. Xu, L.Y., Huang, B.T., Li, V.C., Dai, J.G.: High-strength high-ductility Engineered/Strain-Hardening Cementitious Composites (ECC/SHCC) incorporating geopolymer fine aggregates. Cement Concr. Compos. **125**, 104296 (2022)

16. Xu, L.Y., Qian, L.P., Huang, B.T., Dai, J.G.: Development of artificial one-part geopolymer lightweight aggregates by crushing technique. J. Clean. Prod. **315**, 128200 (2021)
17. Xu, L.Y., Huang, B.T., Lao, J.C., Dai, J.G.: Tailoring strain-hardening behavior of high-strength Engineered Cementitious Composites (ECC) using hybrid silica sand and artificial geopolymer aggregates. Mater. Des. **220**, 110876 (2022)
18. Huang, B.T., Wu, J.Q., Yu, J., Dai, J.G., Leung, C.K.: High-strength seawater sea-sand Engineered Cementitious Composites (SS-ECC): mechanical performance and probabilistic modeling. Cement Concr. Compos. **114**, 103740 (2020)
19. Huang, B.T., Wang, Y.T., Wu, J.Q., Yu, J., Dai, J.G., Leung, C.K.: Effect of fiber content on mechanical performance and cracking characteristics of ultra-high-performance seawater sea-sand concrete (UHP-SSC). Adv. Struct. Eng. **24**(6), 1182–1195 (2021)
20. Huang, B.T., Yu, J., Wu, J.Q., Dai, J.G., Leung, C.K.: Seawater sea-sand Engineered Cementitious Composites (SS-ECC) for marine and coastal applications. Compos. Commun. **20**, 100353 (2020)
21. Ohno, M., Li, V.C.: An integrated design method of engineered geopolymer composite. Cement Concr. Compos. **88**, 73–85 (2018)
22. Peng, K.D., Huang, B.T., Xu, L.Y., Hu, R.L., Dai, J.G.: Flexural strengthening of reinforced concrete beams using geopolymer-bonded small-diameter FRP bars. Eng. Struct. **256**, 113992 (2022)
23. Zhang, S., Li, V.C., Ye, G.: Micromechanics-guided development of a slag/fly ash-based strain-hardening geopolymer composite. Cement Concr. Compos. **109**, 103510 (2020)
24. Kan, L.L., Wang, W.S., Liu, W.D., Wu, M.: Development and characterization of fly ash based PVA fiber reinforced engineered geopolymer composites incorporating metakaolin. Cement Concr. Compos. **108**, 103521 (2020)
25. Nguyễn, H.H., Lương, Q.H., Choi, J.I., Ranade, R., Li, V.C., Lee, B.Y.: Ultra-ductile behavior of fly ash-based engineered geopolymer composites with a tensile strain capacity up to 13.7%. Cement Concr. Compos. **122**, 104133 (2021)
26. Deb, P.S., Nath, P., Sarker, P.K.: The effects of ground granulated blast-furnace slag blending with fly ash and activator content on the workability and strength properties of geopolymer concrete cured at ambient temperature. Mater. Des. **1980–2015**(62), 32–39 (2014)
27. Samantasinghar, S., Singh, S.P.: Effect of synthesis parameters on compressive strength of fly ash-slag blended geopolymer. Constr. Build. Mater. **170**, 225–234 (2018)
28. Ling, Y., Wang, K., Li, W., Shi, G., Lu, P.: Effect of slag on the mechanical properties and bond strength of fly ash-based engineered geopolymer composites. Compos. B Eng. **164**, 747–757 (2019)
29. Huang, B.T., Dai, J.G., Weng, K.F., Zhu, J.X., Shah, S.P.: Flexural performance of UHPC–concrete–ECC composite member reinforced with perforated steel plates. J. Struct. Eng. **147**(6), 04021065 (2021)
30. Huang, B.T., Weng, K.F., Zhu, J.X., Xiang, Y., Dai, J.G., Li, V.C.: Engineered/strain-hardening cementitious composites (ECC/SHCC) with an ultra-high compressive strength over 210 MPa. Compos. Commun. **26**, 100775 (2021)
31. Li, V.C., Leung, C.K.: Steady-state and multiple cracking of short random fiber composites. J. Eng. Mech. **118**(11), 2246–2264 (1992)
32. Lao, J.C., Xu, L.Y., Huang, B.T., Dai, J.G., Shah, S.P.: Development of strain-hardening Ultra-High-Performance Geopolymer Concrete (UHPGC) using steel fibers. Compos. Commun. **30**, 101081 (2022)
33. Lao, J.C., Huang, B.T., Fang, Y., Xu, L.Y., Dai, J.G., Shah, S.P.: Strain-hardening alkali-activated fly ash/slag composites with ultra-high compressive strength and ultra-high tensile ductility. Cement Concr. Res. **165**, 107075 (2023). https://doi.org/10.1016/j.cemconres.2022.107075

Developing CO_2-Sequstrating Strain-Hardening Magnesia-Based Composite (SHMC) with Hybrid Synthetic-Natural Fibers

Bo Wu(✉), Xianjun Su, and Jishen Qiu

Department of Civil and Environmental Engineering, Hong Kong University of Science and Technology, 1 University Road, Clear Water Bay, Hong Kong, China
bwuar@connect.ust.hk

Abstract. Reactive magnesia cement (RMC) is an emerging green cement as it can sequestrate substantial CO_2 to harden itself. However, the penetration of CO_2 in RMC from outside to inside causes a change in microstructure with depth, which influences the fiber/matrix interface bond and fiber-bridging capacity. This work firstly investigated the influence of carbonation degree on interface bond by single fiber pull-out, SEM, FTIR and acid digestion tests, the results demonstrated that the interface bond is positively correlated to the carbonation degree, but high carbonation degree may induce the fiber rupture. Secondly, tensile test was conducted to explore the influence of carbonation degree on tensile performance, the results suggested increase in carbonation degree can significantly improve the tensile performance, and replacing partial PVA fiber with sisal fiber can prominently enhance the tensile performance at early stage. This work is the first time to clarify the relationship between carbonation degree, fiber/matrix interface bond and tensile performance of RMC, which may provide some guidance to the mix design and application of SHMC.

Keywords: Reactive magnesia cement · Carbonation degree · Fiber/matrix interface bond · Hollow natural fiber

1 Introduction

Excessive anthropogenic CO_2 emission has caused significant climate change. Reactive magnesia cement (RMC), which can sequestrate CO_2 by hydration and carbonation, has been acknowledged as a green binder to replace Portland cement (PC) [1]. Different studies have reported using RMC as the sole or primary binder to make concrete, and the effect of mineral additives content [1, 2], hydration agent [3, 4], nucleation seeds [5, 6] and curing condition [7, 8] have been investigated. Due to its intrinsic brittleness, polymer fiber (e.g., polyvinyl alcohol (PVA)) have been added to this composite as reinforcement, which induced tensile ductility comparable to the PC-based SHCCs [9].

Unlike the PC-based composites which are homogenous [10], RMC-based composites have different strength at different depths to the surface and thus its mechanical properties are highly size-dependent. This is because RMC relies on carbonation to form

M. Kunieda et al. (Eds.): SHCC 2022, RILEM Bookseries 39, pp. 43–52, 2023.
https://doi.org/10.1007/978-3-031-15805-6_5

hydrated magnesium carbonates (HMCs), which results in the microstructural refinement and provides the strength [4, 5]. The dense HMCs formed on outside layer would severely slow down diffusion of CO_2 into interior part of specimen [7], resulting in a change in carbonation degree. For strain-hardening magnesia-based composites (SHMC), the reduced carbonation degree would induce decrease in fiber-to-matrix interface bond, and thus could reduce the fiber-bridging capacity based on classic micromechanics-based modeling. However, the effect of carbonation degree on the fiber-to-matrix interface properties have not been studied.

In this work, we investigated the influence of carbonation degree on the fiber/matrix interface bond for the first time. Specifically, single fiber pull-out tests were conducted to specimens cured for different extensions of carbonation time. Mass change under acid digestion and Fourier-transform infrared spectroscopy (FTIR) were used to determine the carbonation degree of the specimens. Scanning electron microscopy (SEM) was used to understand the microstructural failure mode at the interface. Based on the single-fiber tests, we developed SHMC with PVA fibers or hybrid fibers (PVA and sisal fiber); the sisal fibers are a typical hollow natural fiber which can induce faster and deeper CO_2 penetration [11]. The tensile properties of PVA fiber and hybrid fiber SHMC were measured by uniaxial tensile test.

2 Experimental Programs

2.1 Materials

The RMC was provided by Shanghai Yunajiang Chemical Co., Ltd, which is composed of 95 wt.% of reactive MgO and minor impurities like CaO and SiO_2; The sodium hexametaphosphate ($Na(PO_3)_6$) was supplied by Sigma-Aldrich, served as the water reducer to modify the rheology of the fresh paste and improve the fiber dispersion [12]. The concentrated sulfate acid (>97%) was supplied by Sigma-Aldrich. The PVA fibers were acquired from Kuraray Ltd with the tensile strength of 1600 MPa, diameter of about 39 μm, and length of 12 mm (for tensile test) or 10 cm (for single fiber pull-out test). Long sisal fibers were purchased from Zhejiang Rafi Grass Paper Products Co., Ltd, its tensile strength is about 300 MPa, and diameter of 150 to 250 μm. The long sisal fibers were cut into 15-mm-long short fibers before adding into paste. The silica sand with average diameter of 150 μm was used to prepare dumbbell samples. The mix proportions for single fiber pull-out test and tensile test are summarized in Table 1.

2.2 Single Fiber Pull-Out Test

The $Na(PO_3)_6$ was dissolved in water before mixing with RMC for 3 min, then the fresh paste was cast into a mold attached with four parallel long-fibers. The specimen was covered with waterproof membrane and then demoulded after 24 h of ambient curing. The single fiber specimens with the thickness of 1 ± 0.2 mm were acquired through diamond saw (IsoMet 1000) cutting and then moved to a carbonation chamber (temperature of 30 °C, RH of $85 \pm 5\%$ and CO_2 concentration of 10%) for 0, 3, 8 and 24 h. Afterwards, the single fiber pull-out test was conducted by an electrical universal material testing machine (Lyllod, model EZ50) equipped with a 20 N load cell and a displacement-controlled actuator. In this test setup, a X-Y table at the bottom for aiming the fiber tip to the clamper at the top. The loading rate was 0.5 mm/min.

Table 1. Mix proportions for single fiber pull-out test and tensile test

Specimen code	MgO	Water	$Na(PO_3)_6$	Silica sand	PVA fiber (vol. %)	Sisal fiber (vol. %)
Single fiber test	1	0.58	0.058			
PVA (2%)-RMC	1	0.58	0.058	0.2	2	0
PVA (1.5%)-Sisal (1%)-RMC	1	0.58	0.058	0.2	1.5	1

After pulling out, the morphology of fiber tip was characterized by SEM (JEOL-6390). The residual pastes were collected and immersed in isopropanol for 3 days followed by vacuum drying to kill the hydration and carbonation. Afterwards, these pastes were grounded into powder that can pass through the sieve with 0.01 mm-opening. The collected powder was characterized by FTIR (Bruker Vertex 70 Hyperion 1000). Besides, acid digestion was also used to determine the amount of CO_2 absorbed during carbonation [13]. Specifically, 3 g powder was added into a mixed solution consisted of 90 mL deionized water and 10 mL concentrated sulfate acid and then stirring for 30 min to allow the sufficient reaction between sulfate acid and HMCs. The mass change (Δm) before and after reaction was recorded. The mass change (Δm_0) of mixed solution without adding powder was also measured to sever as reference. The mass change ratio (η) was determined by following equation:

$$\eta = \frac{\Delta m - \Delta m_0}{3} \times 100\% \tag{1}$$

2.3 Tensile Test

The mixing procedure for preparing SHMC specimen can refer to [9], then the mixture was cast in a dumbbell shape mold with the cross-section of 13 × 30 mm. After curing in ambient for 24 h and demoulding, the specimens were moved to a carbonation chamber (temperature of 30 °C, RH of 85 ± 5% and CO_2 concentration of 10%) and then kept for 3, 7 and 14 days. The tensile test was carried out by a hydraulic universal material testing machine (MTS, model 810) with the loading rate of 0.3 mm/min. During the tension, a pair of linear variable displacement transducers (LVDT) with an 80-mm gauge length were used to record the displacement. Six specimens were tested for each group.

3 Results and Discussions

Figure 1 shows the representative P-u curves of single fiber specimens after different carbonation times. A sudden load drop from P_a to P_b can be observed for all specimens, this indicates the completion of fiber/matrix debonding [14], then the fiber starts to slip out from the matrix and eventually the load approaches to zero. As the carbonation proceeds, the P_a and P_b continue to increase, especially the slope of the curves after debonding demonstrates a significant improvement. More importantly, after 24 h carbonation, the fibers tend to rupture during slipping stage. With using the characteristic points (e.g., Pa and Pb) and the slop of the curve after debonding, the chemical debonding energy (G_d), frictional bond (τ) and slip-hardening coefficient (β) can be determined [14, 15]:

$$G_d = \frac{2(P_a - P_b)^2}{\pi^2 E_f d_f^3} \left(\mathrm{J/m^2}\right) \tag{2}$$

$$\tau_0 = \frac{P_b}{\pi d_f l_e} \text{ (MPa)} \tag{3}$$

$$\beta = d_f/l_f)[(1/\tau_0 \pi d_f)(\Delta P/\Delta u)_{u' \to 0} + 1] \tag{4}$$

where the E_f, d_f, and L_e, are Young's modulus, diameter, and embedment length of the PVA fiber, respectively. The total work required to completely pull out a fiber can be determined by the area under the curve [16], and the energy absorption of per unit area (G_w) can be calculated by following equation:

$$G_{W\_x} = \frac{\int_0^x P(u)du}{\pi d_f L_e} \left(\mathrm{J/m^2}\right) \tag{5}$$

The calculated parameters are summarized in Table 2. As it can be seen, all the parameters except β exhibit a prominent enhancement ranging from 90% - 197% after 3-h carbonation compared to control group. As the carbonation onwards, the G_d hits its maximum value of 7.85 $\mathrm{J/m^2}$ after 8-h carbonation and then slightly shrink to 5.89 $\mathrm{J/m^2}$ after 24-h carbonation, but it is still 155% higher than control group. The τ presents a continuous increasing and finally reaches 2 MPa, which is about six times that of control group. The β shoots up from 0.1 after 3-h carbonation to 0.428 after 8-h carbonation, and then to 0.489 after 24-h carbonation. The G_w_0.5mm value shows a moderately increase from 686 $\mathrm{J/m^2}$ to 869 $\mathrm{J/m^2}$, and then soars to 1768 $\mathrm{J/m^2}$ after 24-h carbonation. These results directly corroborate that extension in carbonation time can substantially augment the fiber/matrix interface bond, especially the frictional bond and G_w values, this may be due to the continuous carbonation densifies the microstructure near the fiber [4, 17], and the formed HMCs scratch the fiber surface to enhance the friction against the fibers being pulled out.

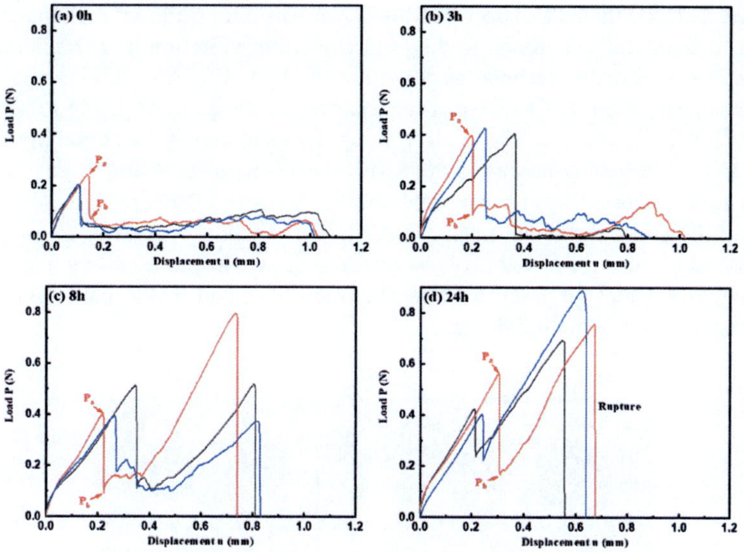

Fig. 1. Representative P-u curves of specimens after (a) 0 h, (b) 3 h, (c) 8 h and (d) 24h carbonation.

Figure 2 compares the surface morphology of fibers pulled out from matrix after different carbonation time. The control group presents intact fiber tip covered with a thick layer of hydration products, and no distinct scratch can be seen, as shown in Fig. 2a. The hydration products are in plate shape, indicating it is likely the brucite [18]. After 3-h carbonation, there is also no visible scratch on fiber surface (Fig. 2b), hence the β value has not been improved in Table 2. However, apart from the plate-like brucite, some needle-like products can be found on fiber surface, implying partial brucites has been carbonized into nesquehonite [5]. With the carbonation time extending to 8 h (Fig. 2c), there are some scratches on fiber surface, and substantial needle-like nesquehonite can be observed, indicating the brucite has been converted into HMCs as the carbonation progressing, thus refining the interfacial transitional zone between fiber and matrix. After 24-h carbonation (Fig. 3d), it can be seen the fiber surface is seriously scratched, this is in line with the fiber rupture in Fig. 1d.

Table 2. Parameters calculated from P-u curves and the results of weight loss ratio

Specimen Code	G_d (J/m^2)	τ (MPa)	β	G_w_0.5mm (J/m^2)	Weight loss ratio
0 h	2.31 ± 0.59	0.33 ± 0.06	0.137 ± 0.013	265.9 ± 43.4	9.3%
3 h	6.85 ± 3.57	0.63 ± 0.31	0.100 ± 0.050	686.6 ± 97.7	19.3%
8 h	7.85 ± 3.13	0.88 ± 0.26	0.428 ± 0.192	869.6 ± 290.1	24.5%
24 h	5.89 ± 3.88	2.00 ± 0.94	0.489 ± 0.182	1768.1 ± 501.1	30.0%

Figure 3 shows the FTIR spectra of the specimens after different carbonation times, the peak at 3700 cm^{-1} belongs to the O-H stretching vibration and shoulder band at 1650 cm^{-1} is assignable to bending vibration of H_2O [19]. Antisymmetric stretching and bending vibration of $CO_3{}^{2-}$ can be observed at the peak of ~ 1450 cm^{-1} and ~ 880 cm^{-1}, respectively [7]. The control group demonstrates weak stretching vibration peak of $CO_3{}^{2-}$, this may be due to the RMC matrix can also be slightly carbonized in ambient atmosphere. After 3-h carbonation, the intensity of this peak exhibits a sharp increase, followed by gradual enhancement as carbonation time extends to 24 h. The mass loss ratio determined by acid digestion test as shown in Table 2 also presents similar trend, suggesting that the interface bond between fiber and RMC matrix is positively correlating to the carbonation degree.

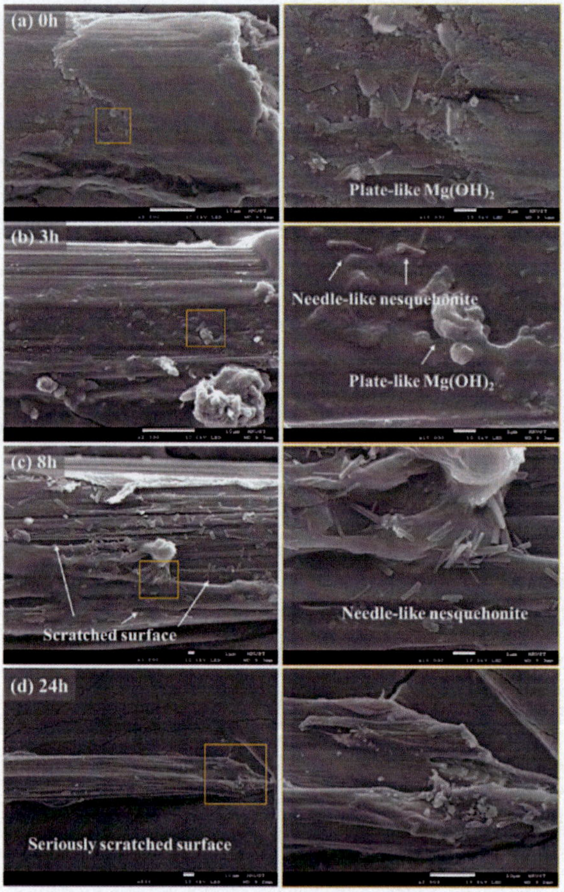

Fig. 2. Surface morphology of fibers pulled out from matrix after (a) 0 h, (b) 3 h, (c) 8 h and (d) 24 h carbonation.

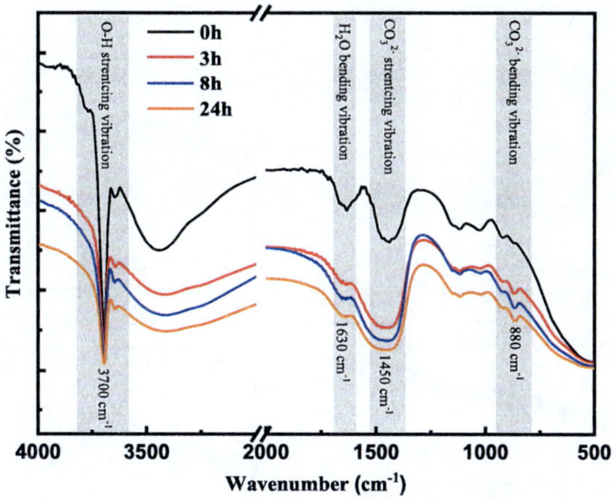

Fig. 3. FTIR spectra of the specimens after different carbonation times.

Figure 4 shows the tensile stress vs. strain behavior of PVA (2%)-RMC and PVA (1.5%)-Sisal (1%)-RMC groups after different carbonation times, and the test results are summarized in Table 3. All the specimens demonstrate strain-hardening behavior, and the first crack strength continuously increases as carbonation progressing. For PVA (2%)-RMC group, it has the smallest first crack strength of 1.18 MPa and the tensile strain of 1.18% after 3-days carbonation, this may be due to the insufficient carbonation leads to the weak fiber/matrix interface bond and mechanical strength of RMC matrix. After 7-days carbonation, the first crack strength increases to 1.41 MPa, and more importantly the tensile strain soars to 4.58%, indicating further carbonation improves the fiber-bridging capacity and the mechanical strength, this is in line with the results of single fiber pull-out test. The first crack strength shows a significant improvement up to 2.11 MPa after 14-days carbonation while tensile strain slightly increases to 4.89%, demonstrating that long-term carbonation can continuously enhance the mechanical strength of RMC matrix but contributes little to the improvement of tensile strain.

With using 1 vol.% sisal fiber to replace 0.5 vol% PVA fiber, the first crack strength can be enhanced from 1.18 MPa of PVA (2%)-RMC group to 1.72 MPa after 3-days carbonation, this verifies the addition of sisal fiber can improve the carbonation degree and mechanical strength of RMC [11], thereby the interface bond and fiber-bridging capacity can also be improved, evidenced by the increase of tensile strain from 1.18% to 3.08%. The first crack strength can be further increased to 2.41 MPa after 14-day carbonation, while the tensile strain decreases to 2.53%, this may be due to the sufficient carbonation excessively modifies the interfacial zone between fiber and matrix, resulting in fiber rupture during tensile loading.

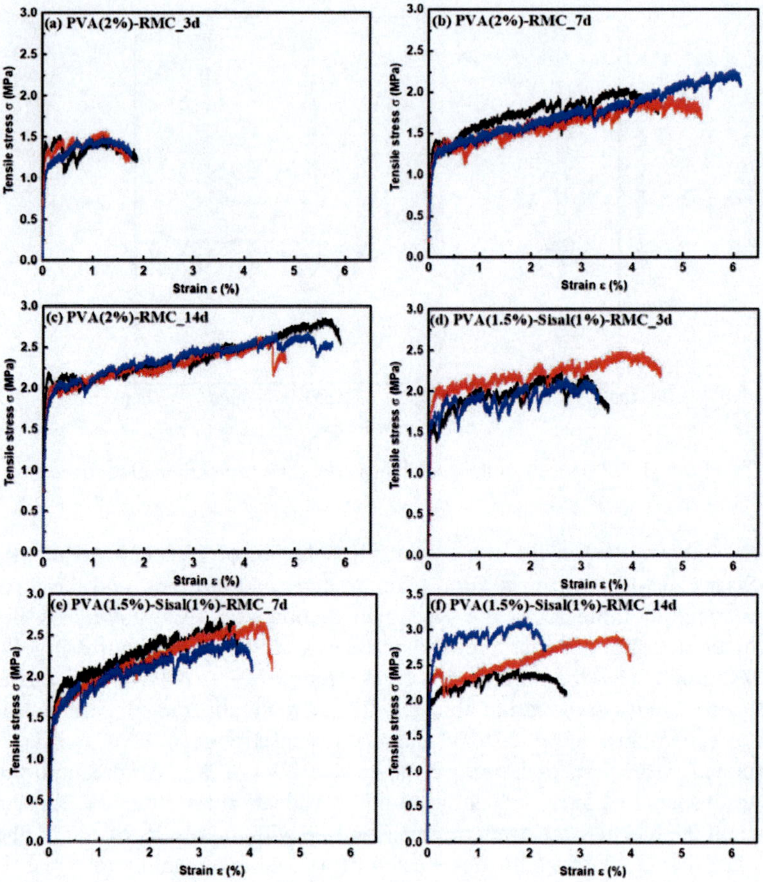

Fig. 4. Representative tensile stress-tensile strain curves of PVA fiber or PVA-Sisal hybrid fiber reinforced RMC.

Table 3. Tensile properties of RMC specimens

Mix code	First crack strength (MPa)	Ultimate tensile strength (MPa)	Tensile strain capacity (%)
PVA (2%)-RMC_3d	1.18 ± 0.17	1.46 ± 0.08	1.18 ± 0.16
PVA (2%)-RMC _7d	1.41 ± 0.06	2.01 ± 0.18	4.58 ± 1.19
PVA (2%)-RMC _14d	2.11 ± 0.08	2.68 ± 0.15	4.93 ± 0.53
PVA (1.5%)-Sisal (1%)-RMC _3d	1.72 ± 0.21	2.11 ± 0.27	3.08 ± 0.73

(*continued*)

Table 3. *(continued)*

Mix code	First crack strength (MPa)	Ultimate tensile strength (MPa)	Tensile strain capacity (%)
PVA (1.5%)-Sisal (1%)-RMC _7d	1.75 ± 0.16	2.55 ± 0.26	3.92 ± 0.88
PVA (1.5%)-Sisal (1%)-RMC _14d	2.41 ± 0.32	2.75 ± 0.33	2.53 ± 0.77

4 Conclusions

This work investigated the influence of carbonation degree on fiber/matrix interface bond and tensile performance in SHMC, the sisal fiber was also utilized to partially replace PVA fiber to increase the carbonation degree. Single fiber pull-out test results suggested that the extension in carbonation time prominently enhances the fiber/matrix interface bond, which however results in fiber rupture after 24-h carbonation. FTIR and acid digestion test results revealed the CO_3^{2-} content in RMC increased with carbonation time, implying the interface bond was positively corelated with carbonation degree. Similarly, extension in carbonation time can improve the tensile strength and tensile strain of PVA (2%)-RMC group. Replacing 0.5 vol.% PVA fiber with 1 vol.% sisal fiber enhanced the tensile strength and tensile strain.

References

1. Wu, H.-L., Zhang, D., Ellis, B.R., Li, V.C.: Development of reactive MgO-based engineered cementitious composite (ECC) through accelerated carbonation curing. Constr. Build. Mater. **191**, 23–31 (2018)
2. Shah, V., Scott, A.: Hydration and microstructural characteristics of MgO in the presence of metakaolin and silica fume. Cement Concr. Compos. **121**, 104068 (2021)
3. Dung, N., Unluer, C.: Development of MgO concrete with enhanced hydration and carbonation mechanisms. Cement Concr. Res. **103**, 160–169 (2018)
4. Dung, N., Unluer, C.: Carbonated MgO concrete with improved performance: The influence of temperature and hydration agent on hydration, carbonation and strength gain. Cement Concr. Compos. **82**, 152–164 (2017)
5. Dung, N., Lesimple, A., Hay, R., Celik, K., Unluer, C.: Formation of carbonate phases and their effect on the performance of reactive MgO cement formulations. Cement Concr. Res. **125** 105894 (2019)
6. Dung, N.T., Unluer, C.: Influence of nucleation seeding on the performance of carbonated MgO formulations. Cement Concr. Compos. **83**, 1–9 (2017). https://doi.org/10.1016/j.cemconcomp.2017.07.005
7. Dung, N., Hay, R., Lesimple, A., Celik, K., Unluer, C.: Influence of CO_2 concentration on the performance of MgO cement mixes. Cement Concr. Compos. **115**, 103826 (2021)
8. Unluer, C., Al-Tabbaa, A.: Enhancing the carbonation of MgO cement porous blocks through improved curing conditions. Cement Concr. Res. **59**, 55–65 (2014)
9. Qiu, J., Ruan, S., Unluer, C., Yang, E.-H.: Autogenous healing of fiber-reinforced reactive magnesia-based tensile strain-hardening composites. Cement Concr. Res. **115**, 401–413 (2019)

10. Bullard, J.W., Jennings, H.M., Livingston, R.A., Nonat, A.: Mechanisms of cement hydration. Cement Concr. Res. **41**(12), 1208–1223 (2011)
11. Wu, B., Qiu, J.: Incorporating hollow natural fiber (HNF) to enhance CO_2 sequestration and mechanical properties of reactive magnesia cement (RMC)-based composites: Feasibility study. J. CO_2 Utilization **57**, 101874 (2022)
12. Ruan, S., Qiu, J., Yang, E.-H., Unluer, C.: Fiber-reinforced reactive magnesia-based tensile strain-hardening composites, Cement and Concrete Composites **89**, 52–61 (2018)
13. Unluer, C., Al-Tabbaa, A.: The role of brucite, ground granulated blastfurnace slag, and magnesium silicates in the carbonation and performance of MgO cements. Constr. Build. Mater. **94**, 629–643 (2015)
14. Yang, E.-H., Wang, S., Yang, Y., Li, V.C.: Fiber-bridging constitutive law of engineered cementitious composites. J. Adv. Concr. Technol. **6**(1), 181–193 (2008)
15. Li, V.C.: Introduction to engineered cementitious composites (ECC). In: Engineered cementitious composites (ECC), pp. 1–10. Springer, Heidelberg (2019). https://doi.org/10.1007/978-3-662-58438-5_1
16. Bo, W., Qiu, J.: Effect of nano-SiO2 coating on the mechanical recovery of debonded fiber-cement interface under water curing. In: Serna, P., Llano-Torre, A., Martí-Vargas, J.R., Navarro-Gregori, J. (eds.) Fibre Reinforced Concrete: Improvements and Innovations II: X RILEM-fib International Symposium on Fibre Reinforced Concrete (BEFIB) 2021, pp. 879–888. Springer International Publishing, Cham (2022). https://doi.org/10.1007/978-3-030-83719-8_75
17. Vandeperre, L., Al-Tabbaa, A.: Accelerated carbonation of reactive MgO cements. Adv. Cement Res. **19**(2), 67–79 (2007)
18. Moreira, F.K., Pedro, D.C., Glenn, G.M..: Brucite nanoplates reinforced starch bionanocomposites. Carbohydrate Polymers **92**(2), 1743–1751 (2013)
19. Li, S., Wang, Z.J., Chang, T.-T.: Temperature oscillation modulated self-assembly of periodic concentric layered magnesium carbonate microparticles. PLoS ONE **9**(2), e88648 (2014)

Fundamental Study on Mechanical Performances of FRCC Using Polypropylene Nanofibers

Miyu Kanri[1], Tomoya Nishiwaki[1](✉), and Masafumi Kitatsuji[2]

[1] Tohoku University, Sendai 9808579, Japan
tomoya.nishiwaki.e8@tohoku.ac.jp
[2] Miyagi University, Sendai 9820215, Japan

Abstract. In recent years, research on nanomaterials has been extensively conducted in various areas, including construction fields. The use of nanofibers, such as carbon nanotubes (CNT) and cellulose nanofibers (CNF), has also been proposed for fiber-reinforced concrete and fiber-reinforced cementitious composites (FRCC). The excellent mechanical properties of nanofibers are very potential to improve the FRCC performances. In many previous works, remarkable results have already been reported on the use of nanofibers in FRCC.However, the use of high-performance nanofibers requires high manufacturing costs. Therefore, it cannot be easily introduced into the concrete that is known as an inexpensive building material. In this study, the application of polypropylene (PP) nanofibers is proposed. The PP nanofibers have nano-order diameters and can be provided in mass production at a low cost. Here, the mechanical performances of FRCC using the PP nanofibers and the combinations with other fibers were investigated. As a result, although the simple mixing procedures had problems with dispersibility, the PP nanofibers can be dispersed well using appropriate admixtures. It has been confirmed that the well dispersed PP nanofibers are improving the mechanical properties of FRCC. In addition, the combination use of PP nanofibers and polyvinyl alcohol (PVA) fibers enhancing the ductility.

Keywords: PP nanofibers · Bending strength · Ductility · Fiber dispersion

1 Introduction

In order to compensate for the weak tensile strength and brittle behavior of cementitious materials, fiber-reinforced cementitious composites (FRCC), which contain reinforcing fibers, have been widely investigated to improve mechanical performance [1]. Especially, hybridization/combination use of different sizes of fibers is expected to enhance further performances of FRCCs [2]. In addition, nanomaterials are also employed to improve the mechanical properties of FRCCs. In previous studies, multi-level fiber reinforcement has been investigated by combining micro-sized steel fibers with cellulose nanofibers (CNF) [3] and carbon nanotubes (CNT) [4], which are the smallest among the applicable fibers. It has been confirmed that the nanofibers suppress and delay the generation of

M. Kunieda et al. (Eds.): SHCC 2022, RILEM Bookseries 39, pp. 53–61, 2023.
https://doi.org/10.1007/978-3-031-15805-6_6

nano-size cracks at the initiation stage. Moreover, the synergistic effect with longer fibers improves tensile/bending strength and ductility [5]. However, although nanofibers such as CNFs and CNTs have excellent mechanical properties such as high strength, their substantial cost is unsuitable especially for construction materials. In this study, polypropylene (PP) nanofibers, as shown in Fig. 1, will be considered as an alternative reinforcing nanofiber. PP nanofibers have characteristics of lightweight, high strength, inexpensive, and readily available due to mass production. On the other hand, the use of PP nanofibers as reinforcement in cementitious materials has not been sufficiently investigated.

This study investigates the reinforcing effect of the PP nanofibers and the synergistic effect of the combined use of other fibers.

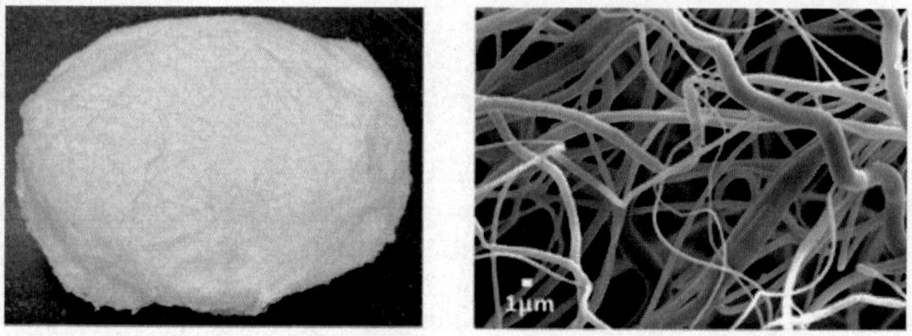

(a) Appearance of PP nanofibers (b) SEM Image

Fig. 1. Employed PP nanofibers

2 Outline of the Experiment

Uniform mixing and dispersion methods of the PP nanofibers were studied. Bending and compressive strength tests were conducted to confirm the mechanical properties with the reinforcing effect of PP nanofibers alone and the combined effect of PP nanofibers with other fibers. Moreover, scanning electron microscope (SEM) observation was conducted to evaluate the dispersion of the PP nanofibers in mortar.

2.1 Materials

The mix proportions are shown in Table 1. High early-strength Portland cement (C, density: 3.14 g/cm^3) as a binder and fine silica sand (S, density: 2.61 g/cm^3, average particle size: 180 μm) as fine aggregate were used. PP nanofiber (PPn, density: 0.92 g/cm^3, shown in Fig. 1), PP fiber (PP, density: 0.92 g/cm^3, diameter: 12 μm, length: 6 mm), and PVA fiber (PVA, density: 1.30 g/cm^3, diameter: 40 μm, length: 12 mm) were used as reinforcing fibers. AE agent (AE, anionic surfactant, density: 1.06 g/cm^3), AE plasticizer (SV, lignin sulfonate and oxycarboxylate, density: 1.09 g/cm^3), AE superplasticvizer (SF, polycarboxylic acid compound, density: 1.06 g/cm^3), foaming agent (FA, alkylsulfate surfactant, density: 1.00 g/cm^3) were used as admixtures (Ad). The anionic surfactant was selected due to its availability and the hydrophobic and hydrophilic end properties that reduce the surface tension and helps to produce a stable structure in the mixture[6]. However, the use of other classifications of surfactants such as nonionic, cationic, and zwitterionic could be considered for further research on its suitability in providing homogenous dispersion of synthetic fibers.

The PP nanofibers were pre-dispersed in the mixing water. For pre-mixing, a juicer-mixer (power: 740 W, 23000 rpm) was used for 3 min as shown in Fig. 2. Because PP nanofibers have high water-repellent properties, they are floated in water before mixing as shown in Fig. 2(a). Also, the PP nanofibers cannot be mixed within simple water. Therefore, the pre-mixing was carried out in water with the above mentioned admixtures.

Table 1. Mix proportions (wt.%)

Series	C	S/C	W/C	Ad/C	PPn (vol.%)	PP (vol.%)	PVA (vol.%)
Control40	100	40	40	–	–	–	–
PPn0.5	100	40	40	–	0.5	–	–
PPn0.5–AE0.001	100	40	40	0.001	0.5	–	–
PPn0.5-AE0.01	100	40	40	0.01	0.5	–	–
PPn0.5-FA0.1	100	40	39.9	0.1	0.5	–	–
PPn0.5-FA1.0	100	40	39.0	1.0	0.5	–	–
PPn0.5-SV2.0	100	40	38.0	2.0	0.5	–	–
Control30	100	40	30	–	–	–	–
PPn0.5-SV0.8	100	40	29.2	0.8	0.5	–	–
PPn0.5-SF0.5	100	40	29.5	0.5	0.5	–	–
PPn0.5-SF0.9	100	40	29.4	0.9	0.5	–	–
PP0.5	100	40	40	–	–	0.5	
PP0.5-AE	100	40	40	0.01	–	0.5	–
PPn0.25-PP0.25	100	40	40	–	0.25	0.25	
PPn0.25-PP0.25-AE	100	40	40	0.01	0.25	0.25	–
PPn0.05-PP0.45	100	40	40	–	0.05	0.45	–
PPn0.05-PP0.45-AE	100	40	40	0.01	0.05	0.45	–
PVA0.5	100	40	40	–	–	–	0.5
PPn0.25-PVA0.25	100	40	40	–	0.25	–	0.25
PPn0.25-PVA0.25-AE	100	40	40	0.01	0.25	–	0.25
PPn0.05-PVA0.45	100	40	40	–	0.05	–	0.45
PPn0.05-PVA0.45-AE	100	40	40	0.01	0.05	–	0.45

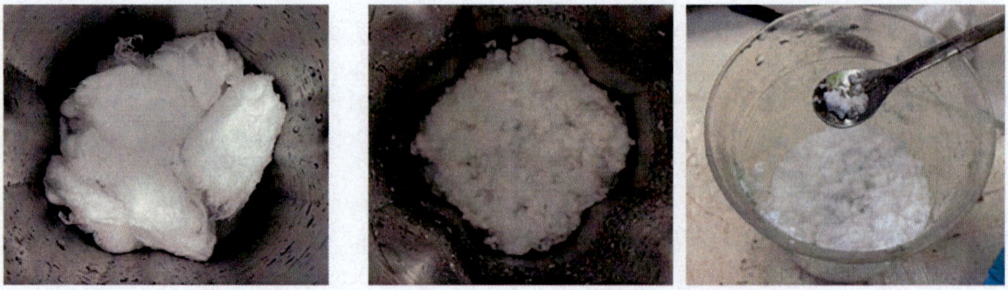

(a) Before premixing (b) After premixing

Fig. 2. Premixing of PP nanofibers using a juicer-mixer

2.2 Specimen Preparation and Curing

An omni-type mixer with a capacity of 5 L was used for FRCC mixing. To mix PP nanofibers alone, after 1 min and 30 s of dry mixing, the water pre-mixed with admixture and PP nanofiber were added and mixed for 2 min and 30 s. In the case of the normal PP and PVA fiber are mixed alone, after the dry mixing, water and admixture were added and mixed for 2 min, and one-third of the fiber was added in three batches and mixed for 1 min each. For the combined use of PP nanofiber and PVA fiber, the same procedures of PP nanofiber were applied, and the PVA fibers were added three times, each for one minute.

Prismatic specimens of 40 × 40 × 160 mm were prepared for each series for the three-point bending test. The cast specimens were stored in a room temperature with 20 °C and more than 95% of relative humidity until demolding at 24 h after casting. Then, the specimens were stored in a water bath at 20 °C until the testing at the age of 7 days.

2.3 Three-Point Bending Test

The specimens for the three-point bending test were notched at the center bottom with a depth of 12 mm (see Fig. 3a). The test was conducted using a universal testing machine (30 kN capacity, Type 5567, Instron Tool Works Inc., USA) in accordance with JCI-S-001–2003 "Test Method for Fracture Energy of Concrete Using Notched Beams" established by the Japan Concrete Institute. The specimen was set up as shown in Fig. 3(a), and the crack mouth opening displacement (CMOD) was measured by a clip gauge bridging the notch. The bending span length was 140 mm, and the loading speed of the cross-head was 0.4 mm/min.

2.4 Compressive Strength Test

The compressive strength test was carried out using a universal testing machine (1000 kN capacity, YU-1000SIV, Tokyo Koki Testing Machine Co.Ltd., Japan) at a loading speed of 0.5 N/sec. A half piece of the specimen after the bending test was used as the specimen, and the test was conducted using the apparatus shown in Fig. 3(b), referring to JIS R 5201.

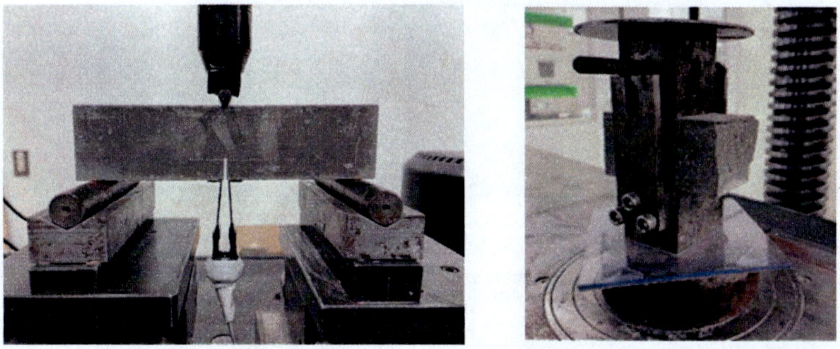

(a) Three-point bending test (b) Compressive strength test

Fig. 3. Testing procedure

2.5 SEM Observation

SEM (JSM 6500F, JEOL Ltd., Japan) observation was conducted to evaluate the dispersion of PP nanofibers in mortar.

3 Results and Discussion

3.1 Mixing and Dispersion Methods of Nanofibers

As shown in Fig. 2, the PP nanofibers have high water repellency. Thus, conventional mixing methods for mortar/concrete are not suitable for obtaining uniform dispersion. Here, the pre-mixing process was conducted by adding the admixtures with surfactant properties to the water. Except for the foaming agent, both the PP nanofibers and the admixtures were simultaneously added into water at this premixing process using the juicer-mixer. The foaming agent was added into water and mixed manually with the PP nanofibers. Figure 4 shows the dispersion conditions after the premixing. Based on the figures, it can be seen that using the AE and FA agents with a cement mass ratio of 0.01% resulted in better dispersion of 0.5PPn when compared to other proportions. In this case, the mixture with FA0.01 was visually observed to have better uniform dispersion. However, as discussed in the next sections, a lot of foam was generated in the FA agent mix, which might cause lower strength performance. In addition, it can also be explained that the use of 0.001% AE showed a little effect on the dispersion level of PP as also found when using water only.

(a) PPn0.5 (b) PPn0.5-AE0.001 (c) PPn0.5-AE0.01 (d) PPn0.5-FA0.1

(e) PPn0.5-SV2.0 (f) PPn0.5-SV0.8 (g) PPn0.5-SF0.5 (h) PPn0.5-SF0.9

Fig. 4. The conditions after the pre-mixing

3.2 Single Use of PP Nanofibers

Bending Test. The results of the bending test are shown in Fig. 5. There was no significant difference between the results of Control40 (without fiber) and PPn0.5 (without admixture). On the other hand, the series with admixtures showed an increase in bending strength, except PPn0.5-FA1.0 (with the foaming agent). Among the admixture series, the AE agent increased the bending strength the most. The higher the amount of AE agent, the greater the strength. However, the strength decreased when the high amount of foaming agent was used due to the generation of the air voids. These results suggest that the use of an appropriate amount of admixture improves the dispersibility of PP nanofibers and increases their bending strength.

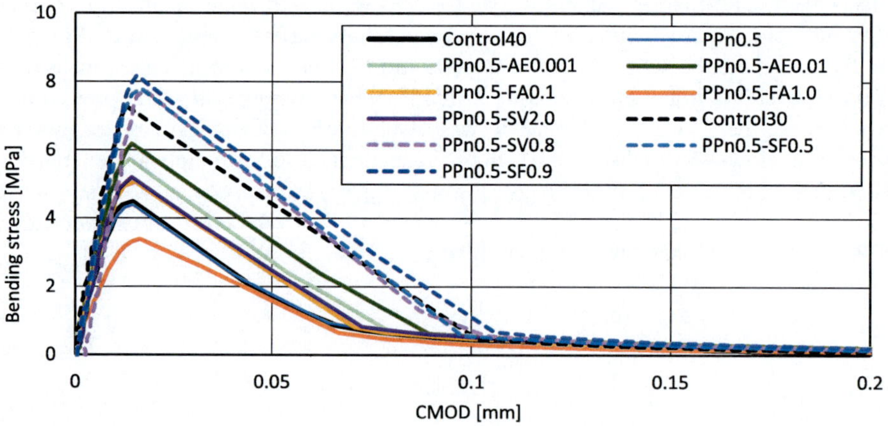

Fig. 5. Bending test results

Compressive Strength Test. Figure 6 shows the compressive strength test results. Among the series of 40% water-cement ratio, the Control40 with compressive strength of 63.2 MPa is higher than PPn0.5 with compressive strength of 54.3 MPa. This is because the PP nanofibers may not be uniformly dispersed. PPn0.5-AE0.001 and PPn0.5-AE0.01 with AE agent showed the same level strength with 62.5 MPa and 63.7 MPa, respectively. For the PPn0.5-FA0.1, PPn0.5-FA0.5, and PPn0.5-FA1.0 series with the foaming agent, the strength decreased from 45.3 MPa to 21.7 MPa. For PPn0.5-SV2.0 and the series of 30% water-cement ratio, the compressive strength increased compared to the control without the PP nanofiber for about 13% and up to 22%, respectively. This can be suggested due to the uniform dispersion of the PP nanofibers, which allows for the reinforcement of fine cracks at the initiation stage.

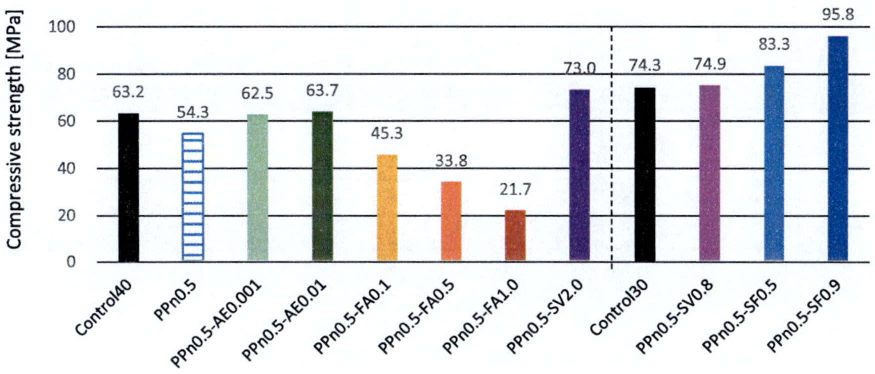

Fig. 6. Compressive strength test results

SEM Observation. Figure 7(a) shows the SEM image of PPn0.5, including agglomeration of the PP nanofiber. On the other hand, Fig. 7(b) shows better dispersion of the fibers in PPn0.5-AE0.01 using the AE agent with surfactant property, which allows the PP nanofibers to perform well resulting in improved mechanical performance.

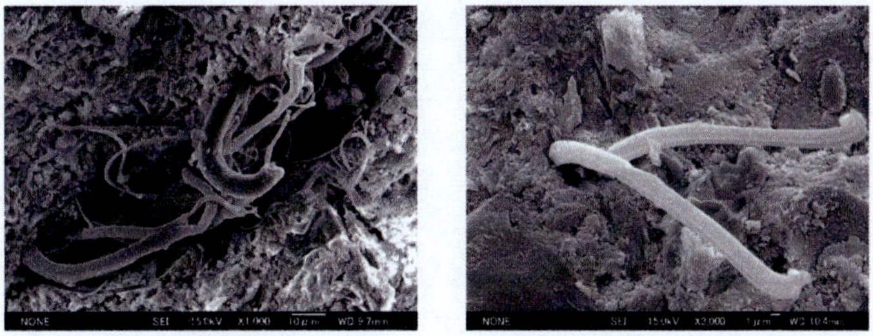

(a) PPn0.5 (b) PPn0.5-AE0.01

Fig. 7. SEM images

3.3 Combined Use of PP Nanofibers with PVA Fibers

Bending Test. The bending test results are shown in Fig. 8. No significant improvement in mechanical properties was observed when PP nanofibers were used in combination with regular PP and PVA fibers. In particular, when 0.25 vol.% of PP nanofibers were added, both strength and ductility decreased. On the other hand, 0.05 vol.% of PP nanofibers can increase the bending stress at the second peak, i.e., the flexural ductility improved when compared to mixtures with PP and PVA fibers alone. In particular, the improvement in flexural ductility can be seen in both cases with AE agents (solid line) and without AE agents (dashed line).

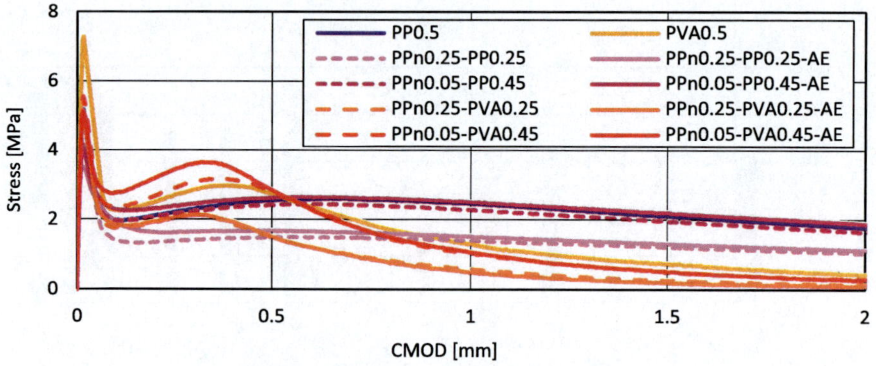

Fig. 8. PP fiber and PVA fiber alone and combined with PP nanofiber

Compressive Strength Test. The test results are shown in Fig. 9. For the combined use of PP fiber and PP nanofiber, the compressive strength was almost the same, ranging from 60.9 MPa to 61.9 MPa, regardless of the amount of PP nanofiber. A little difference on compressive strength is also found when compared to control mortar. On the other hand, for the composite of PVA fiber and PP nanofiber, the compressive strength of the

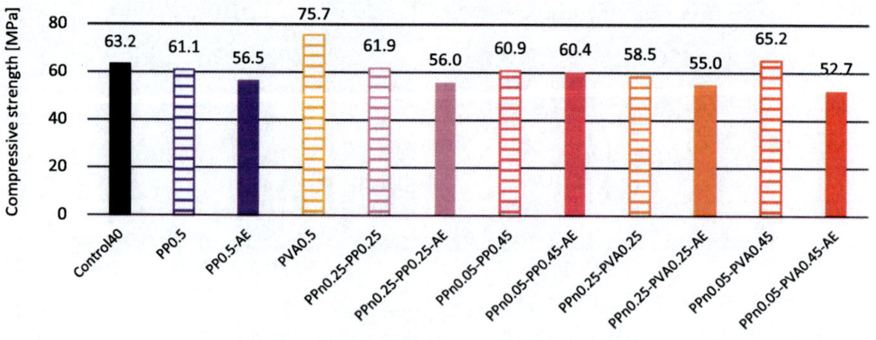

Fig. 9. Compressive strength test results

one mixed with PVA fiber alone was 75.7 MPa, which was higher than that of the ones mixed with PP nanofiber with the strength values 58.5 MPa and 65.2 MPa, respectively, and the compressive strength decreased as the amount of PP nanofiber increased.

4 Conclusion

In this study, the effect of the PP nanofibers on FRCC was investigated. Based on the bending and compressive strength tests, the results showed that the flexural strength of the PP nanofibers mixed alone was improved by the use of the AE agent. In this case, the use of AE agent with anionic surfactant based was found as the most effective admixture in improving the dispersion of PP nanofibers compared to other types of admixtures. This phenomenon was supported by the observation conducted through SEM analysis. It was also observed that the toughness of mortar was improved by combining PP nanofibers with PVA fibers using the AE agent. However, the contribution of strength due to the synthetic fibers addition is not only because of the fiber dispersion but also can be affected by the properties of surfactants that effects the mortar matrix, therefore, the use of other classification of surfactants such as nonionic, cationic and zwitterionic could be considered for further research on its suitability in providing homogenous dispersion of synthetic fibers.

References

1. Yoo, D.Y., Banthia, N.: Mechanical properties of ultra-high-performance fiber-reinforced concrete: A review. Cem. Concr. Compos. **73**, 267–280 (2016). https://doi.org/10.1016/J.CEMCONCOMP.2016.08.001
2. S. Kwon, T. Nishiwaki, T. Kikuta, H. Mihashi, Development of ultra-high-performance hybrid fiber-reinforced cement-based composites, ACI Mater. J. 111 (2014) 309–318. https://doi.org/10.14359/51686890
3. S.W.M. Supit, T. Nishiwaki, Compressive and Flexural Strength Behavior of Ultra-high Performance Mortar Reinforced with Cellulose Nano-fibers, Int. J. Adv. Sci. Eng. Inf. Technol. 9 (2019). https://doi.org/10.18517/ijaseit.9.1.7506
4. Konsta-Gdoutos, M.S., Metaxa, Z.S., Shah, S.P.: Highly dispersed carbon nanotube reinforced cement based materials. Cem. Concr. Res. **40**, 1052–1059 (2010). https://doi.org/10.1016/j.cemconres.2010.02.015
5. Otsuka, K., Mihashi, H., Kiyota, M., Mori, S., Kawamata, A.: Observation of Multiple Cracking in Hybrid FRCC at Micro and Meso Levels. J. Adv. Concr. Technol. **1**, 291–298 (2003). https://doi.org/10.3151/jact.1.291
6. E.S. Negim, L. Kozhamzharova, J. Khatib, L. Bekbayeva, C. Williams, Effects of surfactants on the properties of mortar containing styrene/methacrylate superplasticizer, Sci. World J. 2014 (2014). https://doi.org/10.1155/2014/942978

The Use of Strain Hardening Natural Fabric Reinforced Cement Based Composite Systems for Structural Applications

Felipe Pinheiro Teixeira and Flávio de Andrade Silva(✉)

Department of Civil and Environmental Engineering, Pontifícia Universidade Católica do Rio de Janeiro, Rio de Janeiro, RJ 22451-900, Brazil
fsilva@puc-rio.br

Abstract. It is known that the use of natural fibers as reinforcement for composite materials present economic benefits and eco-friendly appeal when compared to man-made fibers. However, even demonstrating excellent mechanical performance due to the strain-hardening behavior, its use for structural applications still presents a gap in the literature. Therefore, the current work discusses the use of natural fiber cement-based composites as external strengthening for concrete structures. For such, curauá natural fibers were used as reinforcement in a cement composite, which was used as a strengthening material for RC structural beams. The beams were strengthened for flexural and shear. Both externally reinforced specimens presented an increase in loading capacity and deflection decrease compared to their respective references, which was associated with a yielding delay on the rebars in a range between 21% to 34%.

Keywords: Natural fibers · Cement-based composites · Structural applications

1 Introduction

The use of natural fibers as reinforcement in cement matrices has a solid economic and ecological appeal for the new directions of the construction industry. Mechanically, many authors [1–9] described the excellent performance of that kind of composite material, mentioning increases in strength and strain capacity after the first crack due to the multiple-cracking ability. Souza et al. [10] studied cement composites reinforced by curauá long fibers in volume fractions of 4%, 7%, 8%, reaching tensile strength up to 14.7 MPa with a strain capacity of 1.6%. d'Almeida et al. [11] also studied cement composites reinforced by curauá long fibers but under bending testes, which presented flexural hardening behavior with a strength of 27.5 MPa.

However, even showing exceptional mechanical potential, natural fiber composite applications are still limited to elements such as tiles [12–14], paving blocks [15, 16], or non-structural masonry [17–19], while structural applications are commonly associated to man-made fibers, mainly polymeric and steel. As example, Lima et al. [20] evaluated short sisal fiber reinforced concrete (SSFRC) block for one-way precast concrete slabs,

M. Kunieda et al. (Eds.): SHCC 2022, RILEM Bookseries 39, pp. 62–68, 2023.
https://doi.org/10.1007/978-3-031-15805-6_7

and its behavior under the bending test presented a typical flexural hardening. Compared to the commercial blocks (ceramic and EPS), the SSFRC showed more than twice their resistance and non-brittle failure mode, reaching a load capacity 157% higher than the minimum load required by the standard for these types of blocks.

Therefore, to fill this gap, this work presents the use of a cement-based composite reinforced by natural curauá fibers as a strengthening system for RC beams. The composite was externally applied on the surface of the beams, focusing on improving the resistance to shear and bending moments. The beam submitted under flexural tests had its bottom side covered by the composite material, while for shear tests, the composite was applied over the specimen on both lateral sides. For comparison, reference beams were performed during the load tests.

2 Materials and Methods

The curauá fibers were firstly treated with hot water (70 ± 3 °C) for one hour, aiming to eliminate the impurities retained on the fiber surface [21]. Under mechanical tests, the curauá fibers presented a tensile stress equal to 706 MPa at a strain-to-failure of 2.4% with Young's modulus of 32 GPa, as presented by Teixeira et al. [22]. The composite cement matrix was designed for a 1:1:0.4 ratio (cementitious material, quartz sand and water). The cementitious material was composed in mass by 50% of Portland cement type CPV [23], 40% of metakaolin and 10% of fly ash; these pozzolans supplementation aims to produce a calcium hydroxide free matrix [21]. The matrix presented an axial compressive strength equal to 81.0 MPa after 28 days. The composite manufacturing consists of a layering process, in which a matrix layer was placed in a steel mold followed by a layer of curauá fibers, one later at a time. The curauá fibers amount per specimen corresponds to a volume fraction of 5%, divided into three layers longitudinally oriented. This process resulted in composite laminate plates measuring 500 mm length, 60 mm width, and 10 mm thickness. These specimens presented strain-hardening behavior under tensile tests, reaching maximum stress equal to 12.8 MPa at a strain-to-failure of 2.7% with Young‘s modulus of 4.7 GPa, as described by Teixeira e Silva [24].

The concrete mix for structural RC beams is explained in Table 1, and its average axial compressive strength after 28 days reached 33 MPa. The structural beams were designed with a flexural reinforcement ratio of 0.55% (two 8 mm bars), with two distinctions: 1) the beams designed for flexural failure presented shear reinforcement stirrups with 125 mm spacing while 2) the beams designed for shear failure presented no shear reinforcement along the testing region. For the conventional reinforcement, steel rebars with the nominal yield strength of 500 MPa were used, and the beam was cured for 14 days before the composite application. Figure 1 presents the schematic beam details and dimensions.

For the composite application as external strenghening layer, 10 mm thick laminate plates were manufactured over the beams‘ surfaces. The composite plates were fabricated with three curauá fiber layers (volume fraction of 5%). For the external flexural strengthening layer, the composite was applied at the bottom of the beams, while for the shear the application was at both lateral sides. To achieve the desired length, fibers overlaps of 70 mm were adopted for continuity. The tests on structural beams were carried in an MTS servo-hydraulic system (500 kN load capacity) with deflection values acquired

Table 1. Concrete mix proportions.

Materials	Kg/m^3
Portland cement (CPII-F32)	336.0
Natural sand	642.0
Coarse aggregate (9 mm)	441.0
Coarse aggregate (19 mm)	782.0
Water	168.0
Superplasticizer (PLASTOL® 4100)	0.5

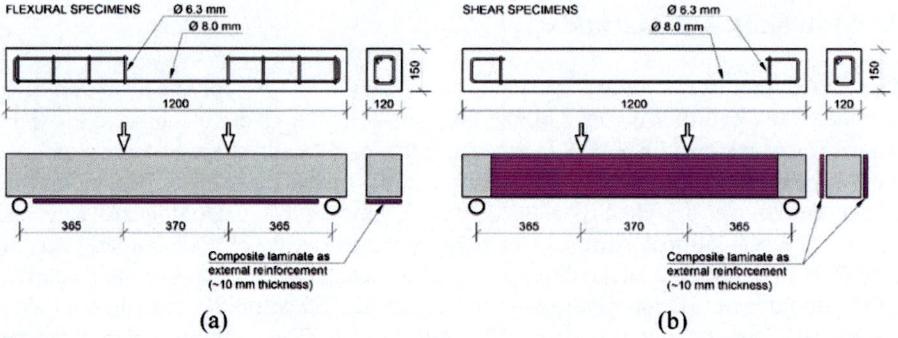

Fig. 1. Schematics of beams details and dimensions: flexural specimens (a) and shear specimens (b).

by three LVDTs arranged at the beam's bottom, aligned with the load points and at its center part. The strain measurements were read by strain gauges on the steel rebars, placed on each rebar. The test displacement rate was 1.0 mm/min over an 1100 mm span between end supports.

3 Results and Discussions

Both structural beams strengthened for flexural and shear presented an increase in loading capacity and a decrease in deflection range compared to their respective references. The flexural reference specimen presented a load peak of 36.0 kN, while its externally reinforced counterpart showed a maximum load capacity of 41.9 kN (16% higher). The same occurred to the shear specimens, which the externally reinforced beam demonstrated a strengthening increase of 28% over its reference (37.5 kN over 29.3 kN, respectively), as well as a higher stiffness. Figure 2 show the failure of both externally reinforced beams and Fig. 3 present the mechanical behavior of flexural and shear specimens.

About the deflection at maximum load, the shear specimens presented a reasonable variation, in which the externally reinforced beam exhibited a decrease of 4.5% compared to its reference (6.6 mm to 6.3 mm). On the other hand, the deflection decrease

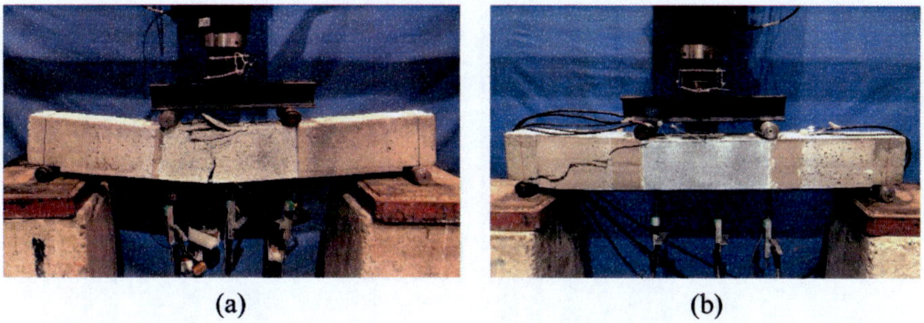

Fig. 2. The externally reinforced beams: flexural failure (a) and the shear failure (b).

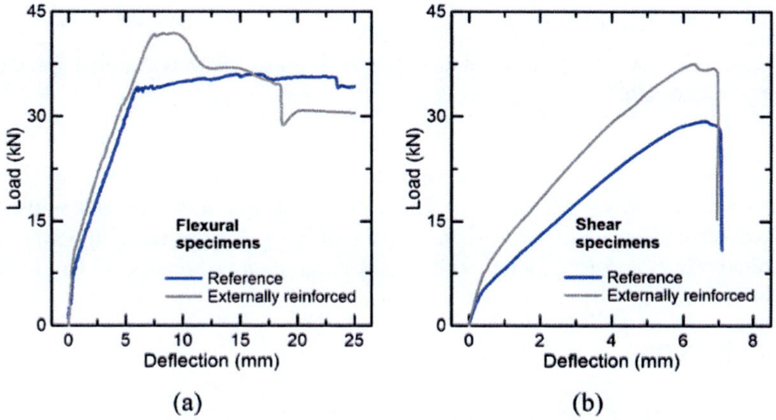

Fig. 3. Mechanical behavior of specimens under flexural tests (a) and shear tests (b).

demonstrated by the externally reinforced beam under the flexural test was expressive, reaching 43.2% (16.2 mm of the reference beam against 9.2 mm of its externally reinforced counterpart). The same load capacity increase with deflection decrease was presented by Schladitz et al. [25] in textile high-performance carbon-composite as concrete slabs reinforcement. It is possible to assume that these decreases in deflection range are associated with a yielding delay on the rebars caused by the composite strengthening contributions, resulting in stiffness gains. Figure 3 shows a comparison of the rebars yielding progress (measured by the strain gauges) at different loading stages up to the maximum strength of each reference specimen, under flexural and shear tests (Fig. 4).

In general, the reference specimens showed higher rebar strains at all known loads, from 5 kN up to each respective maximum strength, which indicates a relevant contribution of the longitudinal continuous curauá fibers to resist the forces at the tension zone. The strain measurements on the rebars were reduced 34% for the flexural reinforced specimen and 21% for the shear reinforced one, compared to their respective references at its maximum loading. It is worth mentioning that, even different from usual

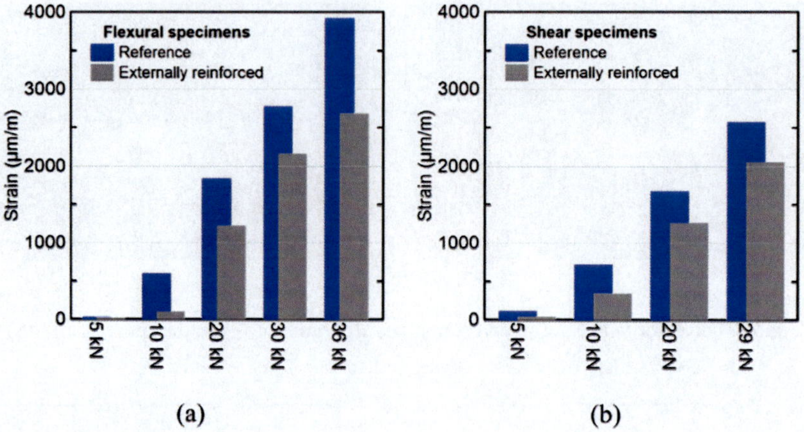

Fig. 4. Correlation between rebars yielding progress and load capacity: flexural specimens (a) and shear specimens (b).

wrapping techniques or transverse external reinforcements, the proposed shear strengthening system (sideways applied over almost the total length of the beam) also had a valuable contribution to flexural resistance and stiffness. Furthermore, in both cases the longitudinal rebars did not reach their nominal yielding at failure.

4 Conclusions

The developed natural curauá fiber-reinforced composite demonstrated an excellent behavior as a structural component, providing a higher load and strain capacity to the externally strengthened RC beam. The following are some highlights of present work:

- The natural fiber composite as a structural reinforcement provided an increase in the ultimate strength of the structural beams before its maximum deflection, about 16% and 28% for flexural and shear specimens, respectively;
- The adopted application method proved to be effective, presenting no signs of failure due to delamination or displacement, providing stiffness enhancement to the beams;
- The stiffness increase and deflection decrease can be associated with the yielding delay on the rebars caused by the composite strengthening contributions.

References

1. Ferreira, S.R., de A. Silva, F., Lima, P.R.L., Toledo Filho, R.D.: Effect of hornification on the structure, tensile behavior and fiber matrix bond of sisal, jute and curauá fiber cement based composite systems. Constr. Build. Mater. **139** (2017) 551–561. https://doi.org/10.1016/j.conbuildmat.2016.10.004

2. Ferreira, S.R., Silva, F.D.A., Lima, P.R.L., Toledo Filho, R.D.: Effect of fiber treatments on the sisal fiber properties and fiber-matrix bond in cement based systems. Constr. Build. Mater. **101**, 730–740 (2015). https://doi.org/10.1016/j.conbuildmat.2015.10.120
3. Alves Fidelis, M.E., Pereira, T.V.C., Gomes, O.D.F.M., de Andrade Silva, F., Toledo Filho, R.D.: The effect of fiber morphology on the tensile strength of natural fibers. J. Mater. Res. Technol. **2**, 149–157 (2013). https://doi.org/10.1016/j.jmrt.2013.02.003
4. de A. Silva, F., Mobasher, B., Filho, R.D.T.: Cracking mechanisms in durable sisal fiber reinforced cement composites. Cem. Concr. Compos. **31**, 721–730 (2009). https://doi.org/10.1016/j.cemconcomp.2009.07.004
5. Silva, F.D.A., Mobasher, B., Soranakom, C., Filho, R.D.T.: Effect of fiber shape and morphology on interfacial bond and cracking behaviors of sisal fiber cement based composites. Cem. Concr. Compos. **33**, 814–823 (2011). https://doi.org/10.1016/j.cemconcomp.2011.05.003
6. de A. Silva, F., Chawla, N., de T. Filho, R.D.: Tensile behavior of high performance natural (sisal) fibers, Compos. Sci. Technol. **68**, 3438–3443 (2008). https://doi.org/10.1016/j.compscitech.2008.10.001
7. Filho, J.D.A.M., Silva, F.D.A., Toledo Filho, R.D.: Degradation kinetics and aging mechanisms on sisal fiber cement composite systems, Cem. Concr. Compos. **40**, 30–39 (2013). https://doi.org/10.1016/j.cemconcomp.2013.04.003
8. Komuraiah, A., Kumar, N.S., Prasad, B.D.: Chemical composition of natural fibers and its influence on their mechanical properties. Mech. Compos. Mater. **50**(3), 359–376 (2014). https://doi.org/10.1007/s11029-014-9422-2
9. Zukowski, B., de Andrade Silva, F., Toledo Filho, R.D.: Design of strain hardening cement-based composites with alkali treated natural curauá fiber. Cem. Concr. Compos. **89**, 150–159 (2018). https://doi.org/10.1016/j.cemconcomp.2018.03.006
10. de Souza, L.O., de Souza, L.M.S., de Andrade Silva, F.: Mechanics of natural curauá textile-reinforced concrete. Mag. Concr. Res. **73**, 135–146 (2021). https://doi.org/10.1680/jmacr.18.00473
11. d'Almeida, A.L.S., Melo Filho, J.A., Toledo Filho, R.D.: Use of curaua fibers as reinforcement in cement composites. Chem. Eng. Trans. **17**, 1717–1722 (2009). https://doi.org/10.3303/CET0917287
12. Savastano, H., Jr., Agopyan, V., Nolasco, A.M., Pimentel, L.: Plant fibre reinforced cement components for roofing. Constr. Build. Mater. **13**, 433–438 (2000). https://doi.org/10.1016/S0950-0618(99)00046-X
13. Roma, L.C., Martello, L.S., Savastano, H.: Evaluation of mechanical, physical and thermal performance of cement-based tiles reinforced with vegetable fibers. Constr. Build. Mater. **22**, 668–674 (2008). https://doi.org/10.1016/j.conbuildmat.2006.10.001
14. Tonoli, G.H.D., Santos, S.F., Savastano, H., Delvasto, S., Mejía De Gutiérrez, R., Lopez De Murphy, M.D.M.: Effects of natural weathering on microstructure and mineral composition of cementitious roofing tiles reinforced with fique fibre, Cem. Concr. Compos. **33**, 225–232 (2011). https://doi.org/10.1016/j.cemconcomp.2010.10.013
15. Zak, P., Ashour, T., Korjenic, A., Korjenic, S., Wu, W.: The influence of natural reinforcement fibers, gypsum and cement on compressive strength of earth bricks materials. Constr. Build. Mater. **106**, 179–188 (2016). https://doi.org/10.1016/j.conbuildmat.2015.12.031
16. Kundu, P.S., Chakraborty, S., Chakraborty, S.: Effectiveness of the surface modified jute fibre as fibre reinforcement in controlling the physical and mechanical properties of concrete paver blocks. Constr. Build. Mater. **191**, 554–563 (2018). https://doi.org/10.1016/j.conbuildmat.2018.10.045
17. Jami, T., Karade, S.R., Singh, L.P.: A review of the properties of hemp concrete for green building applications. J. Clean. Prod. **239**, 117852 (2019). https://doi.org/10.1016/j.jclepro.2019.117852

18. Sassoni, E., Manzi, S., Motori, A., Montecchi, M., Canti, M.: Novel sustainable hemp-based composites for application in the building industry: physical, thermal and mechanical characterization. Energy Build. **77**, 219–226 (2014). https://doi.org/10.1016/j.enbuild.2014.03.033
19. Abdullah, A.C., Lee, C.C.: Effect of treatments on properties of cement-fiber bricks utilizing rice husk, corncob and coconut Coir. Procedia Eng. **180**, 1266–1273 (2017). https://doi.org/10.1016/j.proeng.2017.04.288
20. Lima, P.R.L., Barros, J.A.O., Roque, A.B., Fontes, C.M.A., Lima, J.M.F.: Short sisal fiber reinforced recycled concrete block for one-way precast concrete slabs. Constr. Build. Mater. **187**, 620–634 (2018). https://doi.org/10.1016/j.conbuildmat.2018.07.184
21. de A. Silva, F., Filho, R.D.T., de A.M. Filho, J., de M.R. Fairbairn, E.: Physical and mechanical properties of durable sisal fiber–cement composites. Constr. Build. Mater. **24**, 777–785 (2010). https://doi.org/10.1016/j.conbuildmat.2009.10.030
22. Teixeira, F.P., da F.M. Gomes, O., de A. Silva, F.: Degradation mechanisms of curaua, hemp, and sisal fibers exposed to elevated temperatures, BioResources. **14,** 1494–1511 (2019). https://doi.org/10.15376/biores.14.1.1494-1511
23. ABNT NBR 16697, ABNT NBR 16697 Cimento Portland – Requisitos, Cim. Portl. – Requisitos (2018)
24. Teixeira, F.P., de Andrade Silva, F.: On the use of natural curauá reinforced cement based composites for structural applications. Cem. Concr. Compos. **114**, 103775 (2020). https://doi.org/10.1016/j.cemconcomp.2020.103775
25. Schladitz, F., Frenzel, M., Ehlig, D., Curbach, M.: Bending load capacity of reinforced concrete slabs strengthened with textile reinforced concrete. Eng. Struct. **40**, 317–326 (2012). https://doi.org/10.1016/j.engstruct.2012.02.029

Mechanical Properties of Fiber-Reinforced Cementitious Composites Manufactured Using 3D-Printing Technology

Hiroki Ogura(✉), Shinya Yamamoto, and Hiroyuki Abe

Institute of Technology, Shimizu Corporation, Tokyo, Japan
ogura_h@shimz.co.jp

Abstract. In this study, 3D-printable fiber-reinforced cementitious composites with PE fiber contents of 0.75% were characterized. Specimens were extracted from 3D-printed elements and subjected to compression, splitting tensile, and bending tests to evaluate the mechanical characteristics of the printed elements. The results of the compression tests showed that the compressive strength of the specimens cored in the vertical direction was 109 MPa and the coefficients of variation of the compressive strength and Young's modulus were smaller than those of the mold-cast specimens. Splitting tensile tests were conducted on specimens in which the direction of the interface between each printed layer (layer-to-layer interface) and the direction of crack propagation matched, and the result showed that, at 3.50 MPa, the cracking strength was 30% lower than that of the specimens whose direction of the layer-to-layer interface and crack interface did not match.

Keywords: 3D printing · Additive manufacturing technology · Material extrusion · Fiber-reinforced cementitious composites · Anisotropy

1 Introduction

In recent years, the research and development of construction-scale three-dimensional (3D) printing technology have been pursued aggressively around the world. The technology of 3D printing involves calculating a sliced 2D cross-section shape based on data representing a 3D object on a computer and using the 2D shape to layer and mold materials into a 3D shape. While there are various methods of construction-scale 3D printing, the mainstream technology is the material extrusion method. This involves taking fresh cementitious materials, extruding them through a nozzle, and layering them to create a 3D shape. As this technology enables mechanized construction without the need for molds in concrete works, it not only contributes to labor-saving and productivity improvements but also improves the degree of design freedom, rationalization of layout, and reduction of the environmental burden by eliminating mold wastes.

As it is difficult to add rebars for reinforcement using 3D printing technology, a method of reinforcement that can be an alternative to conventional steel reinforcement may be required. Various reinforcement methods have been suggested so far [1],

M. Kunieda et al. (Eds.): SHCC 2022, RILEM Bookseries 39, pp. 69–78, 2023.
https://doi.org/10.1007/978-3-031-15805-6_8

one of which is reinforcement using short fibers. For example, Le et al. [2], Hambach et al. [3] and Ogura et al. [4] conducted 3D printing using fiber-reinforced cementitious composites and evaluated its effects to increase tensile properties.

The authors also developed the fiber-reinforced cementitious composite suitable for 3D printing; moreover, constructed a 3D printing technology that can print large-scale elements with excellent mechanical characteristics and durability (Fig. 1). To evaluate the hardening properties of 3D-printed elements made using this technique, mechanical tests were conducted on specimens cored and cut from the printed elements in previous research [5]. It was confirmed that the difference between the 3D-printed elements and from the specimens produced using mold-casting was small. However, the integrity and mechanical characteristics of the interface between each printed layer (hereinafter referred to as layer-to-layer interface) and the anisotropy of the printed element were not investigated.

Thus, in this study, to evaluate the mechanical characteristics, including anisotropy in specimens that have been 3D-printed using the fiber-reinforced cementitious composites, mechanical tests were performed on specimens extracted from 3D-printed elements and specimens cast in molds. Then, the effects on the mechanical characteristics of the specimens were investigated by changing the direction of the specimens cored and cut from 3D-printed elements, the loading direction relative to the layer-to-layer interface, and the dimensions of the specimens.

Fig. 1. Image of a material extrusion 3D printing.

2 3D-Printable Fiber-Reinforced Cementitious Composites

When printing the cementitious materials with the material extrusion method, the properties required for the material differ from those required for conventional concrete. Firstly, the property (extrudability) that enables the material to be easily transferred through a thin hose or a print head nozzle is required. If this property is inferior, it is possible that the nozzle may become blocked by the material. It is also necessary for the deformation of the printed material caused by the dead weight of the next layer (upper layer) to be printed so as to keep it to a minimum. This is because the deformation due to the dead weight may cause the collapse of the deposited fresh material. If the deposited fresh material has this resistance, it is said to be highly buildability.

Investigations with both numerical analysis and experiment were conducted to search for a mixture proportion that satisfies both extrudability and buildability and developed the fiber-reinforced cementitious composite called LACTM® (Laminatable Cement-based Tough Material). It satisfies both extrudability and buildability in its fresh state and exhibits superior mechanical characteristics [5].

The state of fabrication of a print with a square cylinder cross-section (500 × 700 mm) is shown in Fig. 2. The printed column can be stacked up to 2 m in height. It maintained a constant shape during and after printing, and no signs of disintegration were observed. The height was limited to 2 m due to the range of motion of the robot arm; however, if the range of motion is not limited, it would be highly possible to print column of over 2 m.

Fig. 2. Printing status of a 2-m high column.

3 Experimental Program

3.1 Mixture Proportions and Raw Materials

The fiber-reinforced cementitious composite used in this study was formulated at a water to powder ratio of 0.24, sand to powder ratio of 0.80, and fiber volume fraction of 0.75 vol.%. For the powder, cement, silica fume, fly ash, and limestone powder were used. The sand had a maximum grain size of 0.85 mm. A polycarboxylic acid-based high-performance water-reducing agent was used as an admixture. The fiber was made of polyethylene with a length of 6 mm and diameter of 0.012 mm.

The material was mixed in a 120-L twin-shaft mixer, and the mixing time was 6 min after loading the materials. The air volume immediately after kneading was 5.8%, and the flow value obtained from the flow test (JIS R 5201:2015) was 125 mm.

3.2 Manufacturing of Specimens

The 3D printing device used in this study was a device that can control the nozzle tip using the robot arm to continuously push cementitious materials to a certain position. The travel speed of the nozzle was set to 100 mm/s, and the print width was set to 70 mm. The nozzle was controlled so that it was raised 7 mm in a vertical direction after printing each layer to move to the next layer.

Figure 3 shows the printed elements created in this study. The shape of the element is a square cylinder with a width of 750 mm, depth of 500 mm, and height of 350 mm. On completion of the prints, the element was left to harden without surface troweling. Using the same mixture, specimens were also prepared through casting (hereinafter referred to as mold-cast specimens).

Fig. 3. 3D-printed elements.

Fig. 4. Specimens extracted from the elements.

The elements printed for the mechanical tests and the mold-cast specimens were left in the laboratory for 1 day under the condition that moisture did not deviate. They were then transferred to a room with a constant temperature and humidity (temperature: 20 ± 2 °C, humidity: 60 ± 5% RH) and sealed and cured for 20 days. The printed elements were removed from the room on the 21st day. They were then, as shown in Fig. 4, cored and cut out, and specimens for mechanical tests (hereinafter referred to as printed specimens) were formed. The printed specimens were then sealed again and stored in the room with a constant temperature and humidity for 27 days.

Multiple printed specimens were cored or cut out from the element with varying direction and diameters. A list of the specimens fabricated is shown in Table 1. For the compression test, five cores with ϕ 50 mm and ϕ 30 mm in the vertical direction and five cores with ϕ 30 mm in the horizontal direction were sampled. For the splitting tensile test, the cores with ϕ 50 mm and ϕ 30 mm in the vertical direction and cores with ϕ 50 mm in the horizontal direction were sampled. Additionally, 10 rectangular specimens measuring 40 × 40 × 160 mm^3 in the horizontal direction were cut as specimens for the three-point bending test.

Table 1. List of specimens prepared for mechanical testing.

	Specimen name	Method of fabrication	Shape (mm)	Core/cut Direction	No. of tests
Compression test	C-mold	Mold-cast	ϕ 50 × 100	–	3
	C-50V	Print	ϕ 50 × 100	Vertical	5
	C-30V	Print	ϕ 30 × 60	Vertical	5
	C-30H	Print	ϕ 30 × 60	Horizontal	5
Splitting tensile test	S-mold	Mold-cast	ϕ 100 × 200	–	3
	S-50V	Print	ϕ 50 × 50	Vertical	5
	S-30V	Print	ϕ 30 × 30	Vertical	5
	S-50Ha	Print	ϕ 50 × 50	Horizontal	5
	S-50Hb	Print	ϕ 50 × 50	Horizontal	5
Bending test	B-mold	Mold-cast	40 × 40 × 160	–	5
	B-40Ha	Print	40 × 40 × 160	Horizontal	5
	B-40Hb	Print	40 × 40 × 160	Horizontal	5

3.3 Method for Mechanical Tests

At the material age of 27 days, the specimens were removed from the room with constant temperature and humidity and prepared for testing. Mechanical tests were performed at the material age of 28 days. The mechanical tests included compression tests, splitting tensile tests, and three-point bending tests; the loading methods were in accordance with JIS A 1107:2012, JIS A 1113:2018, and JIS A 1106:2018 standards, respectively.

For the compression test, Young's modulus and Poisson's ratio were obtained by attaching strain gauges. Strain gauges were also affixed to the specimens for the splitting tensile tests in the direction perpendicular to the loading axis on the bottom of the cylinder (on both sides) to detect cracks. For the 50-mm-diameter specimen, a strain gauge with a measuring length of 30 mm was used, whereas, for the 30-mm-diameter specimens, strain gauges with a measuring length of 10 mm were used. For the three-point bending test, a strain gauge with a measuring length of 60 mm was attached in the longitudinal direction to the central position of the bottom surface (fulcrum side).

Due to the possibility of the high variability of the specimens extracted from the printed element, the number of specimens was set to five. The bottom surface of the 50-mm-diameter cylinder specimen was polished on both sides for smoothness, and for the 30-mm-diameter cylinder specimen, both ends were capped using plaster as it could not be set in the grinder.

Figure 5 shows the loading direction of the specimens. The dashed line in the figure represents the direction of the layer-to-layer interface. As the print-layer width was 7 mm, all specimens consisted of four or more print layers.

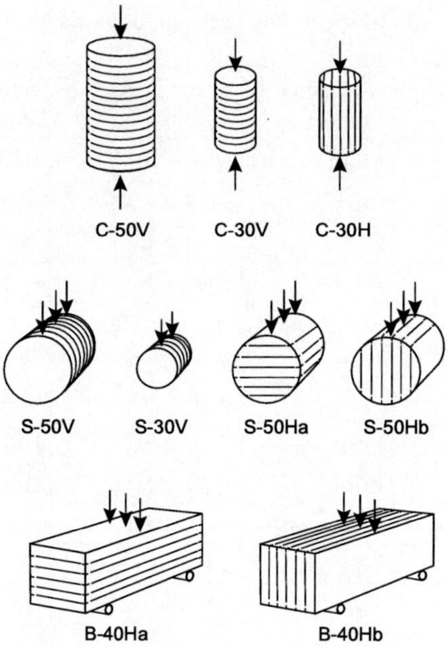

Fig. 5. Loading direction of the specimens (dashed line: layer-to-layer interface).

4 Experiment Results

4.1 Compression Tests

Table 2 lists the results obtained from the compression tests. Each value is the average value of the results obtained from five specimens (3 in case of C-mold cast specimens). The coefficient of variation is shown in parentheses. The printed specimen C-50V had virtually the same compressive strength, Young's modulus, and Poisson's ratio as the mold-cast specimens.

Table 2. Results of the compression tests—mean values: coefficients of variation are given in parentheses.

	Compressive strength [MPa]	Young's modulus [GPa]	Poisson's ratio
C-mold	107 (5.2%)	36.9 (0.80%)	0.221 (1.3%)
C-50V	109 (1.2%)	34.8 (0.34%)	0.209 (0.86%)
C-30V	95.6 (2.5%)	33.7 (2.3%)	—
C-30H	85.0 (3.6%)	36.2 (3.6%)	—

Although the results cannot simply be compared due to different number of sample, the coefficient of variation was smaller for C-50V than for the mold-cast specimens. One reason why the coefficient of variation did not increase for the printed samples is that no significant defects were observed in the interfaces between layers. Figure 6 shows the external appearance of specimens cored from the printed elements. No visible discontinuities between layers were present as well as no voids or cracks were observed in any of the specimens. Based on the above results, it is suggested that there is little disparity in the quality of the 3D-printed elements. The average dimensions of the specimens in the respective cases were ϕ 50.1 mm × 97.4 mm for the C-mold, ϕ 47.9 mm × 98.9 mm for C-50V, ϕ 29.7 mm × 56.5 mm for C-30V, and ϕ 29.7 mm × 56.2 mm for C-30H.

Fig. 6. External appearance of cored specimens.

4.2 Splitting Tensile Tests

Table 3 lists the results obtained from the crack tension tests. The cracking strength was determined based on the stress at the point where the measured strain became discontinuous. Each value is the average value of the results obtained from five specimens (three in case of C-mold cast specimens). The crack initiation strength was generally similar except in the case of S-50Hb. It can be deduced that the higher coefficient of variation for the printed specimens compared to the mold-cast specimen S-mold is related

Table 3. Results of the splitting tensile tests—mean values: coefficients of variation are given in parentheses.

	Cracking strength [MPa]
S-mold	5.40 (8.0%)
S-50V	5.02 (18%)
S-30V	5.75 (15%)
S-50Ha	5.53 (5.7%)
S-50Hb	3.50 (31%)

to the effect of the specimen fabrication accuracy. As the printed specimens are cored from elements, there is a slightly wavy shape on the sides of the cylinder. Contrary to the compression test, as the loading surface for the splitting tensile test is on the side of the cylinder, the loading plate and specimen may not have contacted evenly, which may have resulted in biased stress distribution.

The cracking strength of S-50Hb was 3.50 MPa, which was 70% of that of S-50V. The reason that the cracking strength became smaller is thought to be because the direction of the weak side layer-to-layer interface and the crack propagation direction did not match. The coefficient of variation was a smaller value compared to other cases.

Figure 7 shows the relationship between the stress cracks and strain obtained in the splitting tensile test. For S-50Ha, the stress increased after the cracks were generated and breakage occurred. This is the effect of fibers bridging the crack surface after the cracking. As the direction of the cracking surface and the layer-to-layer interface of S-50Hb coincide, there are few fibers cross-linking the cracked surface, and it is considered that S-50Hb exhibits a behavior in which the strain increases rapidly with the onset of crack.

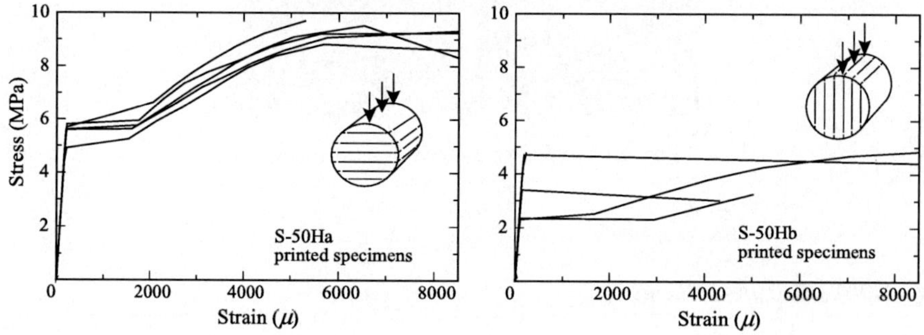

Fig. 7. Relationship between the tensile stress and strain.

4.3 Bending Tests

A list of results obtained from the three-point bending test is shown in Table 4. The bending cracking strength was determined from the stress at the point where the measured strain became discontinuous. Each value is the average of the results obtained from five specimens. In B-40Ha, there are higher values for the cracking strength and bending strength than in mold-cast specimen B-mold. The difference, however, between the mold-cast specimen and the printed specimen was less significant than in the case of the splitting tensile test. This is thought to be that, unlike the splitting tensile test, there were no cases in the bending test where the direction of the crack propagation and layer-to-layer interface matched.

Table 4. Results of the three-point bending tests—mean values: coefficients of variation are given in parentheses.

	Cracking strength [MPa]	Bending strength [MPa]
B-mold	7.89 (3.4%)	12.1 (9.6%)
B-40Ha	8.27 (5.7%)	14.4 (4.6%)
B-40Hb	7.35 (8.1%)	14.0 (9.0%)

Figure 8 shows the stress–strain relationship obtained from the bending test. For all specimens, the specimens did not fail at the same time as cracking; however, stress increased again due to the bridging effect of the fibers, which led to breakage. In the case of the mold-cast specimen, the cracks reached maximum stress at 3000 μ to 6000 μ, whereas in the case of B-40Ha, maximum stress was achieved when they exceeded 6000 μ. This difference in behavior can presumably be explained by examining the changes in the pore structure inside the mortar and the direction of the short fibers depending on how the specimens were fabricated; however, further analysis would be necessary to understand this properly.

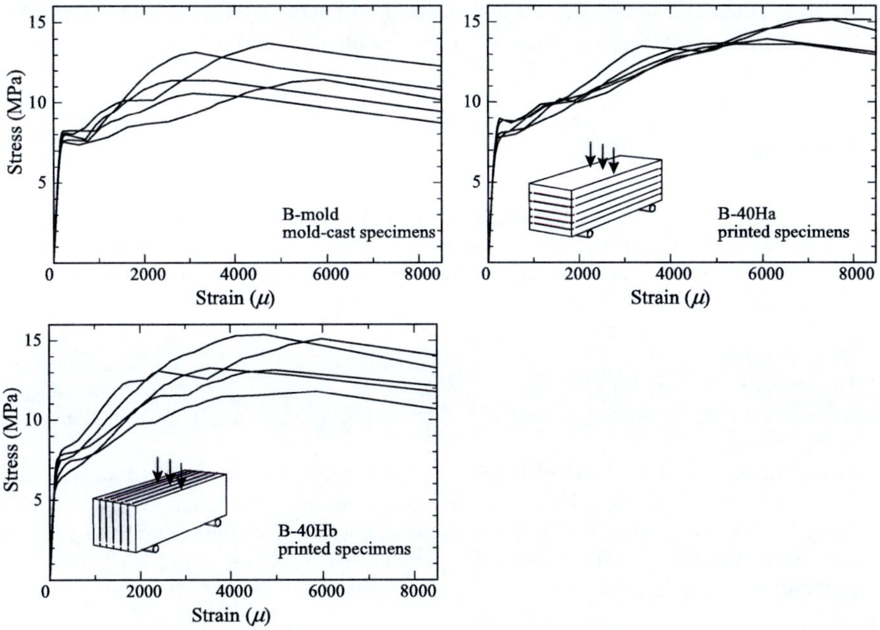

Fig. 8. Relationship between the bending stress and strain.

5 Conclusion

In this study, mechanical tests were performed on specimens that were cored and cut out from printed elements and mold-cast specimens, with the purpose of assessing the basic properties of 3D-printed elements made from the fiber-reinforced cementitious composites. The results obtained are shown below.

(1) Compression tests were performed on specimens cored out of the 3D-printed element in the vertical direction with a diameter of 50 mm. These exhibited behavior equivalent to that of specimens made through casting, yielding a compressive strength of 109 MPa. Compared to the mold-cast specimens, the coefficients of variation of the compressive strength and Young's modulus were smaller, suggesting that there was less disparity in the quality of 3D-printed elements.
(2) Other than the case where the loading direction and the layer-to-layer interface coincided, crack tension test using specimens cored from 3D-printed elements showed that crack initiation strength averaged at or more than 5 MPa. The cracking strength of the specimens, in which the loading direction coincided with the layer-to-layer interface, was 30% lower than that of the other specimens.
(3) In the three-point bending test, it is found that the cracking strength and bending strength were equivalent to those of the mold-cast specimens. All the specimens did not fracture at the same time as the cracking, and the bridging effect of the fibers caused the stress to rise again, leading to breakage. The bending strength of the printed specimens was equal to or greater than 14 MPa.

References

1. Mechtcherine, V., Nerella, V.N., Ogura, H., Grafe, J., Spaniol, E., Hertel, M., et al.: Alternative reinforcements for digital concrete construction. In: Wangler, T., Flatt, R.J. (eds.) First Rilem International Conference on Concrete and Digital Fabrication - Digital Concrete 2018, pp. 167–175. Springer, Dordrecht (2019)
2. Le, T.T., et al.: Hardened properties of high-performance printing concrete. Cem. Concr. Res. **42**(3), 558–566 (2012)
3. Hambach, M., Möller, H., Neumann, T., Volkmer, D.: Portland cement paste with aligned carbon fibers exhibiting exceptionally high flexural strength (>100 MPa). Cem. Concr. Res. **89**, 80–86 (2016)
4. Ogura, H., V., Nerella and Mechtcherine, V.: Developing and testing of Strain-Hardening Cement-Based Composites (SHCC) in the context of 3D-printing. Materials **11**(8), 1375 (2018)
5. Ogura, H., Abe, H., Tanaka, H.: Mechanical properties of fiber reinforced cement-based composites manufactured by 3D-printing. JSCE Annual Meeting Proceedings, V-101 (2019) (in Japanese)

SHCC Reinforced 3D Printed Concrete

Gideon van Zijl[1(✉)], Marchant van den Heever[1,2], and Seung Cho[1,3]

[1] Department of Civil Engineering, Stellenbosch University, Stellenbosch, Republic of South Africa
gvanzijl@sun.ac.za
[2] Harcourt Technologies Limited, Dublin, Ireland
[3] Ulsan National University of Science and Technology, Ulsan, Republic of Korea

Abstract. The paper presents a reinforcement strategy for 3D printed concrete by a thin, bonded strain-hardening cement-based composite (SHCC) overlay. Concrete structures additively manufactured through extrusion-based 3D concrete printing (3DCP) present orthotropic mechanical behaviour. Tensile strength and ductility across layers are a fraction of those in the extruded direction, i.e. parallel to the layers, due to reduced interfacial bond between printed layers. In addition, structural application of 3DCP may require reinforcement to enhance limited tensile capacity of printed concrete. Strategies for automated 3DCP reinforcement are sought that retain the benefits of 3DCP technology. Here, a thin, bonded SHCC reinforcing overlay is proposed. Automation is envisaged by integration of either spraying technology, or multiple nozzle technology for new, composite structural elements. For retrofitting of existing 3DCP, shotcrete or plaster application is appropriate. For the proof of concept here, thin SHCC overlays were simply plastered onto 3DCP elements. Characterising tensile tests are performed on SHCC dumbbells, and SHCC-3DCP interfacial tensile (pull-off) and shearing adhesive (triplet) tests. From these results composite 3DCP-SHCC specimens were designed to exhibit ductile behaviour by multiple cracking of the SHCC overlay, as opposed to abrupt, brittle delamination of the SHCC overlay from 3DCP substrate. Subsequently, reference 3DCP beams and SHCC reinforced 3DCP beams were prepared and subjected to in-plane and out-of-plane actions. The results validate thin bonded SHCC overlay as an appropriate reinforcement strategy for significantly increased resistance and ductility.

Keywords: SHCC · 3DCP · Reinforcement · Freeform · Retrofitting · Automation

1 Introducing SHCC as Novel Freeform Reinforcement

Reinforcement strategies for 3DCP should align with the benefits of this emerging additive manufacturing technology. Bester et al. (2021) reported that the top six advantages of 3DCP are increased design freedom, decreased labor intensity, decreased construction duration, decreased construction waste, and formwork-free construction. Various reinforcement methods have been proposed, broadly categorised as reinforcement before,

M. Kunieda et al. (Eds.): SHCC 2022, RILEM Bookseries 39, pp. 79–88, 2023.
https://doi.org/10.1007/978-3-031-15805-6_9

during and after printing. Both reinforcement before and after printing, respectively pre-manufactured reinforcement cages that are printed over with split nozzles and attachment of external steel, post-tensioning cables, or placement of steel cages in 3DCP permanent formwork, adding reinforcement as an additional step to the 3DCP construction process. Entrained fibres and entrained metal cable reinforcement (see Fig. 1a) are examples of reinforcement during concrete printing that do not compromise the manufacturing benefits. Relatively weak interlayer bond and associated fissility of 3DCP demand reinforcement orthogonal to the interfaces, which can also be inserted during printing (see Fig. 1b). Various forms of nailing and orthogonally inserted steel bars (Marchment and Sanjayan 2020) have been proposed, as reviewed by Bester et al. (2021). This paper presents a novel external reinforcement, by a bonded SHCC overlay (see Fig. 1c), which holds potential to align with the benefits of 3DCP construction. Application during concrete printing or shortly after is foreseen, which requires careful investigation of stability during additive manufacturing, and appropriate bond. Shotcrete application of SHCC to the hardened 3DCP structure is another possibility. Retrofitting, to strengthen existing, damaged or deteriorated unreinforced or reinforced 3DCP structures, similar to SHCC retrofitted masonry walls to enhance resistance and ductility (Van Zijl and De Beer 2019), is another foreseen class of applications of this novel 3DCP reinforcement strategy. In new construction and retrofitting, complex geometrical forms can be accommodated.

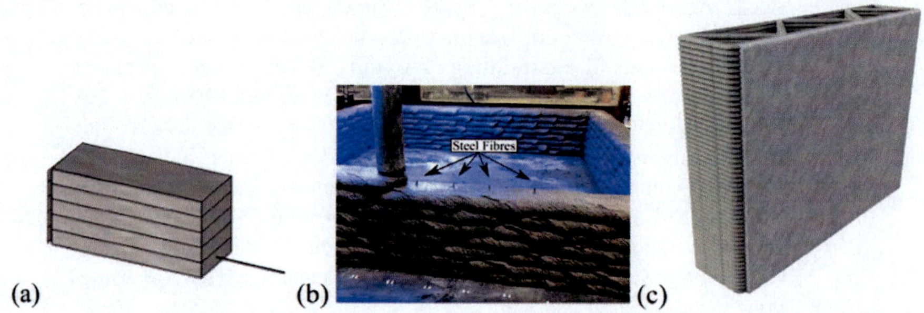

Fig. 1. (a) Entrained reinforcing cable in 3DCP (Bos et al. 2017). (b) 3DCP layer connecting transverse fibres. (c) Illustration of a SHCC reinforced 3DCP wall.

2 Reinforcement Strategy

Depending on the substrate and overlay pairing, objectives of either enhanced resistance, ductility, or both may be achieved with thin SHCC overlays. Van Zijl and De Beer (2019) improved the in-plane shearing resistance of a double leaf masonry wall with a 15 mm thick one-sided bonded SHCC overlay (see Fig. 2). Van Zijl and De Jager (2019) significantly improved the ductility of in-plane shearing resistance on replicas of these 15 mm bonded SHCC overlay retrofitted masonry walls, by adding debonding strips.

Strengthening and/or increased ductility may be designed for by appropriate selection of SHCC reinforcing material and overlay thickness of a bonded overlay in new SHCC-reinforced 3DCP construction, or for retrofitting. This paper contribution presents thin

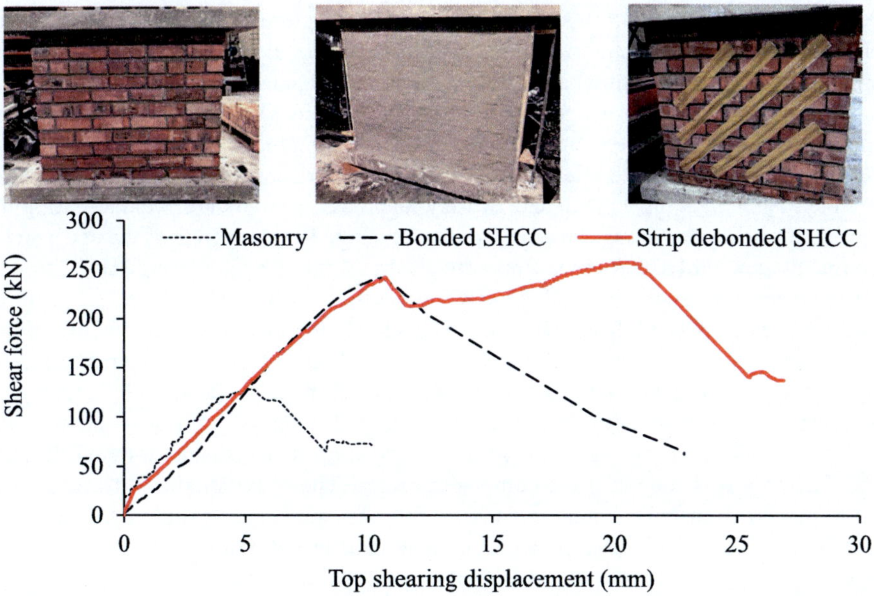

Fig. 2. In-plane shear force-displacement responses of double leaf masonry wall (top left), compared with the enhanced resistance of an identical masonry wall retrofitted with a 15 mm one-sided SHCC bonded overlay (shown top center; Van Zijl and De Beer 2019), and one with a 15 mm one-sided strip-debonded SHCC overlay (top right, showing the debonding strips before spraying on the SHCC; Van Zijl and De Jager 2019).

layer bonded SHCC reinforcement to particularly increase ductility of failure modes in 3DCP involving brittle interfaces between layers.

3 Experimental Program

3.1 Materials and Specimen Preparation

The mixes, mixing protocol, specimen protection and curing established at SU for 3DCP (van den Heever et al. 2021) and SHCC (Van Zijl and de Beer 2019) are selected. Constituents and proportions are presented in Table 1.

SHCC was mixed in a 50 L pan mixer for a predetermined duration at a mixing speed of 25 rpm (Van Zijl and De Beer 2019). For tension, compression and E-modulus mechanical property characterization, the fresh SHCC was cast in dumbbell and cylinder steel moulds respectively, compacted on a laboratory vibration table, left covered in laboratory conditions for 24 h and then stripped and cured in a controlled climate (23 ± 2 °C and 65 ± 5% relative humidity) until the test age of 28 days. From the same freshly mixed batch, SHCC reinforcement layers were plastered onto 3DCP test specimens to the pre-determined thickness (see Fig. 3) with a steel trowel.

The 3DCP was mixed in the same 50 L pan mixer, transported in the pan by trolley indoors over roughly 40 m to the adjacent 3D printing facility to start printing within

10 min after mixing. Figure 3a shows the printed structure on the printer bed of the gantry printer with a 1 m^3 build volume. A 25 mm circular nozzle was used, and printer parameters were set at 10 mm layer height, 3 m print path length, 60 mm/s nozzle speed, resulting in a 50 s pass time and a 40 mm layer width. A total height of 260 mm was printed. From the printed structure, 4 panels of size 260 mm × 400 mm were extracted (Fig. 3c). The 4 panels were left to cure in a same environment as SHCC specimens for 1 day. Then, to control the SHCC overlay thickness, 10 mm thick weather-stripping foam tape was placed along the panel perimeter (Fig. 3b, c) on four of the six panels. The freshly mixed SHCC was subsequently plastered onto the undulating 3DCP surface of these panels, producing the SHCC-3DCP cross-section perpendicular to the printing direction seen in Fig. 3d. The 3DCP and SHCC-3DCP panels were left to cure in the same environment for a further 27 days. At the 3DCP age of 21 days, beam specimens (Fig. 3e) were sawn from the 3DCP and SHCC-3DCP panels, after which the samples remained in this climate until testing age of 28 days. The 6 types of beams had nominal 3DCP cross-section dimensions of 40 mm × 40 mm, to which the nominal 10 mm SHCC overlay was added for the composite beams. The beam lengths allowed a span of 150 mm for four-point bending test with 50 mm equidistant loading cylinders. Six identical specimens of each type were prepared and tested, with the exception of the CP-I-D1 and CP-O-D1 types, of which only 4 each were tested.

Table 1. 3DCP and SHCC mix proportions in kg/m^3 unless indicated otherwise.

Ingredient	SHCC	3DCP
Cement CEM II 42.5N-ALL	420	562
CSA cement, type III with alumina content ≥34%	21	–
Fly-ash (Class F)	620	162
Silica Fume	–	81.4
Natural quarry sand, max particle size 4.75 mm	540	–
Crushed granite sand, max particle size 4.75 mm	–	1144
Water	365	256
Superplasticiser	0.21% of binder mass	0.6% of binder mass
VMA	0.16% of binder mass	0.3%
[a]Polyvinyl Alcohol (PVA) fibre	2% of total volume	–
[b]Polypropylene (PP) fibre	–	1% of total volume

[a]PVA fibres: $L_f = 12$ mm, $d_f = 0.02$ mm, $f_t = 1600$ MPa, $E_f = 40$ GPa.
[b]PP fibres: $L_f = 6$ mm, $d_f = 0.03$ mm, $f_t = 300$ MPa, $E_f = 3$ GPa.

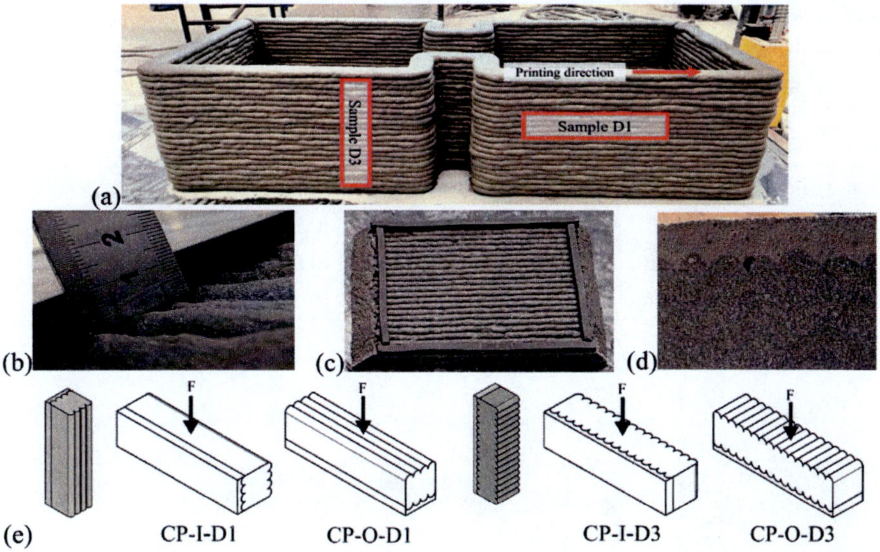

Fig. 3. (a) 3D printed structure from which panels are sawn, (b, c) the perimeter masked to guide the troweled application of a 10 mm SHCC layer. (d) An SHCC-3DCP composite cross-section. (e) Beam test specimens sawn from reference 3DCP and SHCC-3DCP.

3.2 Characterisation Tests

SHCC tensile behaviour was determined as recommended by Van Zijl et al. (2016) on dumbbell specimens nominally 16 mm thick, 30 mm wide and with an 80 mm gauge length. Five specimens were tested in a Zwick Z250 materials testing machine at displacement-controlled rate of 0.05 mm/minute, after applying a pre-load of 200 N at 2 mm/minute. Figure 4 shows the test setup and all five tensile stress-strain responses. The ultimate tensile strain is seen to vary in the range 1 to 2%, with the exception of the specimen labelled 4 in Fig. 4. The ductility might be increased by using finer grained sand than the 4.75 mm maximum particle size sand used here (Table 1), considering the tensile responses of “fine” and “coarse” SHCC categories with sand particle sizes up to 0.3 mm and 1.2 mm respectively reported by Van Zijl et al. (2016). Table 2 summarises the SHCC tensile results, compressive strength and elastic modulus determined on cast cylinder specimens of 100 mm in diameter and 200 mm long.

3DCP compressive and tensile strengths and elastic moduli are also included in Table 2. These characterisations were performed in both material directions, i.e. D1 parallel to the layers, and D3 perpendicular to the layers (see Fig. 3a). The tension and compression specimens were carefully sawn to the nominal dimensions of 40 mm x 40 mm, and the elastic modulus specimens cored from the hardened printed structure to a height to diameter aspect ratio of 2. Note that the 3DCP characterised values reported in Table 2 are taken from van den Heever et al. (2021), hence from a different batch and 3D print, but the same mix (Table 1), printer-pump and settings, as well as curing regimes.

Table 2. SHCC and 3DCP mechanical properties. The values represent averages, with the coefficient of variation and number of specimens given in brackets.

Ingredient	SHCC	3DCP	
		D-1	D-3
Density (kg/m^3)	1822 (0.004;4)	2233 (0.0163;3)	2233 (0.0163;3)
Tensile strength (MPa)	3.80 (0.042;5)	2.45 (0.126;6)	1.25 (0.096;4)
Compressive strength (MPa)	31.4 (0.030; 4)	45.1 (0.057;6)	38.2 (0.071;6)
Elastic modulus (GPa)	11.64 (0.061;4)	21.9 (0.048;3)	21.6 (0.062;3)

a.
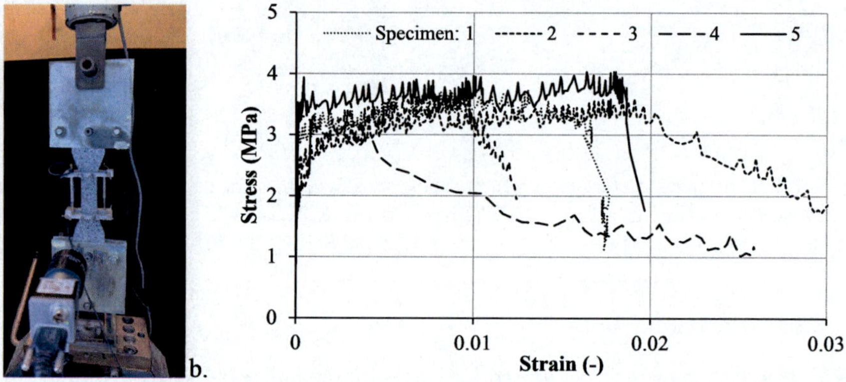
b.

Fig. 4. (a) SHCC tensile test with fixed-pinned boundaries. (b) SHCC dumbbell tensile stress-strain responses.

3.3 Test Protocols

The four-point bending tests of the six beam types (Fig. 3e) were performed under displacement control at a fixed crosshead displacement rate of 0.25 mm/minute in a Zwick Z250 materials testing machine. After each test, the actual cross-section dimensions were measured with a Vernier calliper, and the width and height taken as the average of two measurements each. These measured dimensions were used in calculating equivalent elastic flexural stresses in the cross-section, for both homogenous 3DCP specimens, and the composite (CP) SHCC-3DCP specimens, for simple comparison purposes. Transformation of the section was performed based on the measured dimensions and the elastic modulus ratio of the SHCC and 3DCP in the composite section, as described in the next section.

4 Results

4.1 Flexural Stress Results

From the load-deflection results recorded during the four-point bending test, equivalent elastic maximum tensile stresses were calculated from peak forces. The calculation was based on EN196-1 for the homogeneous 3DCP beams denoted 3DCP-D1 and 3DCP-D3. For the composite sections CP-I-D1, CP-O-D1, CP-I-D3 and CP-O-D3, equivalent 3DCP sections were determined based on the 3DCP and SHCC elastic moduli ratio n from Eq. (1), with moduli taken from Table 2.

$$n = \frac{E_{3DCP}}{E_{SHCC}} \quad (1)$$

Table 3 summarises the peak equivalent 3DCP tensile stress in the beams. Figure 5 presents graphs of these stresses against crosshead travel for the all specimens, grouped per layer direction D1 and D3. A typical response is shown for each beam type, most closely representing the average peak stress and ductility of the 4 to 6 tested specimens for each of the 6 beam types.

From the results it is clear that the bonded SHCC overlay significantly increases flexural ductility in the composite beams compared with the reference unreinforced 3DCP beams. This is true for the D1 and D3 directions, and where the SHCC is loaded in-plane and out-of-plane. In comparison, the equivalent peak tensile stress in the 3DCP is relatively unaffected (Table 3).

Table 3. Equivalent maximum 3DCP tensile stress in the respective unreinforced (3DCP) and reinforced (SHCC-3DCP, denoted CP) beams. The values represent averages, with the coefficient of variation and number of specimens shown in brackets.

Specimen type	Maximum stress (MPa)	Specimen type	Maximum stress (MPa)
3DCP-D1	6.39 (0.063;6)	3DCP-D3	4.27 (0.197;6)
CP-I-D1	7.78 (0.075;4)	CP-I-D3	4.82 (0.079;6)
CP-O-D1	6.36 (0.163;4)	CP-O-D3	3.72 (0.122;5)

4.2 Ductility and Multiple Cracking

Both 3DCP-D1 and 3DCP-D3 specimen types failed in single fracture planes. PP fibres aligned along layers in the print direction bridge the crack in 3DCP-D1 specimens, adding ductility in the softening post-peak response of these specimens seen in Fig. 5a. However, in 3DCP-D3 specimens, the near vertical descending line for these specimens in Fig. 5b depicts abrupt loss of resistance after the peak.

When SHCC reinforcement is added, post-peak ductility is significantly increased for all specimens, whether the loading action is in the SHCC overlay plane, or out of

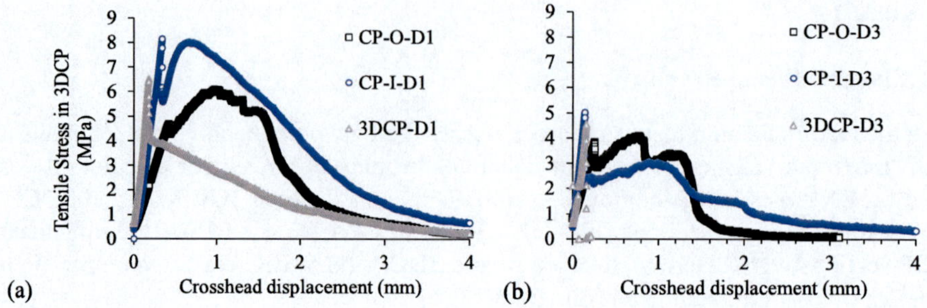

Fig. 5. Four-point bending test results of a representative specimen of each beam type, for (a) D1 direction and (b) D3 direction.

plane (see Fig. 5a, b). Figure 6 shows the failure patterns in a reinforced specimen in each of the CP-O-D1 and CP-O-D3 types, where the SHCC is subjected to in plane and out of plane flexure respectively. As in 3DCP-D1 specimens, PP fibres bridge the crack in the CP-O-D1 specimen (see Fig. 6a). In these specimens, PVA fibres are also seen to bridge the localised, tortuous overlay crack. The photographs of the specimens in Fig. 6 were taken after completion of the tests, when the resistance had reduced to near zero or zero. At that stage the CP-O-D1 specimens were still held in tact by the fibres, but the CP-O-D3 specimens (Fig. 6b) had separated into two parts. In the latter, no fibres are visibly bridging the 3DCP interfaces. A second 3DCP interfacial crack is visible in this specimen. Digital image correlation is under way to visualise and quantify multiple cracking in the SHCC overlay.

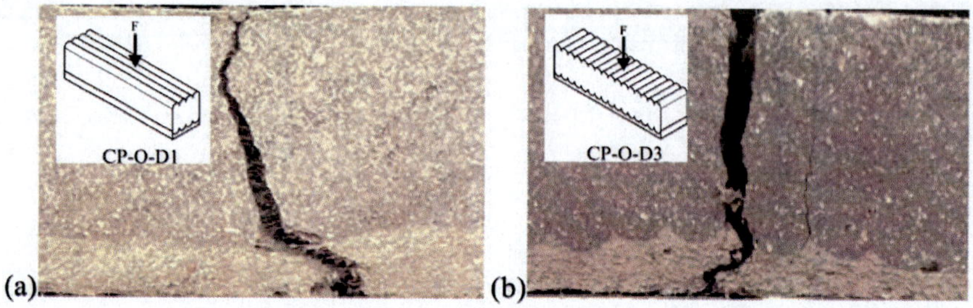

Fig. 6. Fracture zones for SHCC reinforced 3DCP specimens (a) CP-O-D1, showing PP fibres aligned along the layers bridging the 3DCP crack, and in the localised SHCC crack, and (b) CP-O-D3, showing a second 3DCP interface crack.

5 Demonstrative Design Calculations for SHCC-3DCP

As case study, the CP-O-D3 beam is analysed by simple plane cross-sectional analysis, average strength and stiffness material properties from Table 2 and simple assumptions on stress-strain relations. Four (4) stages are analysed to determine the force P_I that causes a 3DCP crack to initiate, P_{II} that causes the first crack in the SHCC, P_{III} that causes a 1% tensile strain in the SHCC, and P_{IV} that results in a 2% tensile strain in the SHCC. Figure 7 illustrates the cross-section flexural strains and stresses at these 4 stages. For these analytical calculations the average dimensions of the 5 successfully tested CP-O-D3 specimen cross-sectional were used, along with the average values of the relevant material properties (Table 2). For these dimensions and material parameters, the compressive stress in the 3DCP remained below the compressive strength (38.3 MPa for D3; Table 3) for all stages up to P_{IV}. However, the linear compressive stress distributions are approximations, as the compressive stress reaches 19.9 and 27.8 MPa for the calculated P_{III} and P_{IV} respectively, which is likely beyond the 3DCP compressive linear stress-strain regime. For simplicity, no refinement was implemented here. The calculated flexural forces are plotted on the graph in Fig. 7 (right). Considering the calculated strain gradients at the 4 stages, and the assumption that the curvature (φ) distribution over the beam span length (L) remains proportional to the elastic curvature, with the maximum (φ_{max}) constant in the constant moment region in the four-point bending setup, the deflection is given by the expression:

$$\delta = 0.051\varphi_{max}L^2 \tag{2}$$

The deflections for the 4 stages calculated from Eq. (2) are marked by the square symbols in Fig. 7 (right). Note, however, that the crosshead displacement was recorded and plotted as the experimental results in Fig. 7. Also, the approximation of the curvature spatial distribution leading to Eq. (2) does not hold once localisation in wide cracks occurs. Nevertheless, the 4 stage sets of analytical results are considered to reasonably depict important stages in the actual, measured response. For reliable design, determination of characteristic values and material factors are recommended.

From these results, the improved ductility of these CP-O-D3 specimens, representing the out-of-plane vertical bending of the 3DCP wall (Fig. 3a), can evidently be designed for with reasonable accuracy. Here, the full compressive capacity of the 3DCP is not exploited by the thin SHCC overlay. However, by selection of other materials and dimensions, this could be possible, in which case the accuracy of using a simplified compressive stress block should be investigated, and bond properties along the SHCC-3DCP interface must be characterised to verify appropriate shear transfer.

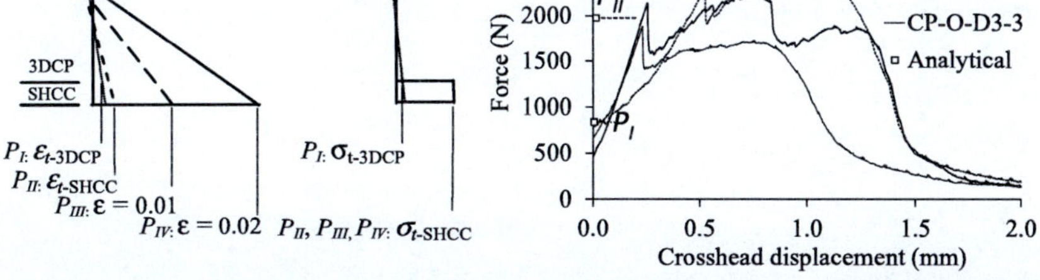

Fig. 7. Illustration of 4 stages in CP-O-D3 deflection response, showing the strains and stresses (left) (not to scale), and comparison of the analytical and experimental results (right).

6 Conclusions

This paper proposes a novel reinforcement strategy for 3DCP, which aligns with the benefits of this emerging construction technology. The design, execution and results of an experimental program to demonstrate the reinforcement of 3DCP with thin, bonded SHCC overlays, are presented. Here, enhanced ductility has been set as objective. The following conclusions are drawn:

- A one-sided, thin, bonded SHCC overlay significantly increases in-plane and out-of-plane flexural ductility of 3DCP.
- For the cases investigated in this research programme, ductility increase is most pronounced for in-plane and out-of-plane bending in the D3 direction, where fracture occurs in the weak interfaces of the layered 3DCP.
- Simple moment-curvature section analysis captures the observed flexural behaviour of the SHCC-3DCP composite with reasonable agreement for design purpose.

Refinement of the design approach to incorporate reliability concepts are recommended. Extension to reinforcement for increased resistance requires careful characterisation of bond between SHCC and 3DCP, and verification whether standardised compressive block concepts for reinforced concrete apply to SHCC reinforced 3DCP.

References

Bester, F.A., van den Heever, M., Kruger, P.J., van Zijl, G.P.A.G.: Reinforcing digitally fabricated concrete: a systems approach review. Addit. Manuf. **37**, 101737 (2021)

Bos, F.P., Ahmed, Z.Y., Jutinov, E., Salet, T.A.M.: Experimental exploration of metal cable as reinforcement in 3D printed concrete. Materials **10**, 1314 (2017)

Marchment, T., Sanjayan, J.: Bond properties of reinforcing bar penetrations in 3D concrete printing. Autom. Constr. **120**, 103394 (2020)

Van den Heever, M., Bester, F.A., Kruger, P.J., van Zijl, G.P.A.G.: Mechanical characterisation for numerical simulation of extrusion-based 3D concrete printing. J. Building Eng. **44**, 102944 (2021)

Van Zijl, G.P.A.G., De Beer, L.: Sprayed SHCC overlay for shear strengthening of unreinforced load bearing masonry. Adv. Struct. Eng. **22**(5), 1121–1135 (2019)

Van Zijl, G.P.A.G., De Jager, D.J.A.: Sprayed SHCC overlay for shear strengthening of unreinforced load bearing masonry. Adv. Struct. Eng. **22**(5), 1121–1135 (2019)

Van Zijl, G.P.A.G., Wittmann, F.H., Toledo Filho, R.D., Slowik, V., Mihashi, H.: Comparative testing of crack formation in SHCC. Mater. Struct. **49**(4), 1175–1189 (2016)

Mechanism and Characterization of Cracking

Influence of Placing Thickness on Fiber Orientation and Bridging Law of FRCC

Hang Zhang[1](✉) and Toshiyuki Kanakubo[2]

[1] Degree Program in Engineering Mechanics and Energy, University of Tsukuba, Ibaraki 305-8577, Japan
s2165296@s.tsukuba.ac.jp
[2] Division of Engineering Mechanics and Energy, University of Tsukuba, Ibaraki 305-8577, Japan
kanakubo@kz.tsukuba.ac.jp

Abstract. It is considered that the different placing thicknesses in casting of fiber-reinforced cementitious composite (FRCC) can be one of the factors affecting the fiber orientation and distribution, which is considered to be one of the most important influence factors of the bridging performance of fibers. In this study, a water glass solution is used to conduct the visualization simulation of the flow patterns of fresh mortar with short discrete fibers. Water glass has high viscosity, and it is colorless and transparent. The rheology of mortar matrix before mixing the fiber has been inspected using the flow time based on the test method for flowability of grout measured by the funnel. The orientation intensity that expresses the fiber orientation tendency for the principal orientation angle is calculated by counting the orientation angles of the black target fibers in the water glass solution of three different placing thicknesses. The effect of different placing thicknesses on the fiber bridging performance is considered in the calculation of the bridging law using the elliptic function characterized by the principal orientation angle and the orientation intensity. The results show that a smaller placing thickness in casting leads to a greater fiber orientation intensity and better tensile performance based on the bridging law.

Keywords: Placing thickness · Fiber orientation · Bridging law · Visualization simulation · FRCC

1 Introduction

Fiber reinforced cementitious composite (FRCC) is a kind of hydraulic cementitious material in which a certain volume fraction of short fibers such as PVA or steel fiber is mixed into mortar or concrete. In recent years, many studies on FRCC have been carried out because the tensile and bending performance of cementitious composite can be improved by the bridging fibers after first cracking, which provides a good tensile strain or deflection hardening and ductile performance. Some previous studies reported that this enhancement by bridging fiber is strongly affected by the fiber orientation intensity in the matrix [1]. The fiber orientation intensity is influenced by a lot of factors such as

M. Kunieda et al. (Eds.): SHCC 2022, RILEM Bookseries 39, pp. 91–97, 2023.
https://doi.org/10.1007/978-3-031-15805-6_10

casting directions, flow and formwork geometry [2]. This study mainly focuses on the effect of different placing thicknesses on fiber orientation and bridging law of FRCC using the visualization simulation which is introduced as following chapter.

2 Visualization Simulation of Placing Thickness on Fiber Orientation

2.1 Materials and Simulation Method for Test

The black-colored nylon fiber is used for visualization simulation to observe each fiber angle easily. The dimensions of nylon fiber used in this study are listed in Table 1. In order to visualize the orientation distribution of fibers in the molds made by acrylic plates, a transparent sodium silicate solution which has high viscosity and usually called water glass is adopted as the matrix. The placing thickness here is defined as the thickness of each layer of target cementitious matrix in the height direction (z-axis direction in Fig. 1) when casting. For example, in Fig. 1, in order to make the experiment comparable, when the thickness was chosen to be 10 mm, the specimen should be casted 10 layers along the height direction and placing thickness of 33 mm should be 3 layers since the standard cross-section is chosen to be 100 × 100 mm. The volume fraction of nylon fiber is set to be 0.05% to distinguish each fiber easily.

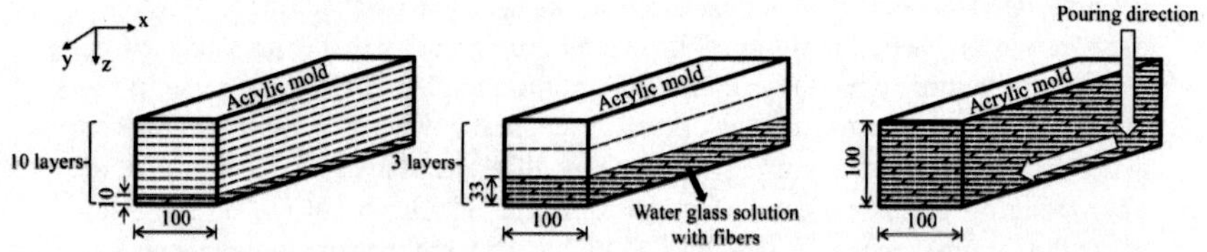

Fig. 1. Instruction of different placing thicknesses in visualization simulation (unit: mm)

Table 1. Dimensions of nylon fiber.

Density (g/cm^3)	Length (mm)	Diameter (mm)	Tensile strength (MPa)
1.14	12	0.24	65

The water glass solution containing nylon fibers is poured into three identical molds along the horizontal casting direction as showed in Fig. 1. Due to the compatibility of the solution, the pouring is stopped when the placing thickness of the water glass solution reached 10 mm, 33 mm and 100 mm respectively in the molds, and the fiber angle distribution of the remaining layers was considered to be the same as this bottom layer. The rheology of mortar matrix before mixing the fiber has been inspected using the flow time based on the test method for flowability of grout measured by the funnel. The flow time of water glass is controlled by adding pure water and the solution temperature in an effort to attain the same flow time of the target mortar matrix. After pouring, the x-y and x-z plane photos of the specimens are taken by digital camera.

2.2 Image Processing and Statistics of Fiber Distribution

Image analysis is conducted to obtain the fiber distribution as following procedures. Firstly, the photo captured by the camera is cropped to include the central part of 100 × 100 mm region as shown in Fig. 2a. Secondly, the photo is converted to black and white after a threshold was determined (Fig. 2b). Thirdly, an approximation line is placed at the location of each fiber. (Fig. 2c) Finally, the projected lengths of the approximation line in the horizontal and vertical directions are recorded to calculate the angle of the fiber. The fiber angle is defined as the angle between the approximation line and the x-axis and ranges from − 90 to +90°. The relative frequency of fiber angle in every 5° can be calculated.

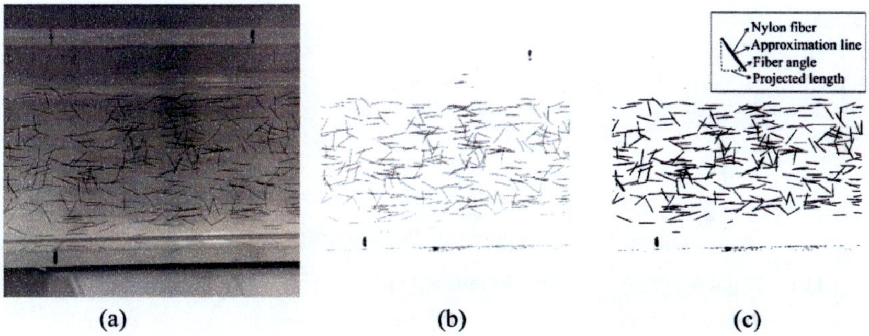

Fig. 2. Process of image analysis (10 mm thickness – x-y plane)

After the image analysis and fiber angle calculation, the quantitative statistics of fiber orientation distribution was conducted. Kanakubo et al. [2] has proposed a probability density function (PDF) using the elliptic function. The fiber orientation distribution is determined by the orientation intensity k (ratio of the two radii of elliptic function) and principal angle θr (argument of one radius of elliptic function), which can be both calculated by the elliptic function. The histogram of fiber angle and calculation result of orientation intensity k and principal angle θr on x-z/x-y plane of each placing thickness are shown in Fig. 3.

The Fig. 3a shows that, as for x-z plane, the relative frequency of placing thickness of 10 mm and 33 mm on the range of fiber angle around 0° (y-axis direction) are much larger which means the fiber angles are mostly around this range. The orientation intensity k of placing thickness of 10 mm and 33 mm are much larger than that of placing thickness of 100 mm, which indicates the fibers show a great directional orientation toward the principal orientation angle for the former two placing thicknesses. As for x-y plane in Fig. 3b, except that the condition of the placing thickness of 10 mm is the same as the previous, the relative frequency distribution of the placing thickness of 33 mm and 100 mm is relatively average and the k values are close to 1.0, which means that the fiber angles of these two specimens are close to randomly distributed.

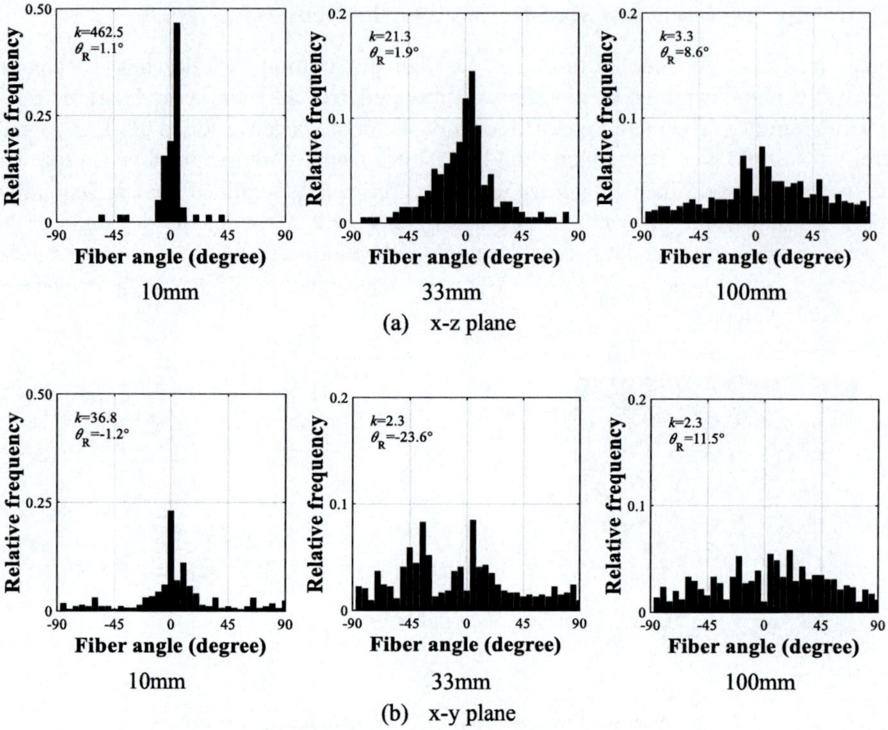

Fig. 3. All fiber angle histogram

3 Calculation of Bridging Law and Section Analysis

3.1 Calculation of Bridging Law

In this section, the calculation of bridging law and section analysis using the models of bridging law are conducted. The target FRCC is PVA-FRCC investigated in the previous study with the fiber volume fraction of 2% [1]. The fiber orientation intensity employed in the calculation is assumed to be the same values obtained from the visualization simulation in the former section for each target thickness. And in this chapter, due to the fiber orientation intensity k of x-z plane is over 100, the specimen of placing thickness of 10 mm is not considered in the calculation. The bridging law is described by the relationship of tensile stress and crack-width. Kanakubo et al. [2] proposed a tri-linear model of the individual fiber for bridging law considering the effect of fiber orientation, which uses the characteristic points in calculating bridging laws considering the corresponding phenomena of the individual fiber pullout properties. The values and instructions of characteristic points in calculating bridging laws are listed in Table 2 and the calculation result of bridging law is shown in Fig. 4.

Table 2. Parameters for calculation of bridging law [1]

Calculation parameter	Input value
Fiber length, l_f	12 mm
Fiber volume fraction, V_f	2.0%
Fiber diameter, *df*	0.10 mm
Snubbing coefficient, *f*	0.5
Fiber effective strength	569 MPa
First peak load, P_a	1.5 N
Crack width at P_a, w_a	0.2 mm
Maximum load, P_{max}	3.0 N
Crack width at P_{max}, w_{max}	0.45 mm
Fiber strength reduction factor, *f'*	0.3

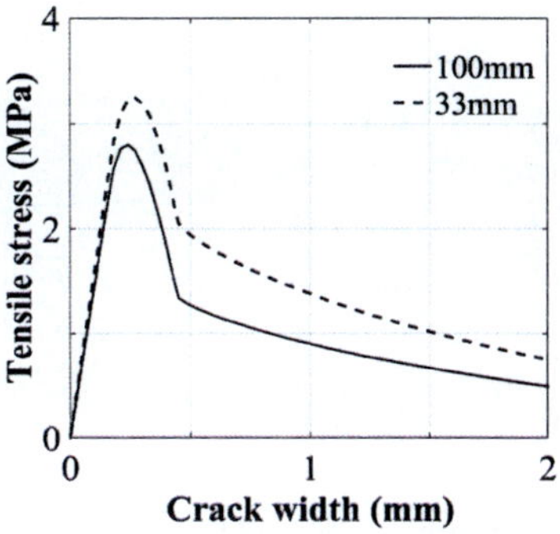

Fig. 4. Calculated bridging law of different placing thicknesses

Figure 4 shows that the bridging law of different placing thicknesses all contain a stress rising branch up to the maximum at first, followed by a softening branch with a larger absolute value of the slope and finally a gentle softening branch that tends to zero. As for the value of maximum tensile stress, the bridging law for the placing thickness of 33 mm is larger than that for the placing thickness of 100 mm.

3.2 Section Analysis

The section analysis based on the tri-linear model of bridging law is conducted to evaluate the bending performance of the specimens using three placing thicknesses for same cross-section (100 mm × 100 mm). Figure 5 shows the whole model of stress-strain relationship including the compression side. The stress-strain model of the compression side is assumed to be a parabola model and the tensile side is the tri-linear model which is derived from bridging law model divided by the target length (=100 mm). The section analysis also complies with the plane-cross section assumption. The section analysis is approached by an approximate method as the following steps. Firstly, the cross-section is divided into a finite number of identical elongated elements from the height direction and an arbitrary curvature was given. Secondly, the strain of each element in cross-section is calculated from linear distribution of strain and stress of each element can be obtained from the stress-strain model in Fig. 5. Finally, neutral axis satisfying equilibrium condition and bending moment is obtained. The values of parameters in the model are expressed as a function of the orientation intensity *k* to simplify the modeling of the bridging law used for section analysis, which are listed in Table 3 as Ozu et al. [1] proposed. For simplification, the *k* value of each specimen used for calculation is the average *k* value of the x-z plane and x-y plane.

As revealed in Fig. 6, the bending moments of all placing thicknesses first increase and then decrease with the increase of curvature. For the maximum bending moment, the value of the specimen with a placing thickness of 33 mm is the largest, followed by the placing thickness of 100 mm. That is, a smaller placing thickness can effectively improve the flexural performance of FRCC.

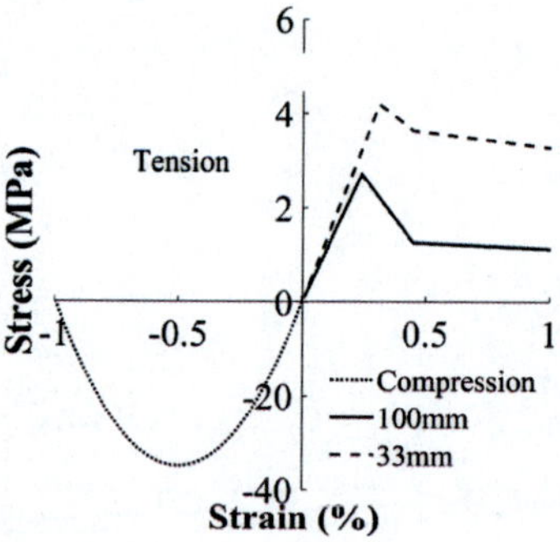

Fig. 5. Stress-strain model used for cross-section analysis

Table 3. Parameters for section analysis [1]

Calculation parameter	Input value
Compressive strength, σ_c	35 MPa
Tensile stress at first snapping point, σ_{cr}	$2.0 \times k^{0.30}$ MPa
Tensile stress at second snapping point, σ_t	$0.60 \times k^{0.73}$ MPa
Compressive strain at σ_c, ε_c	0.005
Tensile strain at first snapping point, ε_{cr}	$0.20 \times k^{0.18}/100$ mm
Tensile strain at second snapping point, ε_t	0.0045
Tensile strain at zero stress, ε_u	0.06

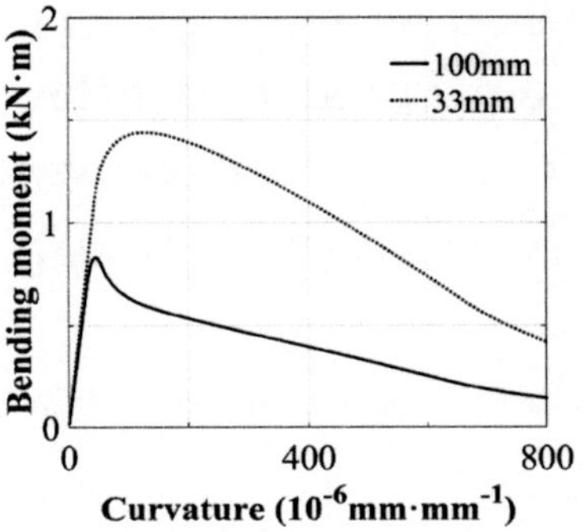

Fig. 6. Bending moment – curvature curve for different placing thicknesses

4 Conclusion

In this study, the visualization simulation using the water glass is conducted to evaluate the influence of placing thickness on fiber orientation distribution and bridging law. The image analysis shows that a smaller placing thickness leads to a greater fiber orientation intensity, which is one of the significant factors on the bridging law of FRCC. From the calculation results of the bridging law, a smaller placing thickness leads to a higher tensile stress due to the centralization of the fiber angles. Based on the simplified tri-linear stress-strain model of bridging law, section analysis was conducted to give strong evidence on that a smaller placing thickness can effectively improve the flexural performance of FRCC.

References

1. Ozu, Y., Miyaguchi, M., Kanakubo, T.: Modeling of bridging law for PVA fiber-reinforced cementitious composite considering fiber orientation. J. Civil Eng. Archit. **12**(9), 651–661 (2018)
2. Kanakubo, T., Miyaguchi, M., Asano, K.: Influence of fiber orientation on bridging performance of polyvinyl alcohol fiber-reinforced cementitious composite. ACI Mater. J. **113**(2), 131–141 (2016)

Comparison Between Experimentally Determined and Theoretical Fiber Orientation Distribution in Strain Hardening Cementitious Composites (SHCC)

Zhenghao Li(✉) and Christopher K. Y. Leung

Department of Civil and Environmental Engineering, HKUST, Clear Water Bay, Kowloon, Hong Kong
zlifb@connect.ust.hk

Abstract. Strain hardening cementitious composites (SHCC) are a class of specially designed fiber-reinforced materials with good crack width control capacity and durability. The crack width of SHCC is mainly governed by the mechanical properties of fibers, interfacial properties between fibers and matrix, the fiber volume fraction as well as fiber orientation distribution. To quantify the effect of these parameters on SHCC behavior, micromechanical models have been developed. In previous studies, the fiber properties and parameters governing interfacial behavior have been experimentally obtained with well-established methods. However, the fiber orientation distributions were simply assumed to be 2D or 3D random (or the average between the two), but such assumptions have never been checked against test results. In this study, the experimentally examined fiber orientation distributions in SHCC tensile samples with PVA fibers were presented and compared with theoretical orientation analysis based on random distribution and consideration of the wall effect. The results show that the real fiber orientation forms a clear peak at around 20° while the pure wall effect analysis cannot reflect this phenomenon. A modified orientation analysis considering the simplified mortar flattening effect was then proposed. Large differences in simulated stress-crack width curves are found between the measured distribution and the wall-effect-induced distribution while the distribution obtained by the modified analytical model can greatly reduce the gap. The measured orientation distribution or proposed analytical approach can be used to replace random distributions in future work to reduce systematic simulation errors.

Keywords: SHCC · Fiber orientation · Flattening effect · Bridging stress

1 Introduction

Strain hardening cementitious composites (SHCC) exhibit tensile strain hardening behavior accompanied by the sequential formation of multiple fine cracks. The sequential formation of cracks is closely related to the cracking strength and bridging stress at various sections. Specifically, when under tensile stress, new cracks can form at the

M. Kunieda et al. (Eds.): SHCC 2022, RILEM Bookseries 39, pp. 98–108, 2023.
https://doi.org/10.1007/978-3-031-15805-6_11

plane where the cracking strength is reached as long as the bridging stress of previously cracked sections can bear the increasing load. As a result, a proper combination of cracking strength distribution and bridging stress distribution is crucial for SHCC behavior.

To guide the design of SHCC with good performance, many micromechanical models have been developed for predicting the cracking strength and bridging stress [1, 2]. Generally, the cracking strength of the composites is believed to be governed by internal flaws [1, 3], and the bridging stress of a plane is affected by fiber distributions [2, 4]. With the help of the non-destructive test method such as X-ray Computed Tomography (X-CT), the full-field mesoscale pore structure inside the SHCC tensile samples (the sections of which are several square centimeters) can be accurately obtained and used for mechanical analysis [3]. However, the fibers used in SHCC are small in radius and difficult to detect with X-CT unless very small specimens (which may not be representative) are used. Currently, the most widely used method for inspecting the fibers in SHCC is sectional analysis [5, 6], which is very time-consuming. As a result, the full-field fiber distribution in tensile samples is seldomly studied, so only very limited information is available. Consequentially, the fiber orientation distributions in mechanical models are usually simplified to 3D random or 2D random (or the average of the two) [2, 7].

In reality, the fiber orientation inside tensile samples is not randomly distributed [4]. In order to minimize the error in deriving the crack bridging stress, various analytical models for predicting the fiber orientation were proposed. In 2006, the approach for analyzing theoretical fiber orientation based on wall effect was proposed for rigid steel fibers and the result was checked with small beam samples [8]. Then in 2017, a model considering the folding effect of synthetic fibers was discussed [9]. The folding effect is a modification for non-stiff fibers by the introduction of a simple length reduction factor. In the same year, the wall effect analysis was applied to PVA fibers in SHCC [4]. The effect of sample thickness was analyzed and the detailed influence of fiber orientation on bridging stress was discussed. However, the simulated results have not been validated by real fiber orientation distribution obtained from tensile samples.

Based on the above discussions, the fiber orientation in SHCC samples is important and not fully studied. In this research, the full field fiber orientation distribution of sections in SHCC tensile samples is experimentally determined. Then a new model for simulating the fiber orientation distribution is introduced. Micromechanical analysis for bridging stress is also conducted. The simulation results with the simplified random fiber orientation distributions, wall effect induced distribution, real distribution, and the distribution derived with the proposed method are compared. This research aims to provide more information on fiber orientation distributions in SHCC tensile samples and propose a new method to model the fiber orientation distributions with higher accuracy.

2 Fiber Inspection Techniques

PVA fibers are one of the most widely used fibers in SHCC. As a kind of synthetic fiber, PVA can emit green to yellow light when excited by ultra-violet light. Based on this property, Torigoe et al. [10] proposed a method for observing PVA fibers with the fluorescence microscope. In this section, the pipeline to obtain fiber orientation distribution with fluorescence microscope based sectional analysis method is presented.

2.1 Sample Preparation and Image Acquisition

SHCC tensile samples measuring 50 mm × 15 mm × 150 mm (width × thickness × length) with six different flowabilities were cast for fiber inspection. The flowability of SHCC is represented by the obtained flow time from the Marsh cone test of the suspending mortar [11]. After proper curing, the coupon samples were cut into pieces to expose cross-section (50 mm × 15 mm) perpendicular to the loading direction. To create a smooth surface for clear images in the microscope, the exposed sections were further polished with #400 and #800 SiC paper (each for 2 min).

After surface preparation, each cross-section was photographed with a digital camera connected to a Nikon SMZ 1270 fluorescence microscope at 30× magnification to capture the shape of each fiber, which will be further used for determining the orientation of each fiber. A sample florescent image after denoising is shown in Fig. 1(a).

Due to the limited receptive field of the camera, each captured image can only cover a small region (2.5 mm × 1.8 mm), so a manual XY stage is employed to enable precise movement and positioning of the samples. A series of images (225 images) were captured following a certain sequence, with about 10% overlap with each other to reform the image of the whole section.

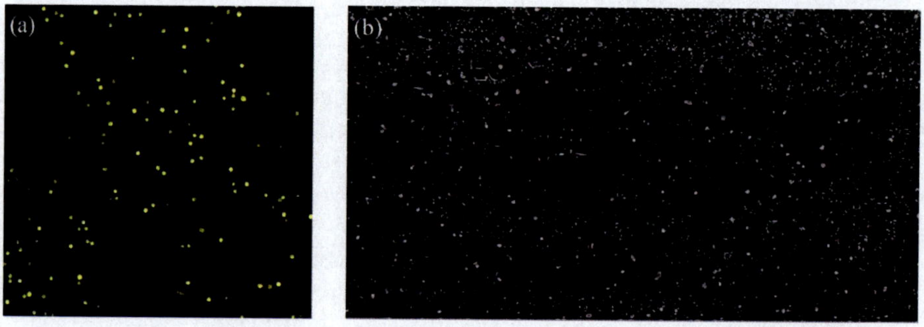

Fig. 1. (a) Sample florescent image after denoising. (b) Binarized cross-section image (partial, fibers in white and background in black).

2.2 Image Processing

225 raw images were first imported into ImageJ to form a whole section with the "stitching" command [12]. Then the images were cropped into the size of 60000 × 16000 pixels (corresponding to 48 mm × 12.8 mm area, around 1 mm of cropping from each side). The edge region was cropped for the reason that the fibers at the edges are more likely to be affected by various random factors, such as troweling effect and rough casting surface, which may bring undesired variability to the results [6].

The cropped images were then processed with the help of the deep learning-based algorithm. The output images are binary images in which the fibers are in white while the background appears in black (Fig. 1(b)). With the watershed algorithm [13], each fiber can be further segmented and identified, and the shape of the fiber cross-sections can be utilized to calculate the orientation.

2.3 Determining the Fiber Orientation

For each section, the intersecting fibers are represented by elliptical dots in the binary images. The major axis and minor axis of each region (which represents a fiber) can be used to calculate the orientation as in Eq. (1).

$$cos\theta = \frac{d}{l} \tag{1}$$

In Eq. (1), l is the major axis of the elliptical regions and d refers to the minor axis as in Fig. 2. With Eq. (1), the orientation of each fiber in the whole section can be determined.

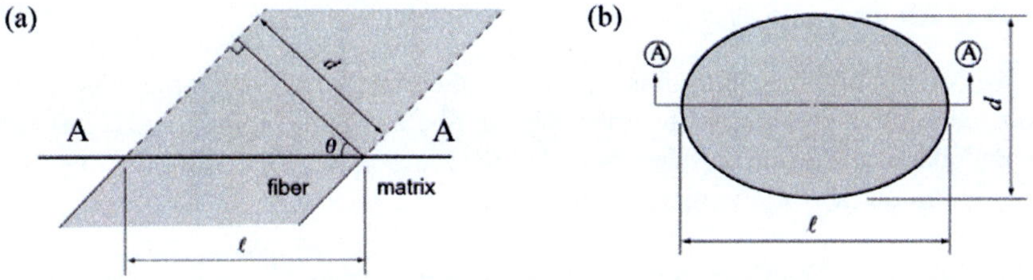

Fig. 2. Dimension of inclined fiber. (a) Pan view on cutting plane and (b) Side view at the section of A-A [5].

3 Comparison Between Experimentally Determined and Theoretical Fiber Orientation Distribution

3.1 Experimental Results

The summarized orientation distributions of all 6 samples with different flowability are plotted as Fig. 3. Each histogram is the summation of 6 cross-sections of a sample and the value in the title denotes the measured flow time of suspending mortar ranging from 15 s (denoted by T15, very flowable case) to 47 s (denoted by T47, very sticky case). From Fig. 3, the fiber orientation distributions of all samples are almost the same, which means that the fiber orientation distribution is barely affected by the flowability of mortars. Also, all samples showed a clear peak at a relatively small angle (around 20°). Over 90% of fibers are within 45°. Although being randomly mixed, fibers are aligned quite well in the coupon samples.

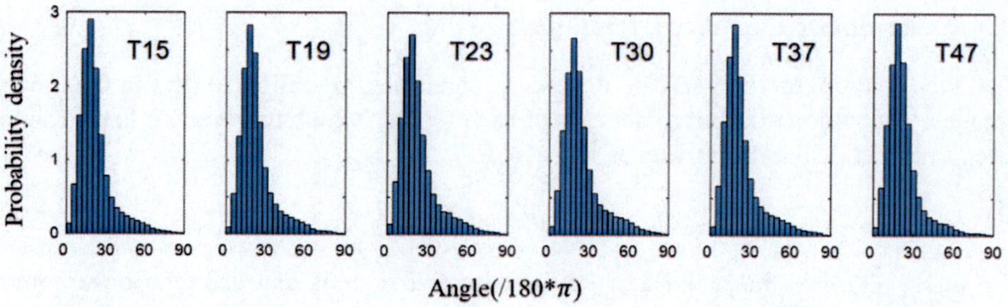

Fig. 3. Measured fiber orientation distribution of samples with different flowabilities.

3.2 Wall Effect Analysis

The fiber orientation distribution is affected by the boundaries of samples. As the fibers cannot reach out of the sample, the orientation of fibers near the boundaries is constrained. This boundary constraint on fibers is called the wall effect. The wall effect analysis is mainly based on the following assumptions [8]:

1. All fibers are stiff enough to stay straight in the matrix.
2. The centroids of fibers are uniformly distributed in the whole sample.
3. The top surface of the section has the same boundary condition as the sides of the mold without considering the effect of smoothening and troweling of the top surface.

For a typical SHCC tensile sample, the cross-section is rectangular. The section can be further separated into 3 regions as in Fig. 4(a): the bulk region (region C), the region with one wall (region B), and the region with two walls (region A).

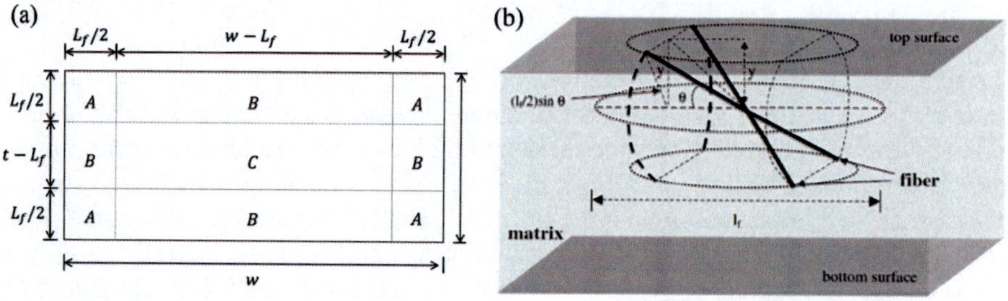

Fig. 4. (a) Cross section and 3 different regions. w is the width of the cross-section and t is the thickness. L_f stands for the length of fibers. (b) 3D view of boundary constraint in region B [4].

The distance between points in region C and boundaries are larger than half of the fiber length, so the fiber orientation is not affected by the boundaries and the distribution should be 3D random [14].

$$p(\phi = \theta_0) = sin(\theta_0) \tag{2}$$

For fibers in region B, the orientation is constrained by the boundary on one side. The orientation distribution can be obtained by geometric analysis. Figure 4(b) is the schematic for a fiber of which the centroid is at the distance y from the boundary while inclined at the angle θ_0 (in which $y < L_f \sin\theta_0/2$). Due to the existence of the top surface, the endpoints of the fiber cannot be a circle as in the bulk region. Instead, the endpoint can only be located on the two arches cut from the circle as the dash lines in Fig. 4(b). As a result, the probability of the fiber orienting at the angle θ_0 is no longer the sine function in Eq. (2) but has to be multiplied by the ratio between the length of arches and the circumference of the circle. While if the angle is small enough so that $y < L_f \sin\theta_0/2$, the circle formed by the endpoints will not intersect with the boundaries which implies the orientation is not affected. With geometric analysis, the probability of fibers in region B orienting at the angle θ_0 can be summarized as

$$p(\phi = \theta_0) = \begin{cases} \frac{sin(\theta_0)}{\pi} * 2\,arcsin\left(\frac{2y}{L_f * sin(\theta_0)}\right), \theta_0 > arcsin(2y/L_f) \\ \sin(\theta_0), \theta_0 < arcsin(2y/L_f) \end{cases} \quad (3)$$

Region A, the corners, is affected by both vertical and horizontal boundaries. The locus of the endpoints of fibers is thus cut by lines from 2 sides. With similar calculation for region B, the length of arches can be computed and the distribution can be obtained.

With the derived distributions of all three types of regions, the probability distribution of fibers inside the sample orienting at θ can be obtained by integrating the p function over the whole section numerically.

3.3 Comparison Between the Test Results and Theoretical Analysis

With the above analysis, the theoretical fiber orientation distributions obtained by wall effect analysis can be plotted with the summarized test results of all 6 samples in Fig. 5. It should be noted that the curve corresponding to the test result in Fig. 5 is not the same as the histogram in Fig. 3. The distributions in Fig. 3 are the orientation distributions of all fibers intersecting with the tested cross-section, which is different from the probability of a fiber orienting at angle θ. The orientation distribution of fibers in a plane should be the multiplication of cosine functions and distribution of fibers, as highly inclined fibers are less likely to intersect with a randomly selected plane (measured intersection) [4]. For better comparison, the widely used random distributions, including 2D random distribution, 3D random distribution, and the average of the two are also plotted.

From Fig. 5, it is found that the theoretical fiber orientation distribution of the coupon sample sections based on wall effect analysis is similar to the 2D random distribution. The reason may be related to the shape of the cross-section. In this research, the width of the section is 4 times of the fibers length while the thickness is only slightly larger than the fiber length. As a result, the thickness constraint is pronounced while the width constraint is less significant, so the distribution is close to the 2D random distribution.

However, the derived distribution based on wall effect analysis is not even close to the experimental results. The significant difference shows that the wall effect analysis may not be accurate enough to model the fiber orientation distributions in tensile samples. The underestimation of the alignment of the fibers implies that there might be other factors causing the fibers to align in the tensile direction.

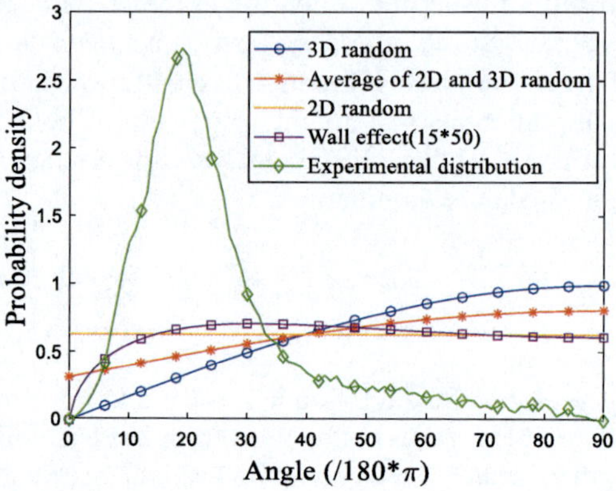

Fig. 5. Probability distribution function of fiber orientation of measured result, wall effect analysis, and simplified random distributions.

4 Flattening Effect Analysis

The large number of fibers orienting at small angles (around 20°) in Fig. 5 cannot be explained by the wall effect. Although the bending of fibers (due to the relatively low stiffness) can affect the fiber angle distributions, this cannot explain the bias of distribution towards small angles, as the bending direction is random spatially. To model the orientation distribution as close to the real case as possible, the flattening effect is introduced. This flattening effect is a simple representation of the reshaping process during sample preparation. With simplifying assumptions, the influence can be evaluated.

4.1 Introduction and Assumptions

Typically, the mixing process within a mixer can be regarded as a random process. Thus, the fibers in well-mixed fresh SHCC mortars should be 3D randomly distributed [14]. When casting tensile samples, the fresh SHCC mortars are usually required to be in one lift to guarantee the continuous fiber bridging effect along the tensile direction. Specifically, if two separate pieces of cast mortar create a contact surface in between, the fibers near the contacting surface cannot penetrate through the surface so a weak section is introduced which is destructive to the tensile performance.

However, it is hard to have a lift of fresh SHCC mortars in a long slender bar shape similar to the test specimen. Usually, the lifted mortar is a bulk shape like a sphere or a cube. Then the viscous bulk mortar is reshaped into the coupon shape either under gravity (for self-compacting SHCC) or under external force (for much viscous SHCC). In either situation, a reshaping process is introduced to turn a bulk shape into a plate shape. This deformation process is considered here as the flattening effect.

When flattening SHCC tensile samples, the viscous mortar is not really flowing but deforms like a viscoelastic material at a small scale. As a simplified analysis to illustrate the effect, the bulk volume is regarded as layers of deformable materials. During the flattening process, the layers are flattened while spreading along other directions without mixing with other layers. With this assumption, the endpoints of fibers will be still in the original layer while the thickness of the layers is reduced as shown in Fig. 6(a).

In a 3D case, the change of fiber orientation corresponding to the deformation is represented by the change in the length of the projection of the fibers in the thickness direction (y values) as shown in Fig. 6(b). Specifically, if the thickness is reduced to half of the original value, the y value of each fiber will also be reduced to half.

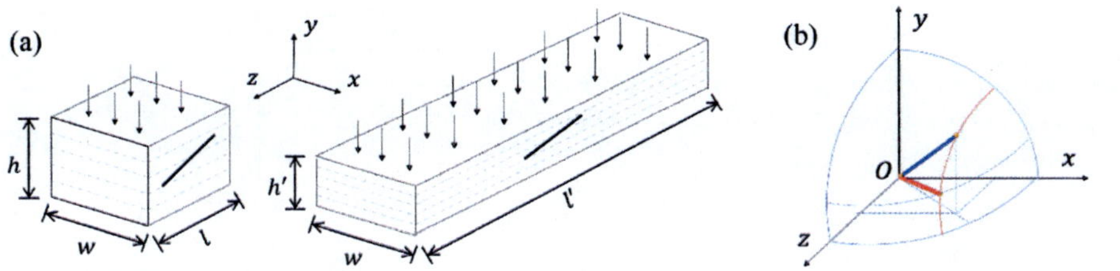

Fig. 6. Schematics for 3D flattening process: (a) the change in shape before and after the flattening process; (b) the rotation path of a fiber, where O represents the gravity center of the fiber.

4.2 Case Studies and Numerical Analysis

For a 3D case as shown in Fig. 6(a), the deformation process is simplified as the volume change from a 45 mm * 50 mm * 50 mm ($h \times w \times l$) solid volume, which is roughly the size of mortar employed to cast the specimens, to a 15 mm * 50 mm * 150 mm ($h \times w \times l$) bar. The thickness is reduced to $h/3$ while the width is not affected. In this case, all fibers are assumed to rotate with their gravity center around the width direction (x axis), and the y values of each fiber will be reduced to 1/3 of the original value.

However, it is not possible that all fibers will rotate with the change in y value exactly being 1/3. In reality, the compacted mortar block does not take the shape of a rectangular prism as Fig. 7(a). When flattening the bulk mortar block, the top layers which are not in contact with the molds are less constrained by friction and can spread to a larger area thus having a higher deformation ratio (the ratio between thickness after and before flattening) than the average value of 1/3 (=15/45) while the bottom layers in contact with mold are less deformed (Fig. 7(b)). This kind of variation in deformation ratio is hard to model and will need further numerical simulation. As a simplified analysis, the deformation ratio is set to be a uniform distribution from 1/4 to 5/12 (1/3 $\pm$ 1/12). The physical meaning of this distribution is that the deformation ratio reduces linearly from 5/12 at the top layers to 1/4 at the bottom layers. By conducting the numerical calculation, the simulated result can be plotted as Fig. 8 together with the experimental result. From the comparison, the flattening effect induced fiber orientation coincides well with the experimental result. While the analysis presented here is greatly simplified, the flattening effect is shown to be a feasible way to explain the fiber distribution in the real member.

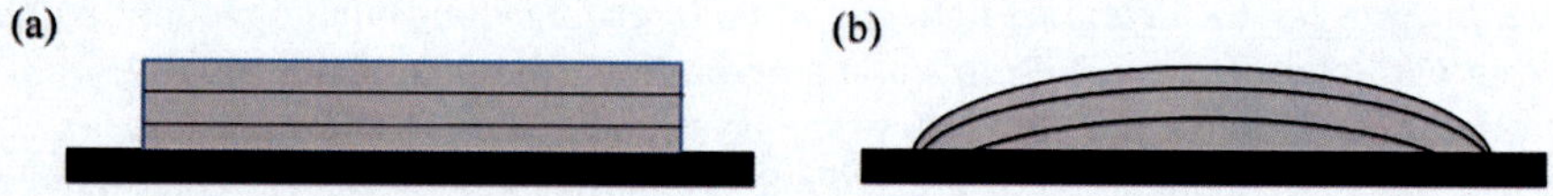

Fig. 7. Side view of flattened mortar in mold: (a) all layers keep rectangular prism shape; (b) due to friction of the mold, top layers spread to a larger area while bottom layers are less deformed.

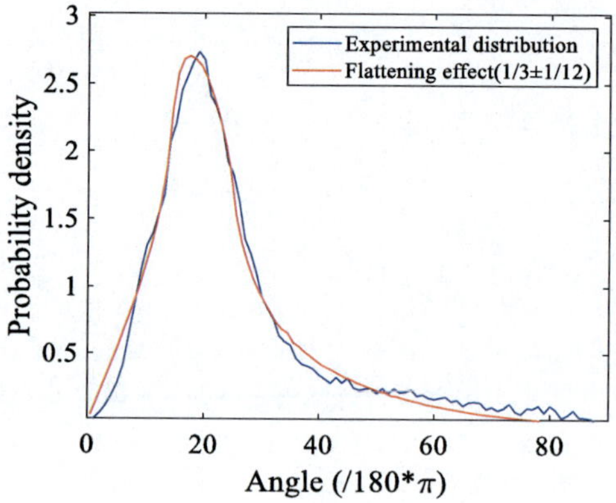

Fig. 8. Comparison between fiber orientation distribution derived with flattening effect and experimentally obtained distribution.

5 Bridging Stress Analysis

The bridging stress of a cross-section in tensile samples is dominated by the fibers crossing the section and interfacial parameters. To evaluate the influence of fiber orientation distribution on the bridging stress, a set of simulations were conducted with all other parameters kept the same.

The simulation procedure and corresponding parameters are following [4]. With different fiber orientation distributions, the different bridging constitutive laws can be plotted in Fig. 9(a). From the comparison, the curve with experimentally obtained distribution shows a much higher peak bridging stress than random distributions and wall effect analysis. Such a large difference can be contributed to two factors. Firstly, when the fibers are more aligned in the tensile direction, there will be more fibers crossing a section. More fibers will lead to higher bridging stress. Secondly, compared with the random distributions, there are fewer ruptured fibers at a certain crack opening (Fig. 9(b)) as highly inclined fibers are more likely to rupture [15]. With a higher percentage of less inclined fibers, more fibers can contribute to the bridging effect so the peak bridging stress is higher [4].

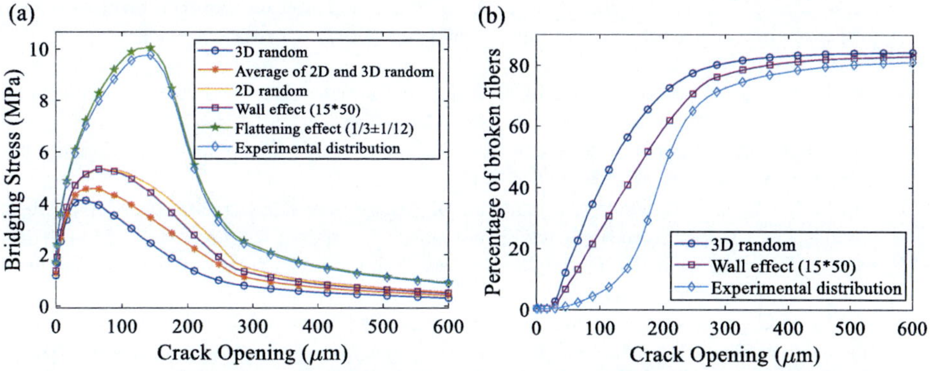

Fig. 9. (a) Comparison between bridging constitutive laws with different orientation distributions; (b) Comparison of percentage of broken fibers with representative orientation distributions.

6 Conclusion

In this paper, the fiber orientation distribution in SHCC tensile samples is experimentally obtained and analyzed. The widely used distributions including simplified random distributions (2D/3D) and wall effect analysis can not accurately predict the real distributions. A new flattening effect analysis is then proposed to model the distribution. In the proposed model, the casting procedure is regarded as flattening the bulk mortar composed of layers to the plate shape with the relative position of the layers kept unchanged. The orientation change can then be calculated according to the movement of the fiber endpoints. The simulation result of the proposed method coincides well with the experimental distribution.

However, as a preliminary study on the effect of casting factor on fiber orientation distribution, the proposed flattening effect analysis is just a highly simplified method. The casting process in reality is highly variable and is hard to model accurately. The flattening parameters selected in this analysis is empirical and may be different with different material properties and sample size. This work aims to bring the casting factor into consideration, especially for modeling the distribution of fibers inclining at large angles. More details should be considered in future studies to make the model more robust and suitable for different composite systems.

References

1. Li, V.C., Leung, C.K.Y.: Steady-state and multiple cracking of short random fiber composites. ASCE J. Eng. Mech. **118**(11), 2246–2264 (1992)
2. Yang, E.H., et al.: Fiber-bridging constitutive law of engineered cementitious composites. J. Adv. Concr. Technol. **6**(1), 181–193 (2008)
3. Lu, C., et al.: Flaw characterization and correlation with cracking strength in Engineered Cementitious Composites (ECC). Cem. Concr. Res. **107**, 64–74 (2018)
4. Lu, C., Leung, C.K.Y.: Theoretical evaluation of fiber orientation and its effects on mechanical properties in Engineered Cementitious Composites (ECC) with various thicknesses. Cem. Concr. Res. **95**, 240–246 (2017)
5. Lee, B.Y., et al.: Quantitative evaluation technique of Polyvinyl Alcohol (PVA) fiber dispersion in engineered cementitious composites. Cem. Conc. Comp. **31**(6), 408–417 (2009)
6. Tosun-Felekoğlu, K., et al.: The role of flaw size and fiber distribution on tensile ductility of PVA-ECC. Compos. B. Eng. **56**, 536–545 (2014)
7. Yao, J., Leung, C.K.Y.: Scaling up modeling of Strain-Hardening Cementitious Composites based on beam theory: from single fiber to composite. Cem. Conc. Comp. **108**, 103534 (2020)
8. Dupont, D., Vandewalle, L.: Distribution of steel fibres in rectangular sections. Cem. Conc. Comp. **27**(3), 391–398 (2005)
9. Alberti, M.G., et al.: On the prediction of the orientation factor and fibre distribution of steel and macro-synthetic fibres for fibre-reinforced concrete. Cem. Conc. Comp. **77**, 29–48 (2017)
10. Torigoe, S., et al.: Study on evaluation method for PVA fiber distribution in engineered cementitious composite. J. Adv. Concr. Technol. **1**(3), 265–268 (2003)
11. Li, M., Li, V.C.: Rheology, fiber dispersion, and robust properties of engineered cementitious composites. Mater. Struct. **46**(3), 405–420 (2013)
12. Preibisch, S., et al.: Globally optimal stitching of tiled 3D microscopic image acquisitions. Bioinformatics **25**(11), 1463–1465 (2009)
13. Vincent, L., Soille, P.: Watersheds in digital spaces: an efficient algorithm based on immersion simulations. IEEE Trans. Pattern Anal. Mach. Intell. **13**(06), 583–598 (1991)
14. Li, V.C., et al.: A micromechanical model of tension-softening and bridging toughening of short random fiber reinforced brittle matrix composites. J. Mech. Phys. Solids **39**(5), 607–625 (1991)
15. Kanda, T., Li, V.C.: Interface property and apparent strength of high-strength hydrophilic fiber in cement matrix. J. Mater. Civ. Eng. **10**(1), 5–13 (1998)

A Depth-Dependent Fiber-Bridging Model to Predict the Tensile Properties Recovery Induced by the Self-healing of Strain-Hardening Cementitious Composites

Yangqing Liu[1(✉)], Bo Wu[2], and Jishen Qiu[2]

[1] Shanghai University, 99 Shangda Road, Shanghai, China
yangqingliu@shu.edu.cn
[2] Hong Kong University of Science and Technology, 1 University Road, Hong Kong, China

Abstract. The self-healing of strain-hardening cementitious composites (SHCCs) relies on the penetration of CO_2 (or dissolved CO_3^{2-}) into the cracks; some SHCCs mixed with special binders, e.g., the reactive magnesia cement (RMC)-based SHCC, even rely on carbonation to harden in the first place. As the carbonation degree decreases with matrix depth, it induces depth-dependent fiber-to-matrix interface properties in these scenarios. In this work, we present a new analytical model that captures the effect of depth-dependent carbonation and self-healing of RMC-based SHCC. In this model, the fiber-bridging *tensile stress* vs. *crack width* curve is formed by summing the *tensile load* vs. *displacement* relationship of individual fibers. On the single-fiber level, the debonding and slip-hardening of the fiber-to-matrix interface induced by a tensile preloading as well as the recovery of the interface properties by self-healing are coherently quantified in a clear kinetic process. On the fiber-bridging level, the experimentally characterized *carbonation* vs. *depth* relationship is added to the model. The modeling results can well capture the single-fiber pullout behavior and the fiber-bridging behavior of the self-healed SHCC specimens.

Keywords: Analytical model · Single-fiber · Carbonation · Self-healing · Fiber-bridging

1 Introduction

Strain-hardening cementitious composites (SHCCs) can engage autogenous crack healings because their crack widths are well controlled by fiber-bridging [1–4]. Recently, it has been noticed that the healing not only happens at the matrix crack but also at the fiber-to-matrix interface, i.e., the fine crack resulting from the fiber debonding from the matrix [5]. It is believed that the healing-induced tensile recovery is mainly attributed to the mechanical recovery of the interface [2], as the matrix crack healing mostly happens at the surface area only [6]. For both the matrix healing and the interfacial healing in a mature SHCC, the healing products to fill the cracks are mainly carbonate precipitates,

M. Kunieda et al. (Eds.): SHCC 2022, RILEM Bookseries 39, pp. 109–118, 2023.
https://doi.org/10.1007/978-3-031-15805-6_12

e.g., $CaCO_3$; their formation relies on the ambient CO_2 penetrating inside, which is depth-dependent. As a result, the autogenous healing of SHCC can be size-dependent, i.e., SHCC with larger cross-sections achieving less tensile recovery as CO_2 cannot penetrate to the deep core. In recent years, SHCCs based on alternative binders have been developed for sustainability. One of them is strain-hardening magnesia-based composites (SHMCs), which adopts zero Portland cement but only reactive magnesia cement (RMC) to bind sands and fibers [7, 8]. As RMC hardens by hydrating and carbonation, i.e., reacting with the penetrating CO_2 to form hydrated magnesium carbonates (HMCs) [9], the SHMCs have strong size-dependency not only in healing but also in its first curing.

The mix proportion design of SHCCs (including SHMCs) are guided by the micromechanics-based analytical model that predicts the fiber-bridging constitutive law [10, 11]. However, there has been no analytical model to quantify the two important effects regarding carbonation, i.e., depth-dependency and post-healing behavior of the fiber-to-matrix interface properties. Therefore, this study modifies the classic two-way pullout fiber-bridging law [11] by adding these effects. First, the key equations in the existing model that adopt interface properties, e.g., chemical bond, frictional bond, and slip-hardening coefficient, are reviewed; second, new equations that quantify the depth-dependency and loading/healing-dependency of these properties are developed, based on our experimental results with SHMC specimens; lastly, a parametric study based on the new model is conducted. The effect of depth and recovered interface property coefficients on the fiber-bridging behavior during the preloading and the reloading is discussed.

2 Review: Existing Fiber-Bridging Analytical Model

Lin et al. [10] derived the tensile load vs. displacement relationship of individual fibers, including Eq. (1) for the debonding stage and Eq. (2) for the slipping stage. In the equations, G_d, τ_0, and β are the chemical bond strength, frictional bond strength, and slip-hardening coefficient, respectively; E_f and d_f are the elastic modulus and diameter of the fibers; η is defined as $V_f\ E_f/V_m E_m$, where V_f and V_m are the volume fractions of fibers and matrix, respectively; L_e is the length of the fiber embedded in the matrix; δ and δ_0 are the fiber pullout displacement and the displacement corresponding to the maximum debonding load, which can be calculated by Eq. (3).

$$P = \sqrt{\frac{\pi^2 \tau_0 E_f d_f^3 (1+\eta)}{2}\delta + \frac{\pi^2 G_d E_f d_f^3}{2}},\ \delta \le \delta_0 \tag{1}$$

$$P = \pi d_f \tau_0 (1 + \beta(\delta - \delta_0)/d_f)(L_e - \delta + \delta_0),\ \delta > \delta_0 \tag{2}$$

$$\delta_0 = \frac{2\tau_0 L_e^2 (1+\eta)}{E_f d_f} + \sqrt{\frac{8 G_d L_e^2 (1+\eta)}{E_f d_f}} \tag{3}$$

Based on the equations above, Yang et al. [11] proposed an analytical model considering the two-way fiber pullout, where the fiber embedment lengths on both sides of the crack are no longer equal. Since the embedment lengths of both sides are different, there are three possible scenarios during the pullout of the fiber. During the derivation process, the fiber tensile force of the short segment is always equal to that of the long segment, as given in Eq. (4).

On the fiber-bridging level, the number of the fibers with the centroidal distance between z_1 and z_2 and the orientation angle between φ_1 and φ_1 can be calculated by Eq. (5), where $p(z)$ and $p(\varphi)$ are the probability density function of fibers with the distance of z and the orientation of φ, respectively. A and A_f are the cross-sectional area of the crack plane and an individual fiber. Further, the total force of the fibers with the corresponding centroidal distance and orientation angle is the product of N_f and the tensile force of a single fiber, as shown in Eq. (6). Any fiber that undergoes a larger load than its tensile strength would be ruptured and not counted in the further analysis. Consequently, the fiber-bridging tensile stress vs. crack width curve can be addressed through Eqs. (1) to (6). However, the previous analytical models cannot consider the effect of depth-dependent carbonation and self-healing.

$$P_s = P_l \tag{4}$$

$$N_f = \frac{AV_f}{A_f} \int_{\varphi_1}^{\varphi_2} \int_{z_1 \cos\varphi}^{z_2 \cos\varphi} p(z)p(\varphi)dzd\varphi \tag{5}$$

$$F = N_f \cdot P \tag{6}$$

3 The Methodology of Depth-Dependent and Self-healing Model

3.1 To Quantify the Depth-Dependency

In this section, an analytical model of fiber-bridging that can quantify depth-dependent carbonation was proposed. Figure 1a shows the relation between carbonation degree and depth from the tests conducted by the authors. The carbonation degree decreased with the depth and got stable somewhere. In the region close to the outer surface of the crack, more CO_2 reacts with $Mg(OH)_2$ to form HMCs that provide high strength for the matrix. On the contrary, as the depth increases, less CO_2 can reach and react with $Mg(OH)_2$ leading to the low strength of the matrix. The interface property coefficients G_d, τ_0, and β are relative to the matrix strength, and therefore relative to the depth. It is assumed that the distribution of the coefficients with the depth is exponential. Taking $G_d(z)$ as an example, the value of the constant coefficients a_1, a_2, and a_3 are determined as 1.7, 0.52, and 0.13 by a nonlinear curve fitting, where the unit of z is millimeter.

$$\begin{aligned} G_d(z) &= G_d \cdot (a_1 \cdot e^{(-z \cdot a_3)} + a_2),\ z \geq 1.5\text{ mm} \\ G_d(z) &= G_d,\ z < 1.5\text{ mm} \end{aligned} \tag{7}$$

Figure 1b shows the fiber-bridging model considering depth-dependent carbonation. The single-fiber pullout behavior at the location of A and B are different because of the depth-dependent interface properties. A small section with the length of dz is taken along the crack depth, as shown on the right of Fig. 1b. Based on Eq. (5), the cross-sectional area of the crack plane A is replaced by $b \cdot$ dz to calculate the number of fibers within the dz-long section, as given in Eq. (8). Further, integrating $P(z) \cdot dN_f$ along the depth of crack can obtain the total force of the fibers with the corresponding centroidal distance and orientation angle, wherein $P(z)$ is the tensile force of a single fiber at the depth of z, which is related to the interface property coefficients. Consequently, the fiber-bridging tensile stress vs. crack width curve considering the depth-dependent carbonation can be addressed through Eqs. (7) to (9).

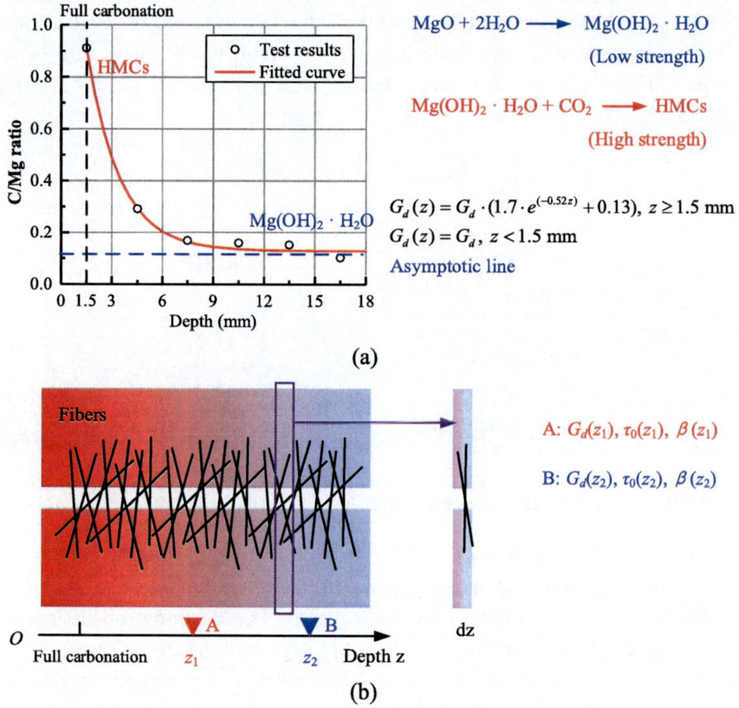

Fig. 1. An analytical model considering the impacts of depth-dependent carbonation. (a) Carbonation degree vs. depth; (b) Fiber-bridging model considering carbonation depth.

$$dN_f = \frac{b \cdot dz \cdot V_f}{A_f} \int_{\varphi_1}^{\varphi_2} \int_{z_1 \cos\varphi}^{z_2 \cos\varphi} p(z)p(\varphi)dzd\varphi \tag{8}$$

$$F = \int_0^l dN_f \cdot P(z) \tag{9}$$

3.2 To Quantify the Effect of Fiber-Matrix Interfacial Healing

Regarding the self-healing behavior, an analytical model of fiber-bridging that considers the impacts of tensile preloading and the interface property recovery was proposed. It is assumed that this effect is independent of depth-dependency.

As discussed in the review, there would be three different scenarios for an individual fiber after the two-way tensile preloading, which are illustrated in Fig. 2. In Scenario 1, the preloading was not significant so both the short and long segments remained at the debonding stage. Based on Eq. (1), the maximum fiber displacements of both sides were the same ($\delta_{s,\max} = \delta_{l,\max}$). In Scenario 2, the preloading level was higher so that the short segment had been fully debonded and entered into the slipping stage, while the long segment remained in the debonding stage. The maximum slip of the short segment $s_{s,\max}$ and the maximum debonding displacement of the long segment $\delta_{l,\max}$ were recorded; In Scenario 3, the preloading level was high and induced full debonding at both sides. The maximum slips of both sides ($s_{s,\max}$, $s_{l,\max}$) were recorded.

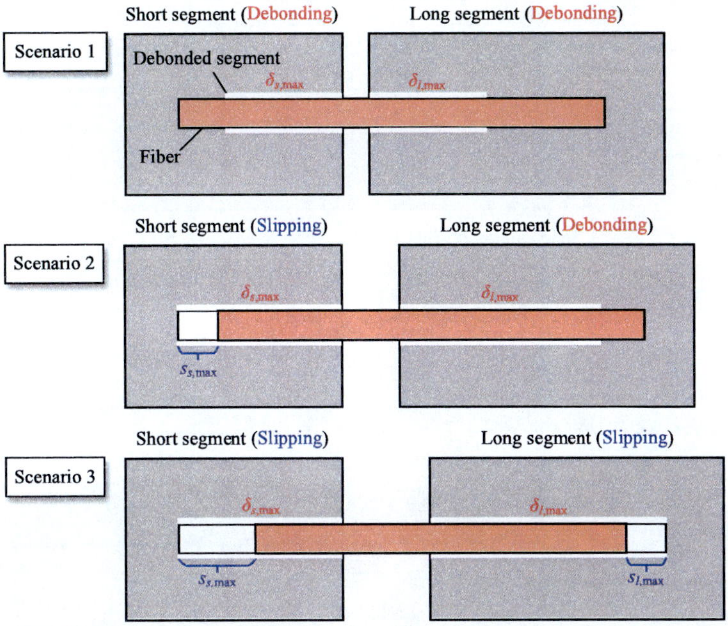

Fig. 2. Different scenarios of fiber-to-matrix after preloading but before healing.

In the three scenarios after preloading, the debonded segments are healed and thus have a new value of chemical bond $G_{d,\text{new}}$ and a new value of frictional bond $\tau_{0,\text{new}}$. Therefore, during reloading, there is a more complex distribution of G_d and τ_0 along with the interface, which not only depends on the preloading scenario (Fig. 2) but also on how far the new debonding crack has propagated. The different scenarios of interfacial properties distributions are summarized in Fig. 3, following the illustration in Fig. 2.

In Fig. 3, L_{s1} and L_{l1} are the length of the short and the long segments that had not experienced debonding during the preloading. By contrast, L_{s2} and L_{l2} are the debonded length of the short and the long segments, which equal the maximum debonding displacements recorded in the preloading. The white region at the fiber end is the residual space after slipping. Their length s_s and s_l are the maximum slip recorded in the preloading. In addition, L'_{es} and L'_{el} are the effective length of the short and the long segments modified by s_s and s_l. The location of the red triangle cursors denotes how far the new debonding has propagated; the location of the blue cursors denotes how far the fiber end, after the full-debonding, has been pulled away from its original location. For a fiber that was only partially debonded and healed, e.g., the left side in Scenario 1, there could be as many as four stages during the reloading, i.e., 1), the debonding frontier is still in the healed segment; 2) the frontier of debonding frontier has entered into the pristine segment; 3) fiber fully debonded and the fiber end is still in the pristine segment; 4) the fiber end has entered into the healed segment. The number in the triangle cursor denotes the current stage of the fiber pullout.

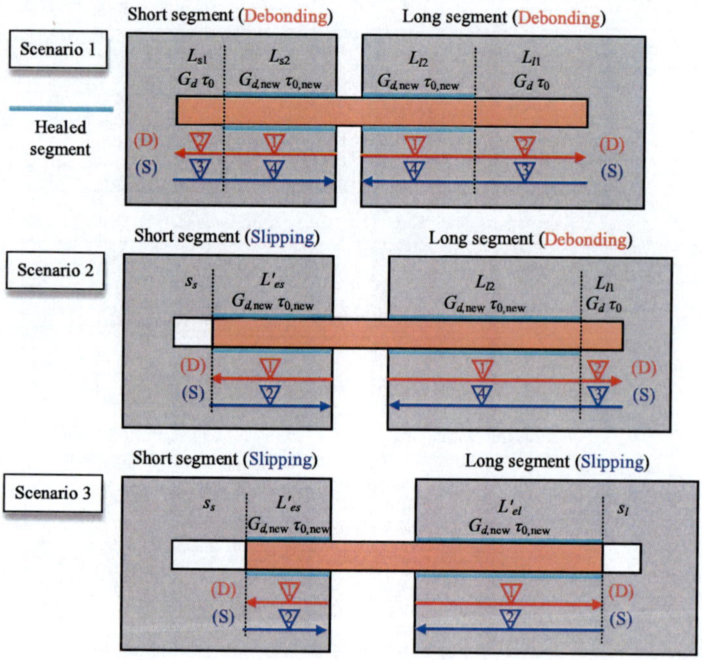

Fig. 3. The kinetic process of the single-fiber pullout after preloading and interfacial healing.

For Scenario 1, taking the short segment as an example, the critical debonding displacement of the healed segment $\delta_{0,s2}$ and the total segment $\delta_{0,s}$ can be obtained by Eqs. (10) to (12) and further used for determining the location of the cursors.

$$\delta_{0,s2} = \frac{2\tau_{0,new}L_{s2}^2(1+\eta)}{E_f d_f} + \sqrt{\frac{8G_{d,new}L_{s2}^2(1+\eta)}{E_f d_f}} \tag{10}$$

$$\delta_{0,s1} = \frac{2\tau_0 L_{s1}^2 (1+\eta)}{E_f d_f} + \sqrt{\frac{8G_d L_{s1}^2 (1+\eta)}{E_f d_f}} \tag{11}$$

$$\delta_{0,s} = \delta_{0,s1} + \delta_{0,s2} \tag{12}$$

When the debonding displacement δ_s is smaller than $\delta_{0,s2}$, corresponding to No. 1 cursor, the fiber force of the short segment P_s can be calculated by Eq. (13). By contrast, when δ_s is larger than $\delta_{0,s2}$ and smaller than $\delta_{0,s}$, corresponding to No. 2 cursor, P_s can be obtained by Eqs. (14) to (16).

$$P_s = P_{s2} = \sqrt{\frac{\pi^2 \tau_{0,new} E_f d_f^3 (1+\eta)}{2}\delta_s + \frac{\pi^2 G_{d,new} E_f d_f^3}{2}},\ \delta_s < \delta_{0,s2} \tag{13}$$

$$P_{s2}(\delta_{0,s2}) = \sqrt{\frac{\pi^2 \tau_{0,new} E_f d_f^3 (1+\eta)}{2}\delta_{0,s2} + \frac{\pi^2 G_{d,new} E_f d_f^3}{2}},\ \delta_s > \delta_{0,s2} \tag{14}$$

$$P_{s1} = \sqrt{\frac{\pi^2 \tau_0 E_f d_f^3 (1+\eta)}{2}(\delta_s - \delta_{0,s2}) + \frac{\pi^2 G_d E_f d_f^3}{2}},\ \delta_s > \delta_{0,s2} \tag{15}$$

$$P_s = P_{s2}(\delta_{0,s2}) + P_{s1},\ \delta_s > \delta_{0,s2} \tag{16}$$

For No. 3 cursor, the fiber end slips within the pristine segment, and δ_s should be larger than $\delta_{0,s}$ as well u_s is smaller than L_{s1} (Eq. (17)). Based on Eq. (2), the fiber force is the sum of the following three components, as given in Eqs. (18) to (21).

$$u_s = \delta_s - \delta_{0,s} \tag{17}$$

$$P_{s1} = \pi d_f \tau_0 (L_{s1} - u_s),\ u_s < L_{s1} \tag{18}$$

$$P_{s2} = \pi d_f \tau_{0,new} L_{s2} \tag{19}$$

$$P_{s3} = \pi d_f \tau_0 \beta u_s / d_f (L_{es} - u_s) \tag{20}$$

$$P_s = P_{s1} + P_{s2} + P_{s3},\ u_s < L_{s1} \tag{21}$$

No. 4 cursor is at the healed segment, where u_s is larger than L_{s1} and smaller than L_{es}. P_s can be calculated by Eqs. (22) to (24).

$$P_{s1} = \pi d_f \tau_{0,new} (L_{es} - u_s) \tag{22}$$

$$P_{s2} = \pi d_f \tau_{0,new} \beta_{new} u_s / d_f (L_{es} - u_s) \tag{23}$$

$$P_s = P_{s1} + P_{s2},\ u_s > L_{s1} \tag{24}$$

The fiber force of the long segment P_l can be analyzed following the derivation process above. Consequently, the fiber forces and the displacements are solved by the force equilibrium condition $P_s = P_l$ and the geometry relationship $\delta = \delta_s + \delta_l$.

For Scenario 2 and 3, the short segments were at the slipping stage and there remained the residual slip s_s that affect the reloading of the short segments. Thus, the length of the short segment is first modified by subtracting the residual slip (Eq. (25)).

$$L'_{es} = L_{es} - s_s \tag{25}$$

Specifically in Scenario 2, the long segment was still at the debonding stage, and therefore the total residual slip is contributed by the short segment. If the displacement is smaller than s_s, the fiber force is equal to 0. Otherwise, the displacement will be modified by subtracting s_s. Subsequently, the fiber force of the short segment P_s can be calculated by Eqs. (1) to (3) with the healed interface property coefficients. And the long segment force can be obtained by the same method mentioned above (Eqs. (10) to (24)).

For Scenario 3, since the long segment was also at the slipping stage, the residual slip s_l is subtracted from the length of the long segment.

$$L'_{el} = L_{el} - s_l \tag{26}$$

At this time, if the displacement is less than the sum of s_s and s_l, the fiber force equals 0. If not, the corrected displacement is the total displacement minus all the residual slips. In this case, both P_s and P_l can be calculated by Eqs. (1) to (3) with the recovery interface properties.

4 Parametric Study with the New Model

Based on the analytical model proposed above, a parametric study to evaluate the effect of depth-dependency and fiber-matrix interfacial healing on fiber-bridging behavior was conducted. Table 1 shows the PVA properties and measured interface property coefficients before and after self-healing from a pullout test of an individual PVA fiber performed by the authors. The related parameters were used in the parametric study. Besides, the width and the elastic modulus of the matrix were 13 mm and 8 GPa, respectively. Referenced from [12] and [11], the snubbing coefficient f and the strength reduction factor f' were set as 0.5 and 0.33, respectively.

Table 1. PVA fiber properties and measured coefficients from the lab tests

Case	L_f (mm)	d_f (μm)	σ_u (MPa)	E_f (GPa)	G_d (J/m^2)	τ_0 (MPa)	β
Preloading	13	200	1000	27	4.28 ± 0.78	0.88 ± 0.12	0.044 ± 0.03
Reloading					–	2.61 ± 0.4	0.045 ± 0.04

Figure 4a shows the preloading and the reloading tensile stress-displacement curves of the models with the depth of 12 mm, 24 mm, and 48 mm. By considering the impacts of depth-dependency carbonation, the fiber-bridging behavior is related to the depth of the specimens. The average tensile stress decreases with the increase of depth because the interface properties are assumed to be lower at the deeper location. Besides, the results demonstrate that the proposed analytical model is capable to take the impacts of depth-dependency and self-healing into account simultaneously.

Figure 4b shows the tensile stress-displacement curves of the models with the $G_{d,\mathrm{new}}$ to G_d ratios of 0, 0.5, and 1. It is found that the recovery chemical bond strength $G_{d,\mathrm{new}}$ significantly affects the peak tensile stress during the reloading, while the magnitude of increment decreases when the ratio rises from 0.5 to 1. By contrast, the residual stress at relatively large displacements is less influenced by the $G_{d,\mathrm{new}}$ to G_d ratio. Figure 4c shows the tensile stress-displacement curves of the models with the $\tau_{0,\mathrm{new}}$ to τ_0 ratios of 1, 3, and 5. It is noted that the residual tensile stress after the peak stress remarkably increases with the enlargement of $\tau_{0,\mathrm{new}}$ during the reloading, while the magnitude of

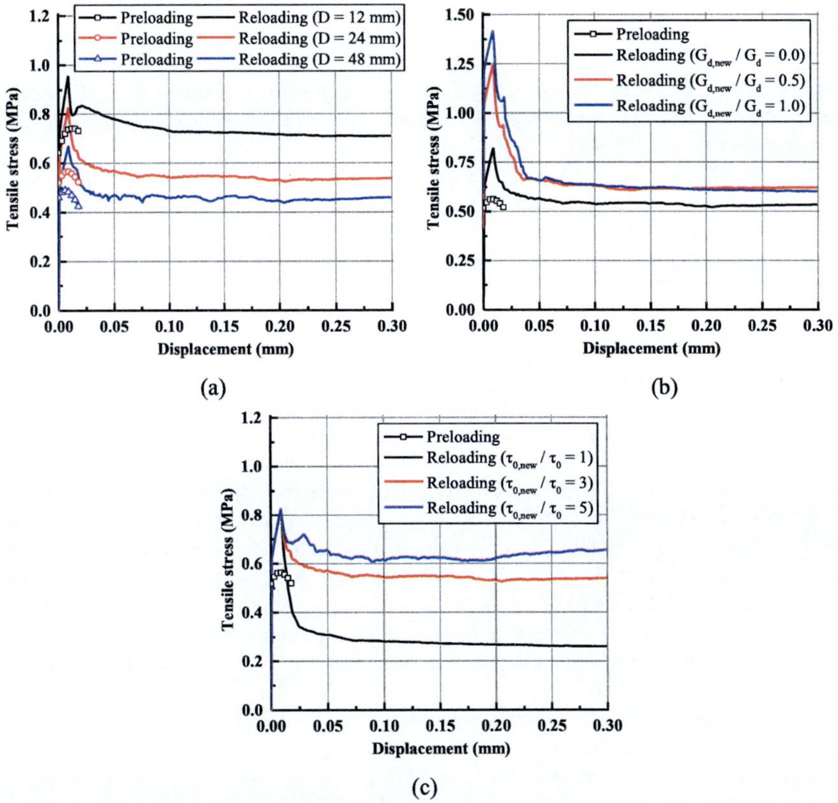

Fig. 4. The comparison of fiber-bridging behavior in the parametric study: (a) Effect of depth; (b) Effect of $G_{d,\mathrm{new}}/G_d$; (c) Effect of $\tau_{0,\mathrm{new}}/\tau_0$.

increment decreases with the increase of the ratio. The results illustrate that the proposed analytical model can capture the effect of interface properties after self-healing on the fiber-bridging behavior.

5 Summary

TO sum up, based on the classic two-way pullout fiber-bridging law, this study developed a new analytical model that can quantify depth-dependent carbonation and the effect of fiber-matrix interfacial healing. The depth-dependency was realized by introducing the carbonation degree vs. depth relationship. A kinetic process of single-fiber pullout considering the different scenarios of preloading was analyzed to evaluate the fiber-bridging behavior after interfacial self-healing. Further, a parametric study with the new model was performed. The effect of depth and recovered interface property coefficients on the fiber-bridging behavior was discussed. The more detailed discussion and model validation will be presented in our upcoming paper.

References

1. Hilloulin, B., Hilloulin, D., Grondin, F., et al.: Mechanical regains due to self-healing in cementitious materials: experimental measurements and micro-mechanical model. Cem. Concr. Res. **80**, 21–32 (2016)
2. Zhang, Z., Zhang, Q., Li, V.: Multiple-scale investigations on self-healing induced mechanical property recovery of ECC. Cement Concr. Compos. **103**, 293–302 (2019)
3. Yang, Y., Lepech, M., Yang, E., et al.: Autogenous healing of engineered cementitious composites under wet–dry cycles. Cem. Concr. Res. **39**(5), 382–390 (2009)
4. Qiu, J., Tan, H., Yang, E.: Coupled effects of crack width, slag content, and conditioning alkalinity on autogenous healing of engineered cementitious composites. Cement Concr. Compos. **73**, 203–212 (2016)
5. Qiu, J., Ruan, S., Unluer, C., et al.: Autogenous healing of fiber-reinforced reactive magnesia-based tensile strain-hardening composites. Cem. Concr. Res. **115**, 401–413 (2019)
6. Fan, S., Li, M.: X-ray computed microtomography of three-dimensional microcracks and self-healing in engineered cementitious composites. Smart Mater. Struct. **24**(1), 015021 (2014)
7. Ruan, S., Qiu, J., Yang, E., et al.: Fiber-reinforced reactive magnesia-based tensile strain-hardening composites. Cement Concr. Compos. **89**, 52–61 (2018)
8. Wu, B., Qiu, J.: Incorporating hollow natural fiber (HNF) to enhance CO_2 sequestration and mechanical properties of reactive magnesia cement (RMC)-based composites: feasibility study. J. CO_2 Util. **57**, 101874 (2022)
9. Unluer, C.: Carbon dioxide sequestration in magnesium-based binders. In: Carbon Dioxide Sequestration in Cementitious Construction Materials, pp. 129–173. Woodhead Publishing (2018)
10. Lin, Z., Kanda, T., Li, V.: On interface property characterization and performance of fiber-reinforced cementitious composites. J. Concr. Sci. Eng. **1**, 173–184 (1999)
11. Yang, E., Wang, S., Yang, Y., et al.: Fiber-bridging constitutive law of engineered cementitious composites. J. Adv. Concr. Technol. **6**(1), 181–193 (2008)
12. Wu, C.: Micromechanical tailoring of PVA-ECC for structural application. Thesis (Ph.D.), University of Michigan (2001)

MicroCT and 3D Image Processing and Analysis to Investigate Strain-Hardening Cement-Based Composites (SHCC)

Renata Lorenzoni[1](✉), Sidnei Paciornik[1], Flavio A. Silva[2], and Giovanni Bruno[3]

[1] Department of Chemical and Materials Engineering, PUC-Rio, Rio de Janeiro, Brazil
renata_lorenzoni@hotmail.com

[2] Department of Civil and Environmental Engineering, PUC-Rio, Rio de Janeiro, Brazil

[3] BAM, Berlin, Germany

Abstract. X-ray micro-computed tomography (microCT) is a non-destructive technique that can provide 3D images of the internal microstructure of a composite material. Optimizing the analysis with modern computational tools leads to a higher precision in quantitative analysis and, consequently, to more accurate results. In this scenario, machine learning has been widely used as solutions for complex image processing and analysis tasks. The SHCC microCT images can be considered complex, given the small scale of analysis and the typical resolution of common microCT, as well as the small differences among the material constituents in terms of density and x-ray absorption. The present work brings innovative solutions for fiber and pore quantification in SHCC using Machine Learning. SHCC were tested in an in-situ testing device coupled to a microCT and the material mechanical response was correlated with microstructure changes through an image sequence. The internal displacement and strain were calculated by Digital Volume Correlation (DVC). The strain results were correlated with the initial quantification of the constituent phases of the material.

Keywords: SHCC · MicroCT · 3D image processing and analysis · Deep learning · DVC

1 Introduction

The microCT, a non-destructive technique that provides information on the material's internal structure, is already widely disseminated for the study of materials [1]. For cementitious materials, this technique is mainly applied for mesoscale analysis. It can provide information such as quantification of constitutive phases and damage assessment when the material is subjected to some mechanical effort or aggressive environment [2]. Also, these 3D images can be used as input or validation of numerical models [3], and for strain determination through the DVC technique [4]. For any these types of analysis, support from the image processing and analysis field is required.

M. Kunieda et al. (Eds.): SHCC 2022, RILEM Bookseries 39, pp. 119–126, 2023.
https://doi.org/10.1007/978-3-031-15805-6_13

Figure 1 shows a typical image processing and analysis procedure of images obtained from microCT. The segmentation step of this procedure is considered the key step for a successful analysis. The segmentation is when the image is no more analyzed voxel by voxel but the sets of these voxels, known as regions of interest (ROIs). Segmentation means defining the voxels that are included in the ROI. The standard method for segmentation is the thresholding method, in which the ROI is determined by a range of gray values in the image histogram. However, when this range includes voxels that should not be part of the ROI, segmentation by thresholding is not efficient and deep learning (DL) method can be a solution. The main feature of DL for segmentation tasks is that feature extraction and pattern recognition and classification are made before the segmentation.

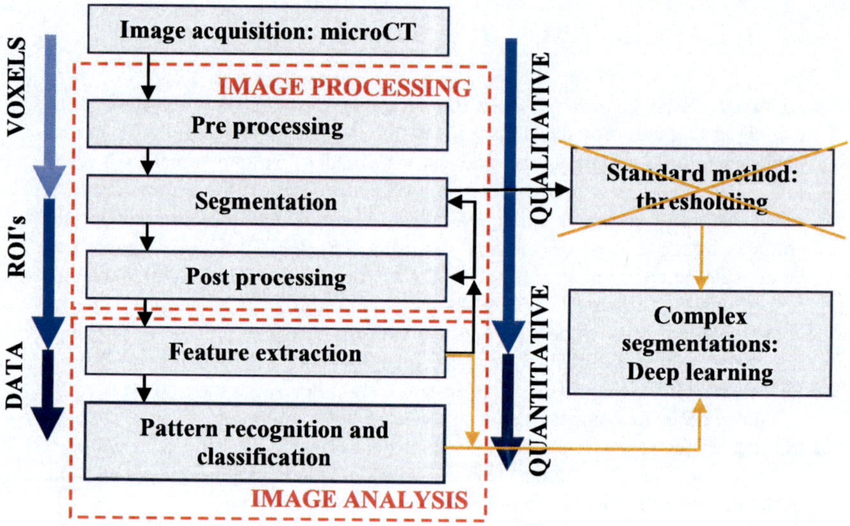

Fig. 1. Image processing and analysis procedure. Replacement of the standard segmentation method (thresholding) by deep learning.

For the Strain-hardening cement-based composites (SHCC), the ROIs would be mainly the phases of pores and fibers since their quantification and distribution influence the material's behavior. SHCC is a cementitious material reinforced with fibers and exhibit the behavior of multiple cracking before failure due to the presence of fibers [5]. To achieve this desired behavior, different materials can be used for its manufacture, both in the mixing the matrix or the type of fiber used [6]. Since the contrast between the phases in the images from microCT is related to the differences of density and x-ray absorption among these phases, the materials used in its manufacture will influence the ease/accuracy of the analysis. For instance, steel fibers are readily distinguished from the matrix, while polymeric fibers present similar gray values of the cementitious matrix, making their distinction complex. For complex segmentation, DL is a recommended method with high potential for accurate and efficient segmentation that cannot be solved by applying a standard method as thresholding.

Besides the quantification of internal phases, microCT can provide valuable information regarding material damage mechanisms. Particularly for the SHCC, it includes the investigation of the multiple cracking. When coupled with an in-situ mechanical test device, it becomes a real-time investigation. Moreover, the analysis of active deformation and damage processes can be complemented by Digital Volume Correlation (DVC), which enabling a detailed assessment of the origin and type of fracture.

2 Experimental Program

This work summarized previous studies carried out in different institutes' facilities. 3D images obtained from microCT of four different specimens of SHCC were reinvestigated and compared. The nomenclature, the base material for manufacture, the microCT used and the voxel size obtained are described in Table 1.

Table 1. SHCC nomenclature, with its base materials, microCT equipment used and voxel size.

	Material	MicroCT	Voxel size
M1-PVA	Normal-strength matrix + polyvinyl alcohol fibers	Nanotom-Baker	$(4.0\ \mu m)^3$
M2-PE	High-strength matrix + polyethylene fibers	Nanotom-Baker	$(4.0\ \mu m)^3$
M1-S + PVA	Normal-strength matrix + steel and PVA fibers	Xradia 510 Versa-Zeiss	$(18.5\ \mu m)^3$
M2-PE_in-situ	High-strength matrix + polyethylene fibers	GE VTomeX	$(13\ \mu m)^3$

The desired voxel size depends, among other parameters, mainly on the size of the cross section of the specimen. The specimens M1-PVA and M2-PE were prepared for a fiber segmentation study, in which a high resolution was required. So, miniature specimens were used, with 100 mm length, 10 mm width, and 3.5 mm thickness [7]. The specimen M1-S + PVA was intended for a mechanical evaluation study, in which the size of the specimen was determined by the mechanical laboratory test: the total length was 200 mm, the width 38 mm, the thickness 9.5 mm, and a notch angle 90° [8]. The specimen M2-PE_in-situ was destined for in-situ compression microCT testing. The miniature specimens had square cross-sections with side dimensions of 5 mm, while their height was 15 mm [9]. The scanner conditions and the reconstruction methods of each specimen are described in the referenced studies [7–9].

All images were pre-processed in FIJI software [10]: first converted to 8-bit, then the contrasts enhanced by histogram stretching, and finally, an edge-preserving low-pass filter called 'non-local means' (NLM) [11] was applied in auto mode. All images were investigated to show the feasibility of the fiber segmentation according to the type of SHCC (matrix and fiber) and image resolution. First, to investigate the feasibility of segmentation by thresholding, were selected gray values ranges in the pre-processed images histograms. The inaccuracy of segmentation by thresholding leads to choosing semantic segmentation by deep learning. M1-PVA and M2-PE images had to be segmented by DL, and the DL network parameters and results of the 3D segmentation are in [7]. The PVA fibers in image M1-S + PVA were hard to distinguish and could not be segmented, the steel fibers were segmented by thresholding. M2-PE_in-situ also needed to be segmented by DL, this work shows a slice of fiber segmentation by DL, and the 3D segmentation and network details are presented in [9].

M2-PE_in-situ images, besides investigated for the feasibility of fiber segmentation, were also analyzed mechanically through the DVC. To map and quantify the strains in the compressed specimen, DVC evaluation was performed with the software VIC-Volume (Correlated Solutions, North Carolina, USA). For this purpose, the software locates a specific sub-volume in two sequences of images and performs image matching with optimal accuracy. The sub-volume size must be set by the user through in a way that each of the sub-volumes contains a pattern of sufficient contrast. Thus, the subset was chosen such that it has one or more phases within it. The amount of analyzed sub-volumes depends on the defined "step size", so that large step sizes result in smaller output data but run faster. This parameter should not be too large so that image information is not lost, and not too small so that the computer can process efficiently. The DVC correlation parameters used in the two subsequent images of M2-PE_in-situ is summarized in [9].

3 Results and Discussions

Attempts to segment the constitutive phases of SHCC by applying the thresholding method were performed on all four images. Figure 2 presents one slice of M1-PVA image and one slice of M2-PE with their respective histograms. Figure 2 (a) and (c) shows an attempt to segment the pores, the range chosen for M1-PVA was from 0 to 65 and for M2-PE it was from 0 to 70. In both, pixels corresponding to fibers are segmented together, highlighted in yellow. Figure 2 (b) and (d) shows an attempt to segment the fiber, the range chosen for M1-PVA was from 60 to 80 and for M2-PE it was from 60 to 85. In both, pixels from the edges of the pores are segmented together, in addition to some sands. Both for pores and fibers segmentation, the M2-PE image presents more complex. This can be traced back to the fact that the PE fiber is smaller and less dense than PVA fibers.

These results show that segmentation by thresholding would not be accurate and efficient, especially regarding fibers, being recommended the segmentation by DL. In [7], a DL network trained with the same training dataset of the M2-PE image was applied in both M2-PE and M1-PVA images, showing satisfactory results.

Figure 3 presents the steel fibers segmentation in one slice of the M1-S + PVA image through a cut-off on the gray value range of the histogram from 155 to 255. Figure 3 also highlights in yellow that before applying the NLM, it's hard but possible to notice the presence of PVA fibers. After the NLM noticing them becomes no longer possible. However, it is not possible to assert that the PVA fibers are distinguishable from the matrix at such resolution. There are two reasons for this. The main reason is the image resolution, which has approximately 2 voxels representing the PVA fiber. Another

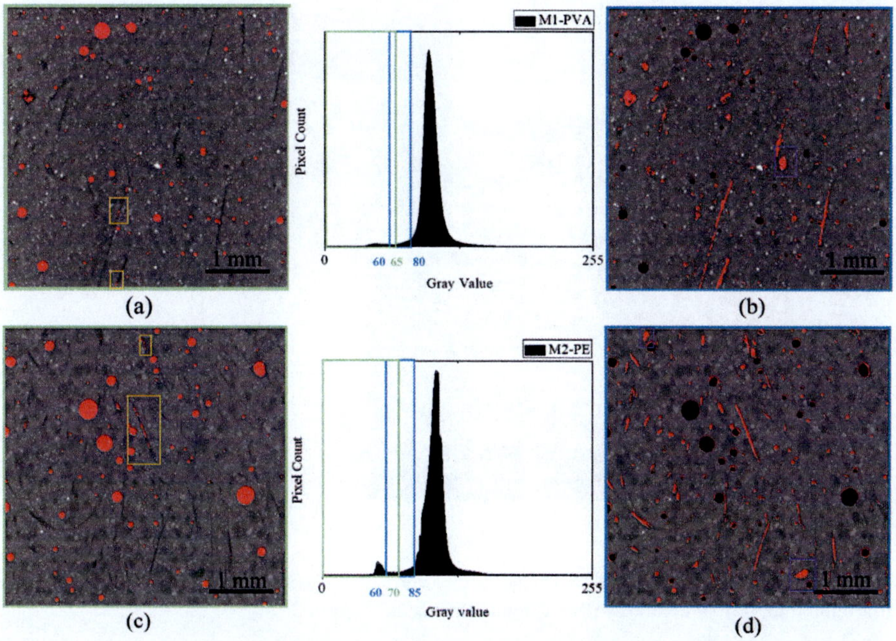

Fig. 2. (a) One slice of M1-PVA image with gray value pixels from 0 to 65 segmented, and (b) the same slice with gray value pixels from 60 to 80 segmented. (c) One slice of M2-PE image with gray value pixels from 0 to 70 segmented, and (d) the same slice with gray value pixels from 60 to 85 segmented. Highlighted in yellow segmented fibers in the same gray value range as pores. Highlighted in purple segmented sands in the same gray value range as fibers.

reason is that the very bright steel fibers lead to a compression of the dynamic range in the dark range of the other phases (matrix, pores, PVA fibers), thus decreasing the contrast between them.

A cross-section of M2-PE_in-situ image is shown in Fig. 4. Figure 4 (b) shows the difficulty of segmenting the fibers by thresholding. Like M2-PE and M1-PVA images, pixels from the edges of the pores are segmented together. Nevertheless, fewer fibers were segmented when this happens sharply. That's because M2-PE_in-situ image has lower resolution. Figure 4 (c) shows the same slice with the fibers segmented by DL.

Regarding the amount of voxels representing a fiber, it is curious that the PVA fibers did not appear to be distinguishable in the M1-S + PVA image and the PE fibers did in M2-PE_in-situ image. This is justified by the contrast between the phases, since M1-S + PVA image presents very bright voxels. Also, it is worth remembering that the microCT scan settings and the image reconstruction influence their quality, being able to generate images with more or less contrasts between the phases.

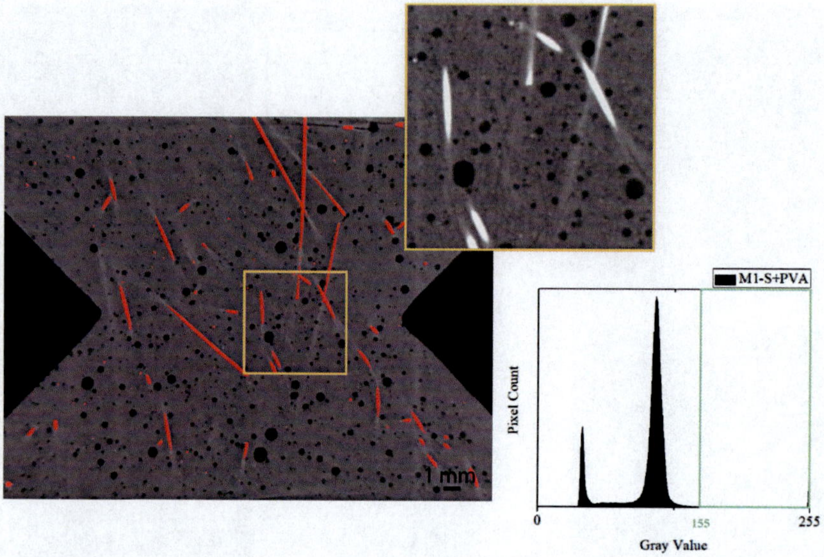

Fig. 3. One slice of M1-S + PVA image with gray value pixels from 155 to 255 segmented. Highlighted in yellow a zoom without 'non-local means' applied.

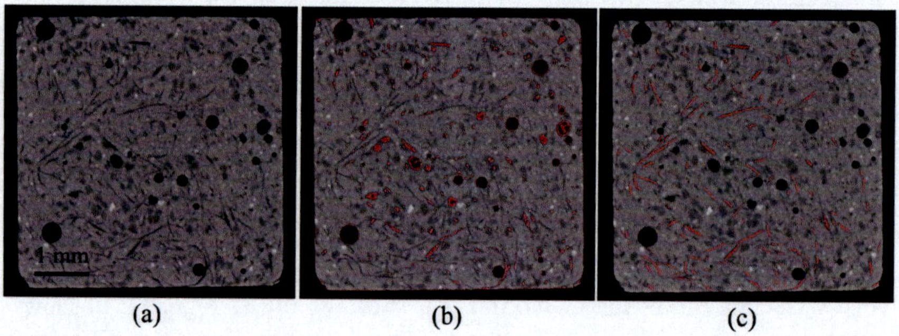

Fig. 4. (a) cross-section M2-PE_in-situ image; (b) fibers segmentation attempt by thresholding; (c) fibers segmentation by DL.

The DVC results applied to two images with subsequent loading of the M2-PE_in-situ specimen is shown in Fig. 5. The compressive strains along the Z-axis are represented in Fig. 5 (b), in which the DVC evaluations captured the apparently eccentric axial loading in the specimen, this being a result of a non-parallelism of the specimen faces. This justifies the occurrence of the crack on the left of the specimen rather than on the right side, where there is a huge pore, see the positive transversal strains in Fig. 5 (a). Strain positive concentrations characterize crack opening. In the subsequent loading step a crack occurs due to this huge pore, see [9].

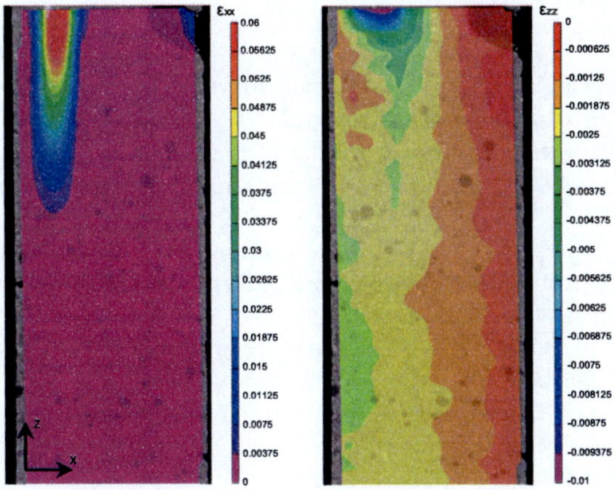

(a) (b)

Fig. 5. DVC analysis of the strain distribution in M2-PE_in-situ: (a) the transversal strains along X-axis and (b) the axial strains along Z-axis are represented.

4 Conclusions

This work showed that PVA and PE fibers need high resolution to be distinguishable in microCT images. For instance, an image with a voxel size of 18,5 μm was not enough to distinguish 40 μm diameter PVA fibers inside a cement. Another image in which the PVA fiber diameter was represented by 5 voxels, presented an inefficient fiber segmentation by the traditional thresholding method, being solved by DL.

The DL technique is a novel method with a high potential for an accurate and efficient segmentation of complex microstructures, which cannot be performed by applying common methods such as the gray-value threshold. When a good training basis is developed, it can be applied in the automatic segmentation of other similar images. This principle was demonstrated in the current work on two different types of SHCC, called M2-PE and M1-PVA.

The in-situ microCT showed potential in monitoring and quantifying active deterioration mechanisms in the specimens subjected to compression loading. The specimens tested in compression were additionally evaluated using Digital Volume Correlation for mapping and quantifying the corresponding strain fields and fracture initiations. The DVC analysis demonstrated the influence of pores on strain concentration and fracture localization, but also highlighted the effect of imperfect specimen geometry on the load distribution in the specimen cross-section.

Acknowledgments. The authors would like to acknowledge the financial support offered by agencies CNPq, CAPES, FAPERJ and the German Academic Exchange Service – DAAD.

References

1. Landis, E.N., Keane, D.T.: X-ray microtomography. Mater. Charact. **61**, 1305–1316 (2010). https://doi.org/10.1016/j.matchar.2010.09.012
2. Lorenzoni, R., Paciornik, S., Silva, F.A.: Characterization by microcomputed tomography of class G oil well cement paste exposed to elevated temperatures. J. Pet. Sci. Eng. **175**, 896–904 (2019). https://doi.org/10.1016/j.petrol.2019.01.022
3. Lu, C., Li, V.C., Leung, C.K.Y.: Flaw characterization and correlation with cracking strength in Engineered Cementitious Composites (ECC). Cem. Concr. Res. **107**, 64–74 (2018). https://doi.org/10.1016/j.cemconres.2018.02.024
4. Lorenzoni, R., et al.: Macro and meso analysis of cement-based materials subjected to triaxial and uniaxial loading using X-ray microtomography and digital volume correlation. Constr. Build. Mater. **323** (2022). https://doi.org/10.1016/j.conbuildmat.2022.126558
5. Van Zijl, G.P.A.G., Wittmann, F.H.: On durability of SHCC. J. Adv. Concr. Technol. **8**, 261–271 (2010). https://doi.org/10.3151/jact.8.261
6. Curosu, I., Mechtcherine, V., Millon, O.: Effect of fiber properties and matrix composition on the tensile behavior of strain-hardening cement-based composites (SHCCs) subject to impact loading. Cem. Concr. Res. **82**, 23–35 (2016). https://doi.org/10.1016/j.cemconres.2015.12.008
7. Lorenzoni, R., Curosu, I., Paciornik, S., Mechtcherine, V., Oppermann, M., Silva, F.: Semantic segmentation of the micro-structure of strain-hardening cement-based composites (SHCC) by applying deep learning on micro-computed tomography scans. Cem. Concr. Compos. **108**, 103551 (2020). https://doi.org/10.1016/j.cemconcomp.2020.103551
8. Lorenzoni, R., Tinoco, M., Paciornik, S., de Andrade Silva, F.: The use of X-ray microtomography to investigate the shear behavior of hybrid fiber reinforced strain hardening cementitious composites. J. Build. Eng. **43**, 103126 (2021). https://doi.org/10.1016/j.jobe.2021.103126
9. Lorenzoni, R., et al.: Combined mechanical and 3D-microstructural analysis of strain-hardening cement-based composites (SHCC) by in-situ X-ray microtomography. Cem. Concr. Res. **136**, 106139 (2020). https://doi.org/10.1016/j.cemconres.2020.106139
10. Schindelin, J., et al.: Fiji: An open-source platform for biological-image analysis. Nat. Meth. **9**, 676–682 (2012). https://doi.org/10.1038/nmeth.2019
11. Wang, X., Shen, S., Shi, G., Xu, Y., Zhang, P.: Iterative non-local means filter for salt and pepper noise removal. J. Vis. Commun. Image Represent. **38**, 440–450 (2016). https://doi.org/10.1016/j.jvcir.2016.03.024

A New Method to Quantitatively Characterize the Porosity of Fiber/Matrix Interfacial Transition Zone (ITZ) via Longitudinal Cross-Sections

Shan He[1](✉), Minfei Liang[1], En-hua Yang[2], and Erik Schlangen[1]

[1] Microlab, Faculty of Civil Engineering and Geosciences, Delft University of Technology, Delft 2628 CN, The Netherlands
s.he-2@tudelft.nl

[2] School of Civil and Environmental Engineering, Nanyang Technological University, Singapore 639797, Singapore

Abstract. The properties of the interfacial transition zone (ITZ) between microfiber and cement-based matrix are of primary significance for the overall behavior of strain hardening cementitious composites (SHCCs). However, due to the relatively small diameter of polymeric microfibers (e.g., PVA fiber), it is technically difficult to obtain quantitative and representative information of the properties of the ITZ. In the current study, a new method that is able to quantitatively characterize the microstructural features of the ITZ surrounding a well-aligned microfiber was reported. With the method, the porosity gradients within the ITZs between PVA fiber and cement paste matrices with different water to cement (w/c) ratios were determined. The results show that the matrix surrounding a microfiber were more porous than the bulk matrix. The thickness of this porous region can extend up to 100 microns away from the fiber surface even at a relatively low water to cement ratio (w/c = 0.3). It is thus believed that the method could facilitate the investigation and modification of fiber/matrix bond properties and also contribute to the development of SHCC with superior properties.

Keywords: ITZ · Porosity · Image analysis · Fiber

1 Introduction

The interfacial transition zone (ITZ) between an inclusion (e.g., aggregate) and bulk matrix in cement-based composites has been the focus of extensive research. Now it is well accepted and documented that the ITZ is quite different in its microstructure, composition and mechanical properties than the bulk cement matrix as a result of the disruption created by the inclusion to the packing of its surrounding cement particles [1, 2]. Because of the loose packing, the ITZ is usually more porous and weaker compared to the bulk matrix, and in many cases governs the mechanical properties and durability of cement-based materials. This phenomenon is usually addressed as the "wall" effect

M. Kunieda et al. (Eds.): SHCC 2022, RILEM Bookseries 39, pp. 127–134, 2023.
https://doi.org/10.1007/978-3-031-15805-6_14

[3, 4]. It is believed that as long as an inclusion is sufficiently large to act as a wall against the cement particles, an ITZ will form around the inclusion.

The question arises as to whether an ITZ will form around a microfiber that has only one dimension (length) being much larger than cement particles, while its another dimension (diameter) is of the same order of magnitude as that of the cement grains. Previous studies on the characterization of the ITZ between microfibers and cement paste also reported contradictory results. Based on scanning electron microscopy (SEM) examination, it has been reported that microstructure of interface between microfiber and cement matrix is dense and no obvious ITZ can be observed, which differs from that around bigger inclusion, such as aggregates and macro-fibers [5]. This has been attributed to the fact that diameter of microfibers is of the same order of magnitude as that of cement particles, and thus the wall effect that gives rise to the formation of ITZ may be largely minimized. However, Chan and Li [6] reported that porous structure were observed around a polyethylene fiber with a diameter of 38 μm in cement paste and showed that the porosity decreases when the water-to-cement ratio (w/c) is reduced. Recently, He et al. [7, 8] reported that porous ITZ could still be observed even when the w/c of the matrix is as low as 0.2.

We argue that disagreement among published literatures on the subject matter may be due to 2 reasons. Firstly, microstructural observation made by SEM is qualitative by nature. It is possible to compare 2 microstructures and decide which one contains more pores. But determining whether a microstructure is porous or not would be largely subjective. Secondly, established methods for studying the ITZ around a fiber often perform analyses on cross-sectional planes that are either perpendicular or intersect with the fiber axis at a random angle. This way of analysis can lead to errors and uncertainties for that the microstructure of the ITZ could be quite different from place to place due to the heterogeneity along its longitudinal direction.

To address the issues mentioned above, this paper reports a new method to quantitatively characterize the ITZ between microfiber and cement matrix. Instead of analyzing the porosity of ITZ on sections perpendicular to the fiber axis, we prepared longitudinal section passing through the central axis of the fiber for ITZ characterization by backscattered electron (BSE) imaging. With the method, the porosity gradients within the ITZs between PVA fiber and cement paste matrices with different water to cement (w/c) ratios were determined.

2 Materials and Methods

2.1 Materials

In this study, CEM I 42.5N Portland cement was used to prepare cement pastes with varying water to cement (w/c) ratio from 0.3 to 0.5. The microfiber used to produce the ITZ is a PVA microfiber with a diameter of 39 μm. Table 1 summarize the physical properties and geometry of the PVA fiber used in this study. All the specimens were cured in a climate room (20 °C and ≥ 98% RH) and have an age of 14 days on the day of analysis.

Table 1. Physical and mechanical properties of PVA fibers.

Diameter (μm)	Density (kg/m3)	Nominal tensile strength (MPa)	Young's modulus (GPa)	Surface oil-content (wt.%)
39	1300	1640	41.1	1.2

2.2 Detailed Sample Preparation Method

In the current study, characterization of the porosity follows the well-established method of quantitative backscattered electron (BSE) image analysis [9], which usually comprises: 1) section a specimen to expose the targeted microstructure; 2) use epoxy resin to impregnate the pores; 3) grind away excessive epoxy on the surface and polish the section; 4) examine the polished section under BSE; and 5) BSE image analysis. The challenge of applying this method to characterize the ITZ surrounding a microfiber lies on making a well-controlled cut near the fiber. This is because that only by making a cut as close and parallel to the embedded fiber as possible can the pores of the ITZ be exposed. As the diameter of a microfiber is only several tens of microns, the positional accuracy of the cutting need to be at micron level. For this reason, in the current study we adopted an automatic dicing machine, which is originally used in the semiconductor industry to perform cutting of silicon wafers. A detailed sample preparation procedure is given below.

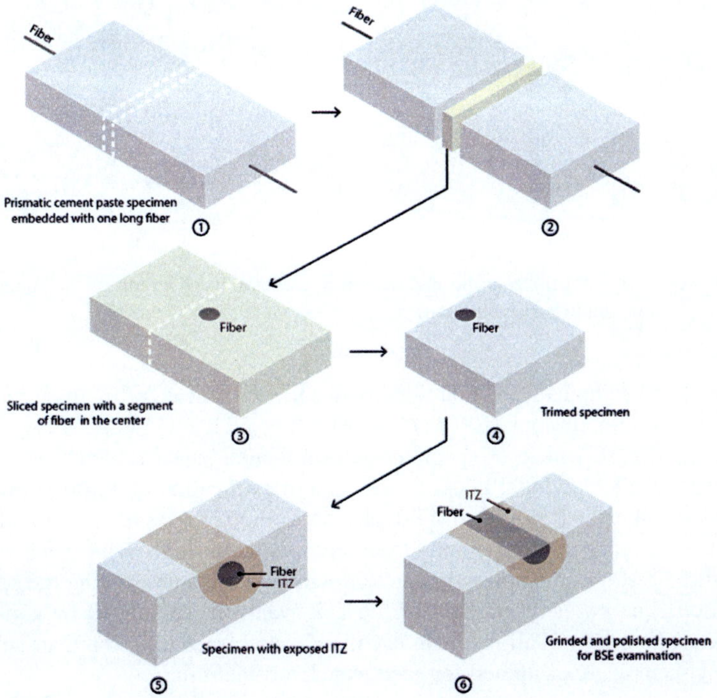

Fig. 1. Schematic illustrations of the sample preparation procedure.

Figure 1 illustrates the main step of preparation procedure. To prepare the specimen, a long PVA fiber was first cut into about 150 mm in length and fixed at the center of a prismatic mold, which will then be filled with fresh cement paste on a vibration table. After demolding, the cement paste prism embedded with one continuous fiber was sliced into thin plate-like specimens with the help of a diamond saw. The thickness of the specimens is between 1 to 2 mm. Each resulting specimens would have a segment of fiber embedded in the center perpendicular to the base surface. The position of the fiber was then identified with the use of an optical microscope and a precise cut was made by a dicing saw roughly 10 microns away from the fiber. Right after the cutting, the specimens were impregnated with epoxy resin and cured in an oven at 40 °C for 24 h. Figure 2 shows the fluorescence and electron microscope images of the specimen after epoxy impregnation. As can be seen, the epoxy successfully impregnated into the pores and a relatively wide porous region can be observed in the middle of the images.

The next step of the preparation is the progressive grinding of the specimen from the cutting plane to remove excessive epoxy and to expose the ITZ. The grinding was stopped exactly at the moment when the grinding surface reaches the center of the fiber. This is to avoid having an angled slice which may lead to exaggerated measurement of the ITZ thickness. The last step is to carefully polish the surface for optimum imaging under the BSE. The grinding and polishing were performed by using an in-house designed system. The details of the set-up can be found in a previous publication by the authors [8].

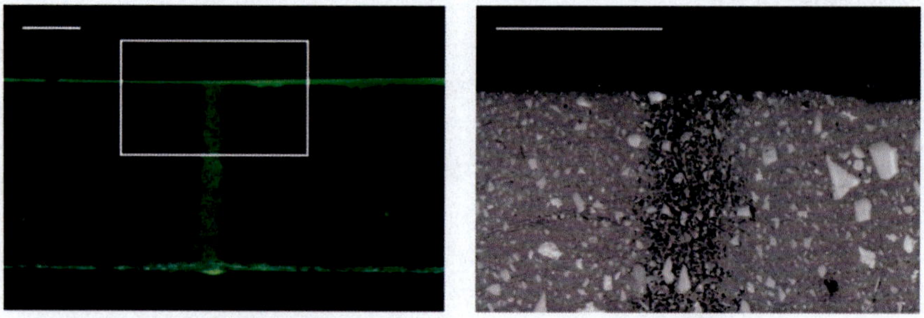

(a) (b)

Fig. 2. Fluorescence light micrograph and scanning electron micrograph of the specimen after epoxy impregnation (scale bar = 500 μm)

The polished samples were examined under the BSE detector in an environmental scanning electron microscope (ESEM) at a voltage of 15 kV without coating. Figure 3a shows a typical BSE image of specimen with polished surface, which was made by digitally stitching 8 individual images taken at a magnification of ×500 (1 pixel = 0.3 micron). Contrast in the BSE image yields clear definition of constituents, e.g., the region in light grey corresponds to the anhydrous cement while regions in black resembles pores. Figure 3b shows the binary image with white pixels representing the pores after segmentation. The porosity gradient of the ITZ was then calculated by counting the fraction of white pixels to all pixels in each horizontal lines from the fiber surface. At least 2 specimens were examined for each water to cement ratio.

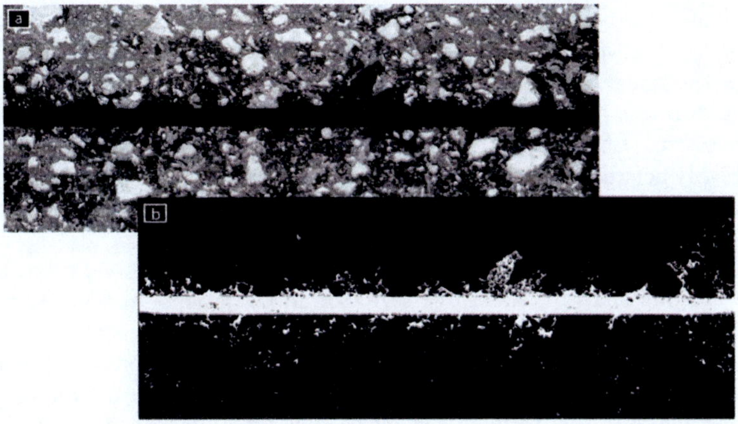

Fig. 3. (a) Typical BSE image of a polished specimen and (b) binary image converted from the BSE image (length of embedded fiber = 2.5 mm).

3 Results and Discussion

3.1 Porosity Gradient

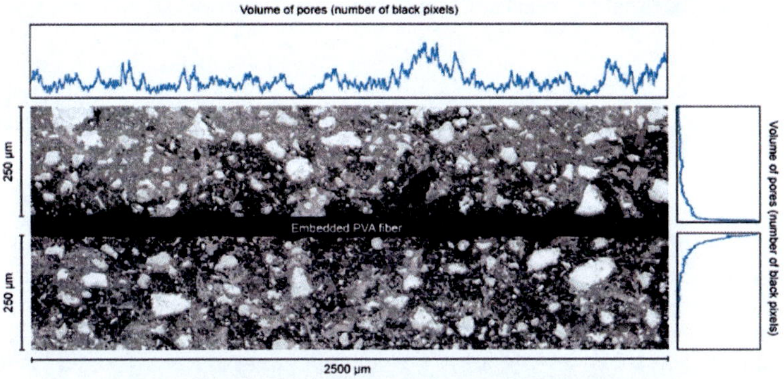

Fig. 4. BSE image of a polished ITZ specimen.

Figure 4 shows a typical stitched BSE image of a longitudinal section of the ITZ surrounding a microfiber, in which anhydrous cement and hydrated phase can be distinguished on the basis of their grey level. The fiber can be easily identified in the middle of the graph as a black rectangular for that PVA fiber contains mainly the element of carbon and therefore appears as dark black under BSE. Also in black color, epoxy filled pores in the size of several tens of micron could be seen in the vicinity of the fiber. The line figures above and next to the BSE image shows the number of black pixels per each vertical and horizontal lines, respectively, representing the change of the volume of the pores along the longitudinal and radial directions. As can be seen form the figure on the right of the BSE image, the zone in the vicinity of the fiber contains a much higher number of pores as compared to the bulk matrix far away from the fiber, suggests existence of a porous transition zone between micro-fiber and cement matrix. Furthermore, from the figure about the BSE image, it can be found that porosity in the ITZ is not homogeneous along the length direction of the fiber. Concentration of pores can be seen at some locations, while some areas remain dense. This reveals that the ITZ between microfiber and cement paste is highly heterogenous along its length direction. As a result, existing ITZ analysis methods performed on sections intersecting with fiber at a random angle can lead to errors and uncertainties.

Figure 5 shows the porosity gradient in the ITZ for different w/c ratios based on the BSE imaging analysis. As can be seen, for all the groups the porosity within 5–100 μm is significantly higher than the bulk matrix. Volume fraction of pores decreases with increasing distance from fiber/matrix interface. The reduction rate is high within 5–50 μm from the interface and turns into moderate beyond 50 μm. This suggest that the microstructure of the hydrated cement paste is highly modified in the vicinity of microfibers for all the tested w/c ratios. The porous zone can extend up to 100 μm from the interface into the matrix with the most porous region being in the region of within 50 μm adjacent to the fiber. Surprisingly, the porosity gradient was found not to be influenced by the w/c ratios. This may be because the age of the specimens used in the current study is only 14 days. It is possible the effect of w/c may be more pronounced when the hydration degree is higher. It should be noted that in Fig. 5 the porosity measurement starts from 5 microns away from fiber surface but not from the exact physical fiber/matrix interface. This is because we consider our measurement of

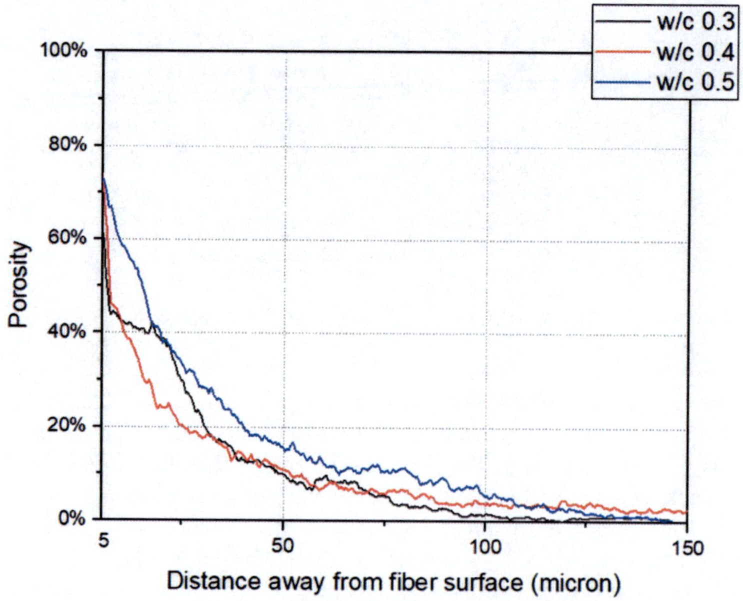

Fig. 5. Volume fraction of pores at different distances from the fiber/matrix interface.

the ITZ porosity in the very close vicinity of the fiber surface (less than 5 μm) will be influenced by the peeling of the fiber and therefore is prone to error.

3.2 Unhydrated Clinker Gradient

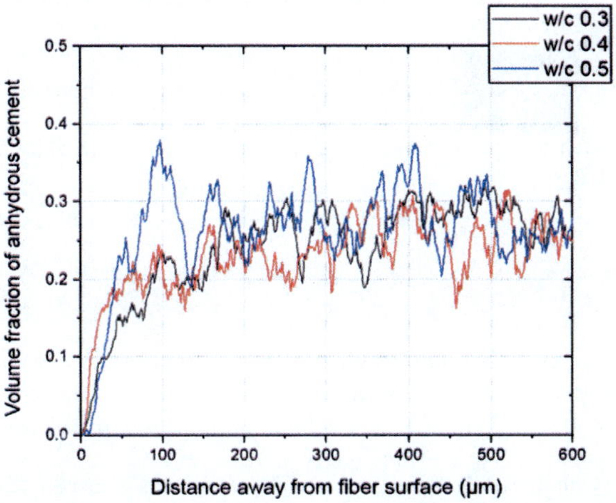

Fig. 6. Volume fraction of anhydrous cement at different distances from the fiber/matrix interface.

By using the same image analysis method, the variation in volume fraction of anhydrous cement adjacent to microfibers can also be obtained based on the BSE imaging analysis. As can be seen from Fig. 6, volume fraction of anhydrous cement is substantially less within the 0–100 μm than the regions far away from the fiber surface. The fact that the zone in the vicinity of the fiber contains significantly less unhydrated cement suggests that there were less cement particles and more water close to the fiber. The particles there thus hydrated faster and easier, resulting in less unhydrated cement. This demonstrates that the packing of the cement grains has indeed been perturbed and becomes rather loose in the region between the fiber and the bulk matrix. This inefficient packing of cement grains may have directly contributed to the high porosity as shown in Fig. 4. These features, including a deficit in anhydrous cement grain and a corresponding high porosity, closely resembles the ITZ between aggregate and cement matrix, indicating that although the diameter of microfibers is small, the axial dimension of microfibers is large enough to act effectively as a 'wall' to perturb the packing of the cement grains surrounding it, resulting in the formation of the ITZ between microfiber and cement matrix.

4 Conclusions

A new approach was proposed to quantitatively characterize the ITZ between microfiber and cement matrix and to reveal its porous microstructure. With the new method, the porosity of the ITZ between PVA microfiber and cement pastes with different w/c ratios was characterized. Results show that a porous transition zone between micro-fiber and cement matrix exists; and the average thickness of this porous zone is roughly 100 μm. It is also found that the microstructure of ITZ is highly heterogenous along its length direction. At an age of 14 days, the porosity of the ITZ is insensitive to the w/c ratio. It is thus believed that the current method could aid the investigation of fiber/matrix bond performance and also contribute to the development of SHCCs with superior properties.

Acknowledgement. This project has received funding from the European Union's Horizon 2020 research and innovation programme under the Marie Sklodowska-Curie grant agreement No 860006.

References

1. Gao, Y., de Schutter, G., Ye, G., Tan, Z., Wu, K.: The ITZ microstructure, thickness and porosity in blended cementitious composite: effects of curing age, water to binder ratio and aggregate content. Compos. B Eng. **60**, 1–13 (2014). https://doi.org/10.1016/J.COMPOSITESB.2013.12.021
2. Scrivener, K.L., Crumbie, A.K., Laugesen, P.: The Interfacial Transition Zone (ITZ) between cement paste and aggregate in concrete. Interface Sci. **12**, 411–421 (2004). https://doi.org/10.1023/B:INTS.0000042339.92990.4C
3. Monteiro, P.J.M., Maso, J.C., Ollivier, J.P.: The aggregate-mortar interface. Cem. Concr. Res. **15**, 953–958 (1985). https://doi.org/10.1016/0008-8846(85)90084-5
4. Scrivener, K.L., Gartner, E.M.: Microstructural gradients in cement paste around aggregate particles. MRS Online Proc. Libr. **114**, 77–85 (1987). https://doi.org/10.1557/PROC-114-77
5. Katz, A., Bentur, A.: Mechanical properties and pore structure of carbon fiber reinforced cementitious composites. Cem. Concr. Res. **24**, 214–220 (1994). https://doi.org/10.1016/0008-8846(94)90046-9
6. Chan, Y.W., Li, V.C.: Effects of transition zone densification on fiber/cement paste bond strength improvement. Adv. Cem. Based Mater. **5**, 8–17 (1997). https://doi.org/10.1016/S1065-7355(97)90010-9
7. Sonat, C., He, S., Li, J., Unluer, C., Yang, E.-H.: Strain hardening magnesium-silicate-hydrate composites (SHMSHC) reinforced with short and randomly oriented polyvinyl alcohol microfibers. Cem. Concr. Res. **142**, 106354 (2021). https://doi.org/10.1016/j.cemconres.2021.106354
8. He, S., Li, Z., Yang, E.-H.: Quantitative characterization of anisotropic properties of the interfacial transition zone (ITZ) between microfiber and cement paste. Cem. Concr. Res. **122**, 136–146 (2019). https://doi.org/10.1016/j.cemconres.2019.05.007
9. Scrivener, K.L.: Backscattered electron imaging of cementitious microstructures: understanding and quantification. Cement Concr. Compos. **26**, 935–945 (2004). https://doi.org/10.1016/J.CEMCONCOMP.2004.02.029

Pull-Out Behavior of Single Fiber Embedded in Porosity Free Concrete(PFC)

Koki Banno[1(✉)], Minoru Kunieda[1], Eiki Yasuda[2], and Katsuya Kono[2]

[1] Gifu University, 1-1 Yanagido, Gifu 501-1193, Japan
a4523029@edu.gifu-u.ac.jp

[2] Central Research Laboratory, Taiheiyo Cement Corporation, 2-4-2 Osaku, Sakura-Shi, Chiba 285-8655, Japan

Abstract. Pull-out behavior of single fiber embedded in Porosity Free Concrete (PFC), which is a non-porous ultra high strength matrix, was experimentally investigated. Four types of fibers were used: steel fiber, stainless steel fiber, bundled aramid fiber, and bundled PBO fiber. Two kinds of embedment length (2 mm and 5 mm) were adopted in the test. Regardless of the difference in embedment length, the maximum load was in the order of stainless steel fiber, steel fiber, aramid fiber, and PBO fiber. Regarding the shape of the load-displacement curve, steel fibers, aramid fibers, and PBO fibers showed softening behavior after the maximum load, whereas stainless fibers showed yielding behavior that caused displacement while maintaining the load. It seems that the stainless fiber has good bonding with the ultra-high strength matrix.

Keywords: PFC · Pull-out behavior · Fiber reinforced concrete

1 Introduction

Porosity Free Concrete (hereinafter referred to as PFC) [1] has an ultra-high strength matrix with a compressive strength of about 400 N/mm^2, and the addition of short fibers is effective in suppressing brittle fracture. By optimizing the combination of matrix and fibers, it is possible to develop even higher performance fiber reinforced composites. Although one of the findings necessary for the material development is the bond characteristics between matrix and fibers, there is little knowledge about the bond between PFC matrix and fibers. Furthermore, the numerical analysis method [2] proposed by the authors for discretizing the short fibers makes it possible to evaluate the mechanical behavior of the fiber reinforced composite.

In this study, pull-out tests of single fiber embedded in high-strength and ultra-high-strength matrices (PFC) were conducted on steel fibers, stainless fibers, bundled aramid fibers, and bundled PBO fibers, and the bond properties of each fiber were clarified experimentally.

M. Kunieda et al. (Eds.): SHCC 2022, RILEM Bookseries 39, pp. 135–145, 2023.
https://doi.org/10.1007/978-3-031-15805-6_15

2 Outline of Experiments

2.1 Test Series

Test method using a specimen with a notch in the matrix was adopted, in which a fiber was arranged inside of the matrix. Tensile force was applied to both specimen ends. The test series are shown in Table 1. The test series name consists of water to binder ratio (15%, 30%) -fiber type (FM, SUS, AR, PBO) -embedment length (5 mm, 2 mm) -curing conditions (D, S). For the matrix, both PFC mix (W/B = 15%) and a high-strength mortar mix (W/B = 30%) were used. There are two types of embedment length, 2 mm and 5 mm. Four types of fibers were used: steel fiber, stainless steel fiber, bundled aramid fiber, and bundled PBO fiber (see Table 2).

Table 1. Test series.

No	Serie name	W/B	Type of fiber	Embedment length	Curing condition
1	15-FM-5-D	15%	Steel	5 mm	Full-spec (D)
2	15-FM-2-D	15%	Steel	2 mm	Full-spec (D)
3	15-SUS-5-D	15%	Stainless steel	5 mm	Full-spec (D)
4	15-SUS-2-D	15%	Stainless steel	2 mm	Full-spec (D)
5	30-FM-5-D	30%	Steel	5 mm	Full-spec (D)
6	30-FM-2-D	30%	Steel	2 mm	Full-spec (D)
7	30-SUS-5-D	30%	Stainless steel	5 mm	Full-spec (D)
8	30-SUS-2-D	30%	Stainless steel	2 mm	Full-spec (D)
9	15-FM-5-S	15%	Steel	5 mm	Steam (S)
10	15-FM-2-S	15%	Steel	2 mm	Steam (S)
11	15-SUS-5-S	15%	Stainless steel	5 mm	Steam (S)
12	15-SUS-2-S	15%	Stainless steel	2 mm	Steam (S)
13	15-AR-5-S	15%	Bundled aramid	5 mm	Steam (S)
14	15-AR-2-S	15%	Bundled aramid	2 mm	Steam (S)
15	15-PBO-5-S	15%	Bundled PBO	5 mm	Steam (S)
16	15-PBO-2-S	15%	Bundled PBO	2 mm	Steam (S)

For curing, from No. 1 to No. 8, degassing and water absorption treatment (30 min), steam curing (maximum temperature 90 °C Celsius, maximum temperature holding time 48 h), heating curing (maximum temperature 180 °C, maximum temperature holding time 48 h) were conducted. In this study, this curing method is called full spec (D). In addition, for synthetic fibers, there is concern about deterioration of the fibers due to water absorption treatment and heat treatment. Therefore, for the curing of test series No. 9 to No. 16, only steam curing after demolding (90 °C, 48 h) (S) was applied.

Table 2. Mechanical properties of fibers (in catalogue).

Type of fiber	Diameter (mm)	Length(mm)	Elastic modulus (GPa)	Strength of rapture (MPa)
Steel	0.2	15	200	2800
Stainless steel	0.3	15	193	600
Bundled aramid	0.21	15	81	2958
Bundled PBO	0.23	15	140	3500

Table 3. Mix proportions of mortar.

Name	W/B (%)	Unit content (kg/m^3)			Chemical admixture(B × %)	
		W	B	S	SP	DF
PFC	15	200	1333	934	2.5	0.2
High strength mortar	30	305	1017	934	0.65	0.2

B: Premixed powder for PFC. S: Silica sand, SP: Superplasticizer, DF: Air reducing agent

2.2 Specimens

The mix proportions of PFC and high-strength mortar are shown in Table 3. An omnimixer, in which inside pressure can be reduced, was used for mixing. Table 4 shows the material test results of PFC and high-strength mortar used in the experiment. A wooden formwork with a fiber fixing unit placed inside was used to prepare the specimen (Fig. 1). In the fiber fixing unit made of styrene board, a groove for fixing the fiber was made, and the fiber was sandwiched so as to have a predetermined embedment lengths of 2 mm and 5 mm, and placed in a wooden formwork. After that, the fibers were placed in the mortar by pouring the mortar into the first layer. After curing the first layer (material age 48 h), the fixing unit was taken out, the mortar was poured into the second layer, and demolding was performed after curing (material age 48 h). After the completion of the demolding of the second layer, the prescribed curing was performed. After curing is completed, cut it to a width of 20 mm using a micro cutter and have dimensions of 20 × 20 × 25 mm. Notches were provided on the four sides so that the area of the ligament was approximately 25 mm^2. Figure 2 shows an example of the prepared specimen. The number of specimens was 5 for each test series.

2.3 Experimental Setup

The mechanical jack shown in Fig. 3 was used for the pull-out test. The specimen and apparatus were fixed using an epoxy resin adhesive, and tensile force was applied.

Table 4. Test results of mortar.

Specimen No	W/B (%)	Curing	Flow (mm)	Air content (%)	f_c (N/mm^2)	f_t (N/mm^2)	f_b (N/mm^2)
No.1–4 (1st layer)	15	D	276	1.4	372	13.2	22.8
No.1–4 (2nd layer)	15	D	278	1.2	371	13.8	20.1
No.5–8 (1st layer)	30	D	191	0.3	142	4.0	11.0
No.5–8 (2nd layer)	30	D	192	0.2	154	3.8	12.9
No.9–16 (1st layer)	15	S	262	0.6	298	10.9	15.6
No.9–16 (2nd layer)	15	S	260	0.7	304	10.9	16.7

f_c: Compressive strength (φ50 × 100 mm), f_t: Splitting tensile strength (φ100 × 100 mm), f_b: Flexural strength (40 × 40 × 160 mm)

Rate of loading was approximately 1 mm/1 min. The tensile load was measured with a loadcell having a capacity of 1 kN and an accuracy of 0.25 N. The stroke displacement of the jack was measured with a displacement transducer (accuracy 1/500 mm).

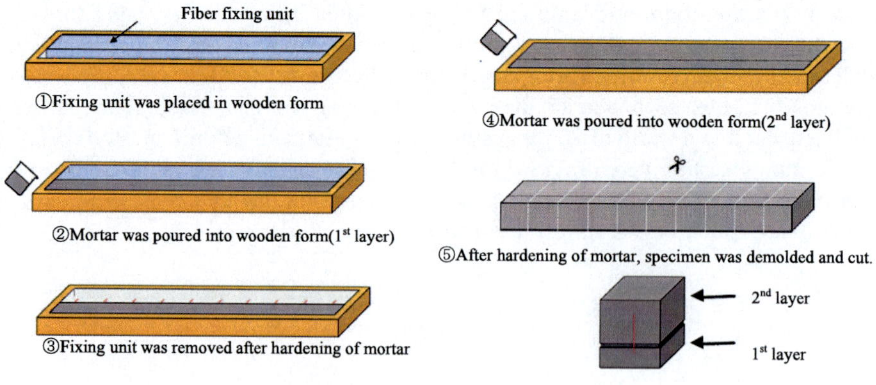

Fig. 1. Preparation of specimen.

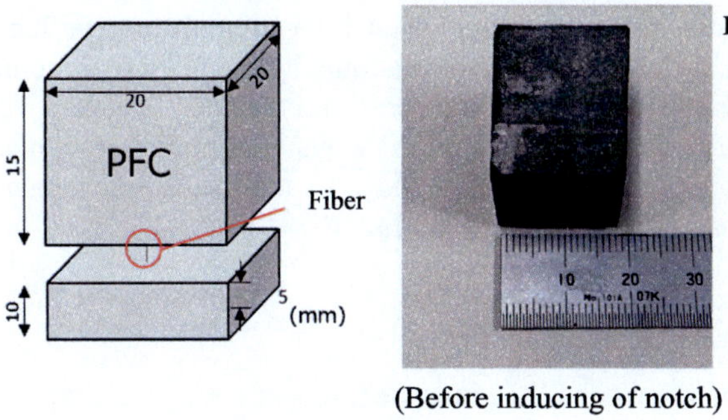

Fig. 2. Specimen.

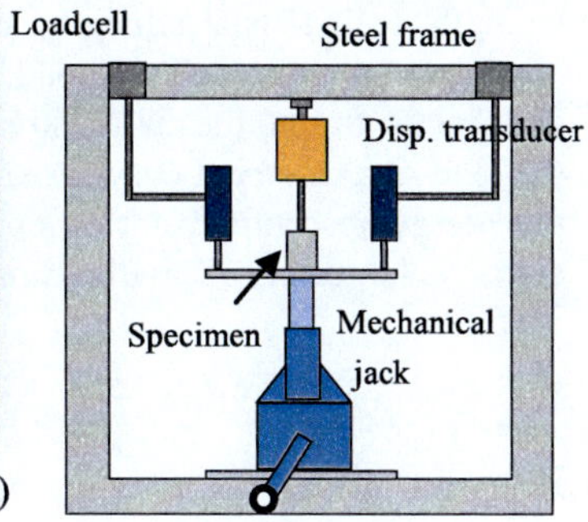

Fig. 3. Test setup.

3 Experimental Results

3.1 Matrix Strength

The load-displacement curves obtained by the experiment are shown in Figs. 4, 5, 6 and 7. Figures 8 and 9 show the averaged value and standard deviation of the maximum load at each test series. As shown in Figs. 4 and 5, PFC greatly exceeded the maximum load of high-strength mortar regardless of the type of fiber and the difference in embedment length. From the results of the strength test in Table 4, PFC has higher compressive

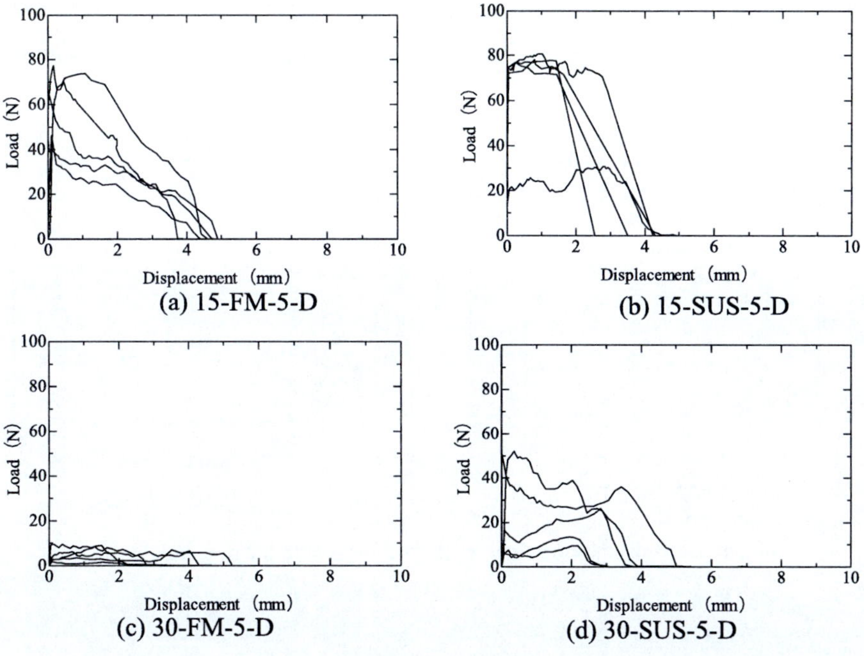

Fig. 4. Load-displacement curves (Embedment length 5 mm, Full-spec).

strength, split tensile strength, and bending strength than high-strength mortar. It is considered that the pull-out load of the fiber has a correlation with the matrix strength, and that the higher the matrix strength, the stronger the bond with the fiber. The shape of the load-displacement curve greatly differs depending on the water to binder ratio, and the energy absorption (corresponding to the area surrounded by the load-displacement curve) is larger in PFC with a smaller water to binder ratio.

3.2 Curing Condition

Regarding the effects of curing conditions on PFC, since the full-spec curing has a higher matrix strength than the case of steam curing, for steel fiber (FM), the maximum load of full-spec curing was larger than that of steam curing (Fig. 4 (a) and Fig. 6 (a), Fig. 5(a) and Fig. 7(a)). The maximum load of full-spec curing was larger than that of steam curing. However, there is no difference in the maximum load of stainless steel fiber only for full-spec curing and steam curing (Fig. 4 (b) and Fig. 6 (b), Fig. 5(b) and Fig. 7(b)). It seems that the maximum load did not increase because the pull-out load was near the strength of the stainless steel fiber and the plastic deformation of the fiber was dominant.

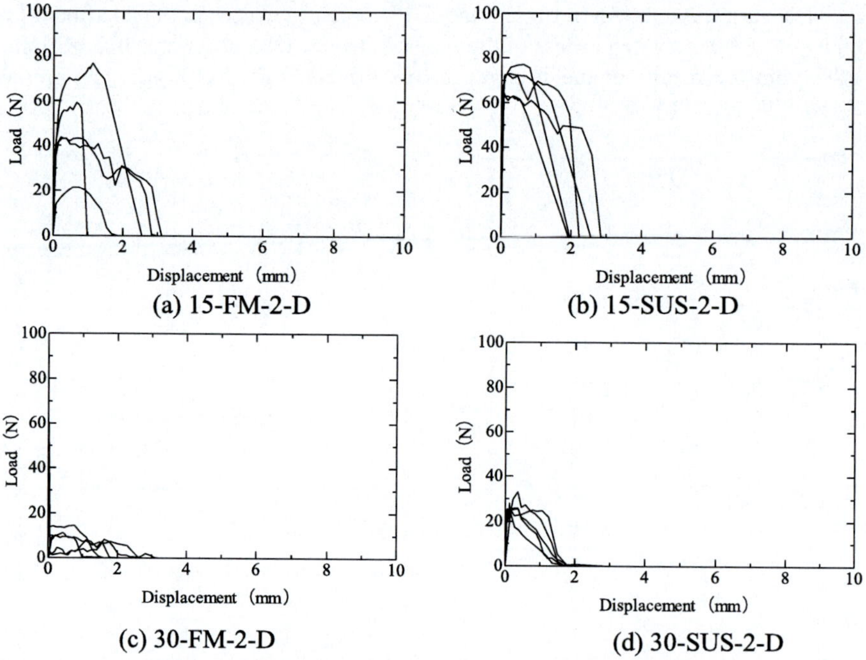

Fig. 5. Load-displacement curves (Embedment length 2 mm, Full-spec).

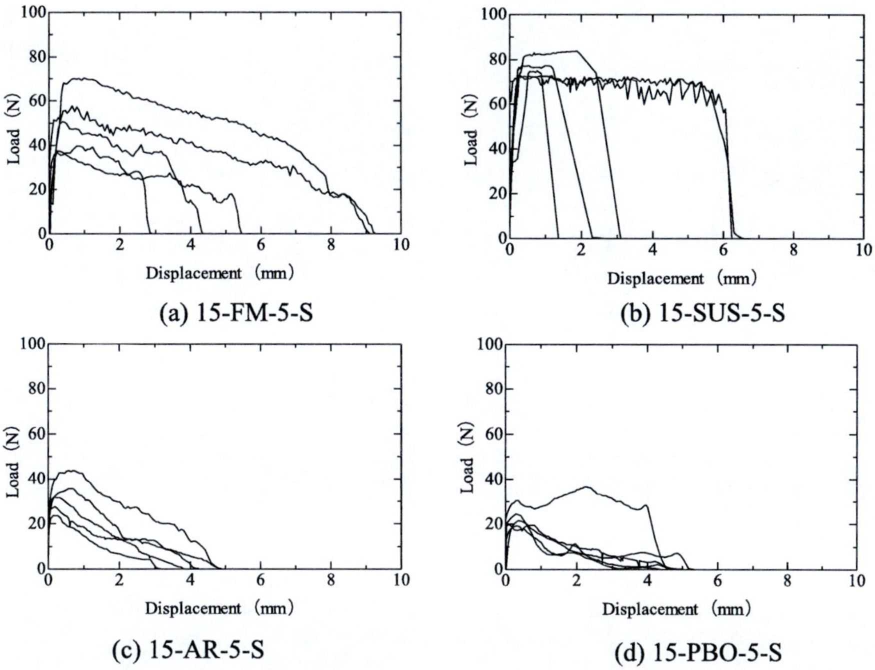

Fig. 6. Load-displacement curves (Embedment length 5 mm, steam).

3.3 Type of Fiber

The maximum load was in the order of stainless steel fiber, steel fiber, focused aramid fiber, and focused PBO fiber. As mentioned above, it was confirmed that the fiber was broken when the tensile force was applied in the some specimen with stainless steel fiber. From the mechanical properties (nominal values) of the fibers in Table 2, it can be seen that the breaking load of the stainless steel fibers is about 42N. It seems that plastic deformation of fiber itself was dominant in the series of stainless steel fiber. Note that, yielding strength over about 70N represented by Figs. 4, 5, 6 and 7 was observed in stainless steel fiber. It seems that the catalogue value of yielding strength in Table 2 is smaller than that of experimental one. Figures 8 and 9 show the normalized load and displacement of each specimen divided by the maximum load and the displacement where the softened load becomes zero, respectively. It was confirmed that the shapes of the load-displacement curves were similar in the specimens of the same series, and the shapes were characteristic depending on the type of fiber. In other words, steel fibers, focused aramid fibers, and focused PBO fibers showed a behavior in which the load gradually decreased as the displacement increased when the maximum load was reached, whereas stainless fibers showed a load even after the maximum load was reached. Although the influence of the difference in fiber diameter may be considered, it was clarified that the stainless steel fiber had good bond property to the matrix (Figs. 10(b) and 11(b)).

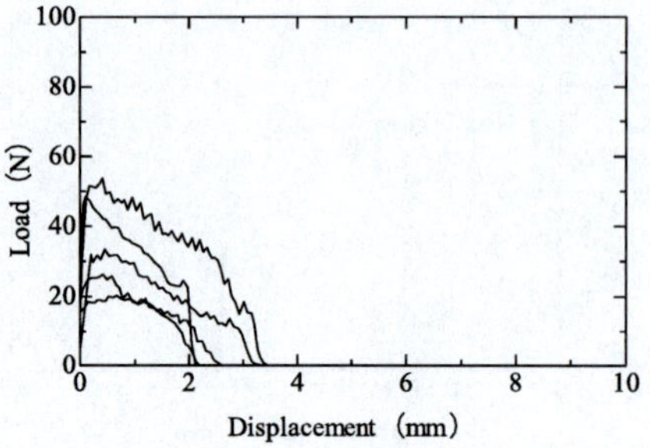

(a) 15-FM-2-S

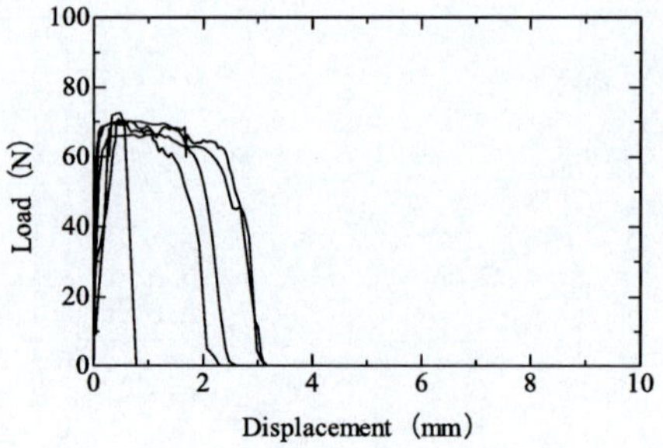

(b) 15-SUS-2-S

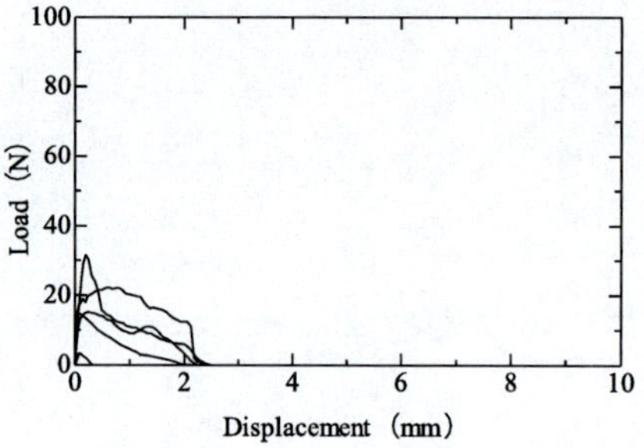

(c) 15-AR-2-S

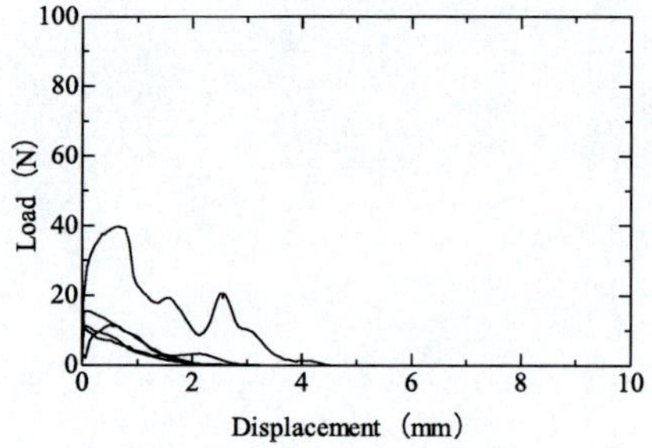

(d) 15-PBO-2-S

Fig. 7. Load-displacement curves (Embedment length 2 mm, steam).

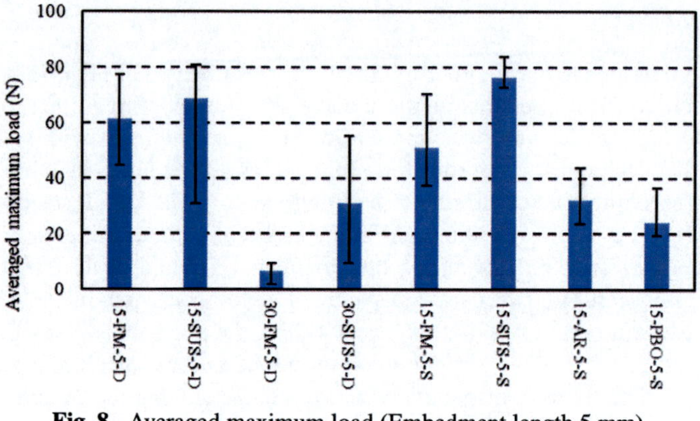

Fig. 8. Averaged maximum load (Embedment length 5 mm).

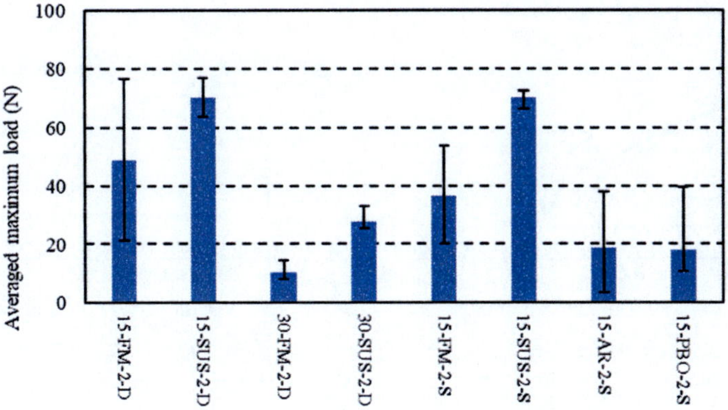

Fig. 9. Averaged maximum load (Embedment length 2 mm).

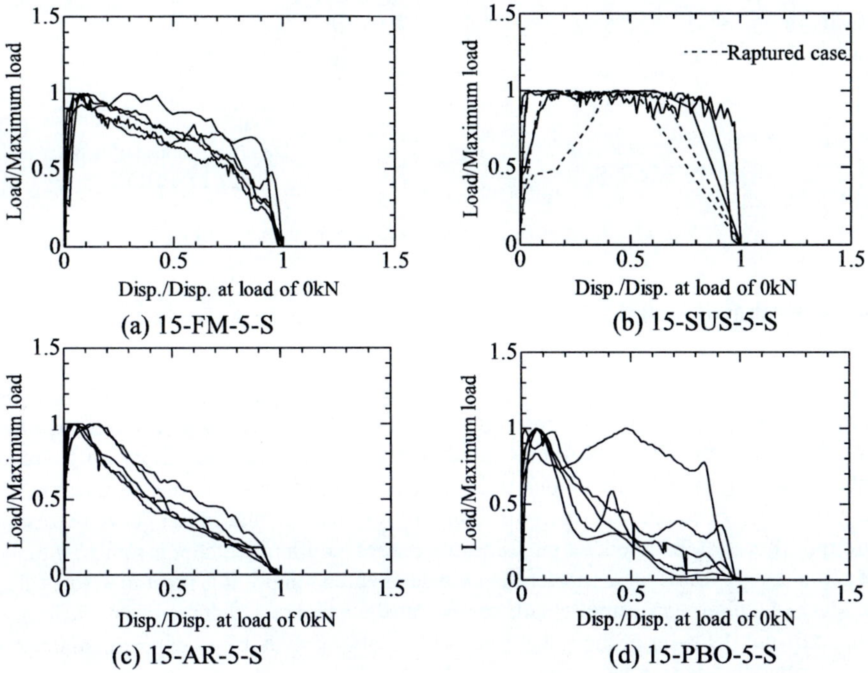

Fig. 10. Normalized curves (Embedment length 5 mm).

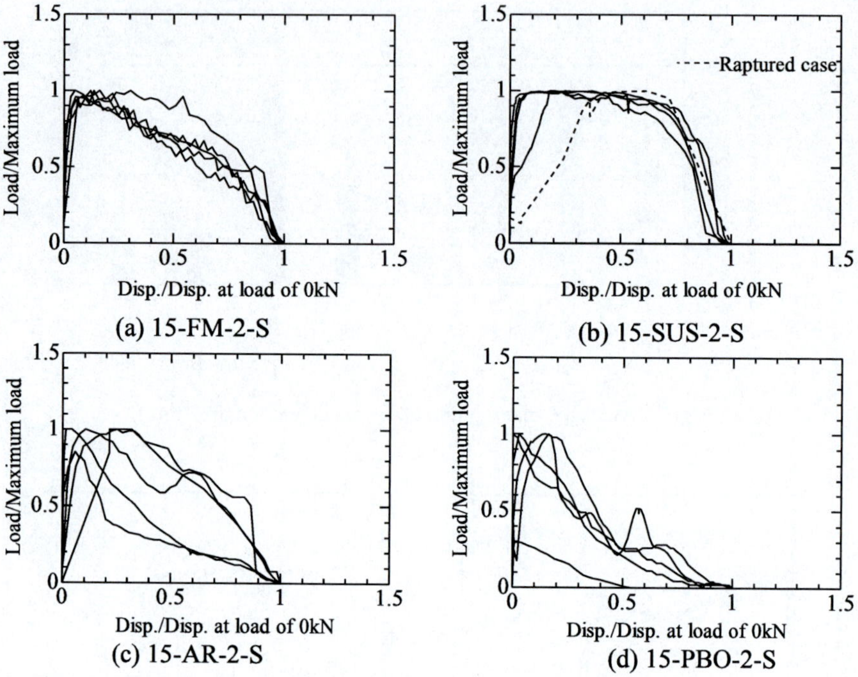

Fig. 11. Normalized curves (Embedment length 2 mm).

3.4 Embedment Length

Comparing the effects of the embedment length from Figs. 4, 5, 6 and 7, the maximum load with an embedment length of 5 mm tends to be larger than the maximum load with an embedment length of 2 mm, except for stainless steel fibers. For aramid fibers and PBO fibers, the results with an embedment length of 5 mm and the results with an embedment length of 2 mm tended to be similar. Considering that the displacement at which the load becomes zero after softening of the load-displacement curve represents the approximate embedment length, the embedment length of the fiber is set to 5 mm, but the actual length tends to be shorter than the designated length. It is presumed that this is the cause of the same maximum load. In this way, it can be seen that there is no difference in the pull-out behavior of the same type of fiber, although there are experimental errors in the embedding length.

4 Conclusions

In this study, the pull-out behavior of short fibers embedded in PFC was experimentally investigated, and following conclusions were obtained.

(1) The bond between the matrix and the fibers in PFC was evaluated. It was confirmed that there is a correlation between the strength of the matrix and the bond of the

fibers, and that the higher the matrix strength gives the higher bond strength with the fibers.

(2) The order of maximum load was stainless steel fiber, steel fiber, aramid fiber, and PBO fiber. Regarding pulling behavior, steel fibers, aramid fibers, and PBO fibers showed softening behavior after the maximum load, while stainless fibers showed yielding behavior that caused displacement while maintaining the load.

References

1. Kurihashi, Y., Kono, K., Komuro, M.: Response characteristics of a steel fiber-reinforced porosity-free concrete beam under an impact load. Int. J. Civil Eng. **18**, 673–684 (2020)
2. Kunieda, M., Ogura, H., Ueda, N., Nakamura, H.: Tensile fracture process of Strain Hardening cementitious composites by means of three-dimensional meso-scale analysis. Cement Concr. Compos. **33**(9), 956–965 (2011)

Experimental Study on Bond-Slip Behavior of Steel Reinforcement in High-Strength Strain-Hardening Cementitious Composites (SHCC) Under Direct Tension

Haroon Younas[1](✉), Jing Yu[2], and Christopher K. Y. Leung[1]

[1] Department of Civil and Environmental Engineering, Hong Kong University of Science and Technology, Hong Kong SAR, People's Republic of China
hyounas@connect.ust.hk

[2] School of Civil Engineering, Sun Yat-Sen University, Guangzhou 510275, People's Republic of China

Abstract. Strain-hardening cementitious composites (SHCC) are pseudo-ductile materials with remarkably high tensile ductility and excellent crack control capability. In some promising applications of SHCC (e.g., construction of structural joint between pre-cast reinforced concrete components), good bonding between SHCC and rebars is essential to ensure sufficient stress transfer. However, the present understanding on the bond between high-strength SHCC and steel rebars is limited. This paper presents an experimental investigation on the bond property between deformed rebars and high-strength SHCC under direct tension. High-strength SHCC with compressive strength of 112 MPa, tensile strength of 8.6 MPa, and tensile strain capacity of 5.5% was used. Steel rebars with three different diameters of 20, 25 and 32 mm, as well as three different cover-to-rebar diameter (C/D) ratios of 1, 1.5, and 2 were examined. The effects of rebar diameter and cover thickness on the bond-slip behavior of steel rebars in high-strength SHCC were discussed. The results showed a significantly improved bonding characteristics between SHCC and deformed steel rebars in comparison to concrete. The bond strength generally increases with an increasing C/D ratio but may stay constant beyond a certain cover thickness. The findings of this study can hopefully improve the current understanding of bonding characteristics of SHCC with deformed steel rebars and can facilitate the structural design of reinforced SHCC members.

Keywords: Strain-hardening Cementitious Composites (SHCC) · Engineered Cementitious Composites (ECC) · Reinforced SHCC members · Stress transfer · Cover thickness · Pull-out test

1 Introduction

Strain-hardening cementitious composites (SHCC), also known as Engineered Cementitious Composites (ECC), were developed in the 1990s under the guidance of fracture mechanics and micro-mechanics by Li and Leung [1]. SHCC exhibit high ductility

M. Kunieda et al. (Eds.): SHCC 2022, RILEM Bookseries 39, pp. 146–155, 2023.
https://doi.org/10.1007/978-3-031-15805-6_16

and superior strain-hardening performance under tension, and therefore many studies reported that SHCC is attractive for joining the pre-cast reinforced concrete components [2] and structural repairing [3]. When used with conventional steel rebars, the effectiveness of SHCC can be compromised if there is insufficient bond length for stress transfer. For example, in Chen et al. [4], a deteriorated reinforced concrete flexural beam with significantly corroded rebar was repaired with SHCC. For the SHCC patch to fully compensate for the loss in load-carrying capacity due to the reduction in steel area, sufficient bond length needs to be provided. For the construction of a joint between two pre-cast reinforced concrete components with steel rebars protruded from both sides, the required bond length is also a key design parameter.

To investigate the bond properties between steel rebars and concrete, different test methods have been proposed by various investigators [5, 6], with the pull-out test being the most commonly used method due to the ease of fabrication [7–11]. In [7–11], the steel rebar was pulled-out from one side of the concrete block which was being pushed. However, the compressive stress in the concrete block may not reflect the actual stress conditions in many practical applications, and the resulting increase in crack resistance in the concrete cover would lead to over-estimation of the bond strength. In [11–13], a direct tension test was carried out on a concrete block where two pieces of steel rebar were embedded on both sides with good alignment. The bond behavior without compression effect was measured from the pull-out of the rebar with shorter embedment length. Therefore, the direct tension approach can evaluate the bond properties more accurately. In the literature, the bond behavior between normal-strength SHCC and steel rebar were reported extensively [14–16], but findings on the bond property for high-strength SHCC (tensile strength > 10MPa) are very limited. While some results have been reported in [11], the effects of covering thickness and rebar size on steel-SHCC bond is yet to be fully understood. This study aims to investigate more extensively the bond behavior between steel rebars and high-strength SHCC under direct tension through an experimental program. The bond effects on block material (includes normal concrete and high-strength SHCC) with different rebar diameters (D = 20, 25, and 32 mm), and different cover-to-rebar diameter ratios (1.0D, 1.5D, and 2.0D) were evaluated with the steel embedment length fixed at 4D.

2 Experimental Program

2.1 Materials

The concrete and SHCC used in this experimental study was adopted from the literature [4]. The mix composition of SHCC is listed in Table 1. Type I cement according to standard (BS EN 197–1:2011) was acquired. The binder ratio of 20% silica fume and 80% cement was used. A low quartz silica sand content was adopted to ease the fiber dispersion. A super low water-to-binder ratio of 0.145 was used, and a high dosage of polycarboxylate-type superplasticizer (SP) was added to maintain sufficient workability. Polyethylene (PE) fiber, 12mm in length and 24 μm in diameter was added at volume fraction of 2.2%. The mix ratio of concrete was also listed in Table 1 with a target compressive strength of 50 MPa similar to normal strength concrete. The rebars were all deformed rebars made with high yield steel, with a yield strength of more than 460 MPa.

Table 1. Mix compositions of SHCC and plain concrete [4]

Material	Cement	Silica Fume	Sand	Aggregate	Water	Super-plasticizer	PE Fiber (vol. %)
Concrete	1	–	1.3	2	0.45	–	–
SHCC	0.8	0.2	0.3	–	0.145	0.025	2.2

2.2 Specimen Preparation and Testing Procedure

In parallel to the bond tests, the direct tensile performance of the adopted SHCC should also be experimentally determined. Dog-bone shape specimens were cast for direct tensile tests under the guidance of the Japan Society of Civil Engineering (JSCE) [17] and four cubes measuring 40x40x40 mm were prepared for compressive strength. For the measurement of bond behavior, the geometric configuration of the pull-out test specimen is shown in Fig. 1 and the various parameters are summarized in Table 2. In the specimen ID, the first alphabetical letter indicates the block material type, either concrete (C) or SHCC (S). The first two digits representing the size of rebar (in mm) being pullet-out and the following last two digits represents the cover thickness (in mm). For each specimen, two major rebars with the same diameter (D) and four minor rebars with 0.5D diameter were placed. Two major rebars were embedded in the block with different embedment lengths. The rebar with shorter embedment length (4D) was expected to show bond failure, while the one with longer embedment length (6D) acts as anchorage end. For each specimen, three cover thicknesses (1D, 1.5D and 2D) were investigated. The minor rebars were placed through the whole specimen block to avoid tensile splitting failure of the block, and they are at a distance from the major rebars to minimize their effect on bond property. For each specimen, at least two blocks were fabricated for testing.

Table 2. Various parameters of the pull-out test specimens

Series	Specimen ID	Major Rebar Diameter (mm)	Minor Rebar Diameter (mm)	Block Material	Cover (mm)	Embedment Length
Control	C2020	20	10	Concrete	20	4D
	C2030	20	10	Concrete	30	4D
	C2040	20	10	Concrete	40	4D
Y20	S2020	20	10	SHCC	20	4D
	S2030	20	10	SHCC	30	4D
	S2040	20	10	SHCC	40	4D

(continued)

Table 2. (*continued*)

Series	Specimen ID	Major Rebar Diameter (mm)	Minor Rebar Diameter (mm)	Block Material	Cover (mm)	Embedment Length
Y25	S2525	25	12	SHCC	25	4D
	S2537.5	25	12	SHCC	37.5	4D
	S2550	25	12	SHCC	50	4D
Y32	S3232	32	16	SHCC	32	4D
	S3248	32	16	SHCC	48	4D
	S3264	32	16	SHCC	64	4D

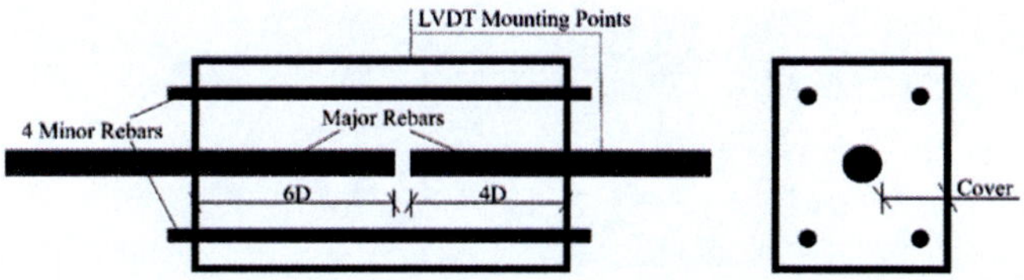

Fig. 1. Specimen geometry of pull-out test

Prior to casting, the major rebars for SHCC blocks were placed in long wooden box and were kept in good alignment by passing them through closely matched holes in two parallel wood plates (Fig. 2). And for concrete specimens, a stiff aligning fixture was used for aligning the rebars (Fig. 3). After placing the major rebars, the minor rebars were installed. The dry materials of SHCC were mixed together for two minutes in the mixer. Tap water as well as SP were then added, and mixing was continued until the mix becomes flowable. Afterward, PE fibers were added and mixed until the fibers dispersed. The fresh mixture was then cast into pull-out molds along with four dong-bone specimens, three cubes (40 × 40 × 40) and three cylinders (50 × 100 mm) for acquiring their tensile performance, compressive strength, elastic modulus, and Poisson's ratio. The specimens were demolded after 48 h of casting time and placed into curing room at a temperature of 23 ± 2^0C and relative humidity of 95 ± 5% for 14 days as per a previous study [4]. Similarly, for concrete, the dry materials were mixed together first and then water was added and mixed. Fresh concrete was then poured into pull-out molds along with three cubes (100 × 100 × 100) and three cylinders (100 × 200) for obtaining their compressive strength, elastic modulus, and Poisson's ratio. These specimens were demolded and cured similar to above mentioned condition for 28 days. Servo-hydraulic machine were used to perform the direct tensile pull-out test (Fig. 4). Two linear variable displacement transducers (LVDTs) were installed on the side with shorter embedded rebar for measuring the pull-out displacement. The major rebars were then secured into the pressurized grip of the testing machine and the initial displacement control rate of 0.001667 mm/sec was adopted. The displacement control rate was increased to 0.025 mm/sec when the load showed a decreasing trend.

Fig. 2. Wooden mold for SHCC blocks

Fig. 3. Aligning fixture for concrete specimens

Fig. 4. Pull-out test setup

3 Results and Discussions

The tensile performance of SHCC at 14 days shows a high tensile stress capacity of over 8 MPa and strain capacity of more than 5% (Fig. 5). The summary of material properties

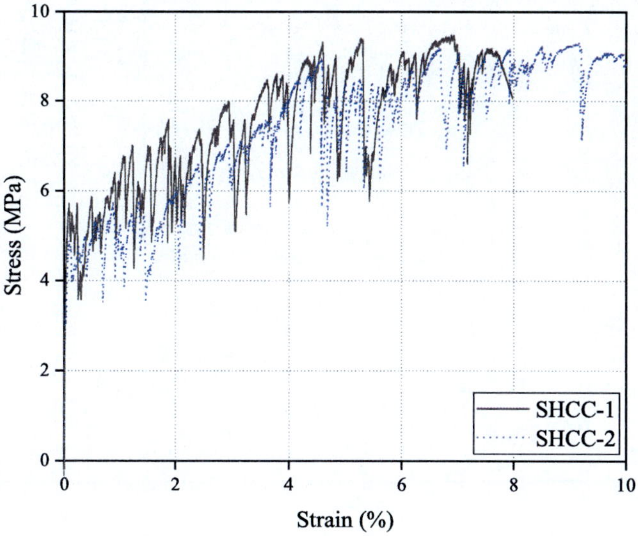

Fig. 5. Tensile stress vs strain curves of SHCC

(concrete and SHCC) is listed in Table 3, and the pull-out test results are summarized in Table 4. Figure 6(a) and 6(b) presents the pull-out load vs slip curves of different specimens. In Table 4, average bond strength was calculated by directly dividing the peak load with the cylindrical contact area of the shorter embedded major rebar.

Table 3. Summary of material properties

	Compressive Strength (MPa)	Elastic Modulus	Poison's Ratio	Tensile Strength	
				Stress (MPa)	Strain (%)
Concrete (28d)	49.9	30.3	0.206	–	–
SHCC (14d)	112	41.5	0.18	8.66	5.56

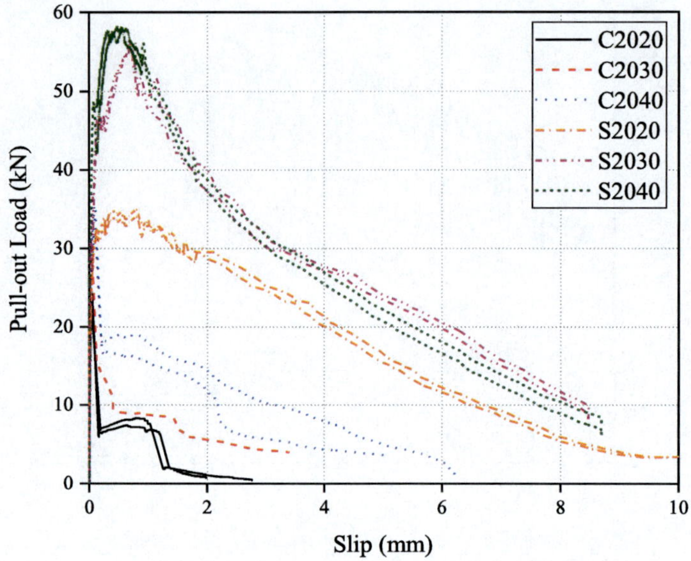

(a)

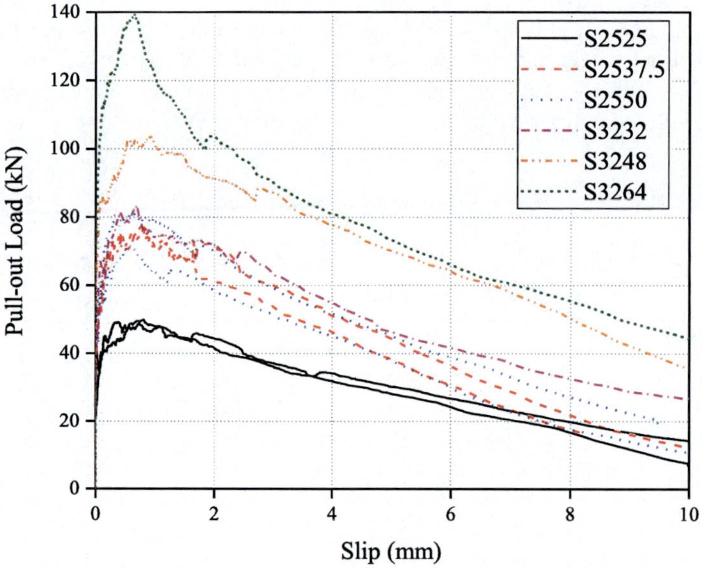

(b)

Fig. 6. Pull-out test curves; (a) Y20-Concrete, Y20-SHCC and (b) Y25-SHCC, Y32-SHCC

According to results shown in Fig. 6, it is clearly observed that the bond behavior between concrete block and SHCC block was very different in terms of load vs displacement curve and bond strength. Figure 7 shows the failure modes of pull-out test specimens. A sudden drop was observed in concrete specimens after the peak load was reached. The drop corresponds to the splitting of concrete cover during the test, which significantly degrades the bond between steel rebar and concrete, leading to bond failure (see Fig. 7(a)). On the other hand, the SHCC specimens shows much higher bond strength and gradual decrease in load after the peak which clearly reflects the ductile and strain hardening behavior of SHCC. Although final failure is also due to cover splitting, it occurs at a late stage with large pull-out displacement, preceded by the formation of multiple cracks indicating high ductility due to the effective crack bridging effect of the PE fibers (Fig. 7 (b), (c), and (d)). For the effect of cover-to-rebar diameter (C/D) ratio, there was a general trend that the higher the cover, the higher the resistance to the splitting failure so bond failure occurs at a higher load. However, for Y20 and Y25 rebars, a plateau value was reached when C/D is beyond 1.5. On the other hand, the bond strength continues to increase with cover thickness for the Y32 rebar. This may be due to the different surface deformation texture on the different rebars and will be further studied.

Table 4. Pull-out test results

Rebar Size	Specimen ID	Peak Load (kN)	Average Bond Strength (MPa)	Failure Mode
Y20	C2020	24.40	4.86	Splitting
	C2030	32.28	6.42	Splitting
	C2040	34.81	6.93	Splitting
Y20	S2020	34.6	6.90	Splitting
	S2030	55.87	11.12	Splitting
	S2040	58.25	11.59	Splitting
Y25	S2525	49.28	6.27	Splitting
	S2537.5	77.45	9.87	Splitting
	S2550	76.52	9.74	Splitting
Y32	S3232	83.4	6.48	Splitting
	S3248	103.65	8.05	Splitting
	S3264	139.4	10.83	Splitting

Regarding the effect of rebar sizes, the test results showed decrease in bond strength for all C/D ratios when the rebar size increases from 20 to 25 mm. However, when the results for 32 mm rebar were included, the trend becomes unclear, as the bond strength may increase or decrease when compared to 25 mm rebar with the same C/D ratio. As we only have one successful test for each case with the 32mm rebar, additional tests have to be performed to clarify the trend.

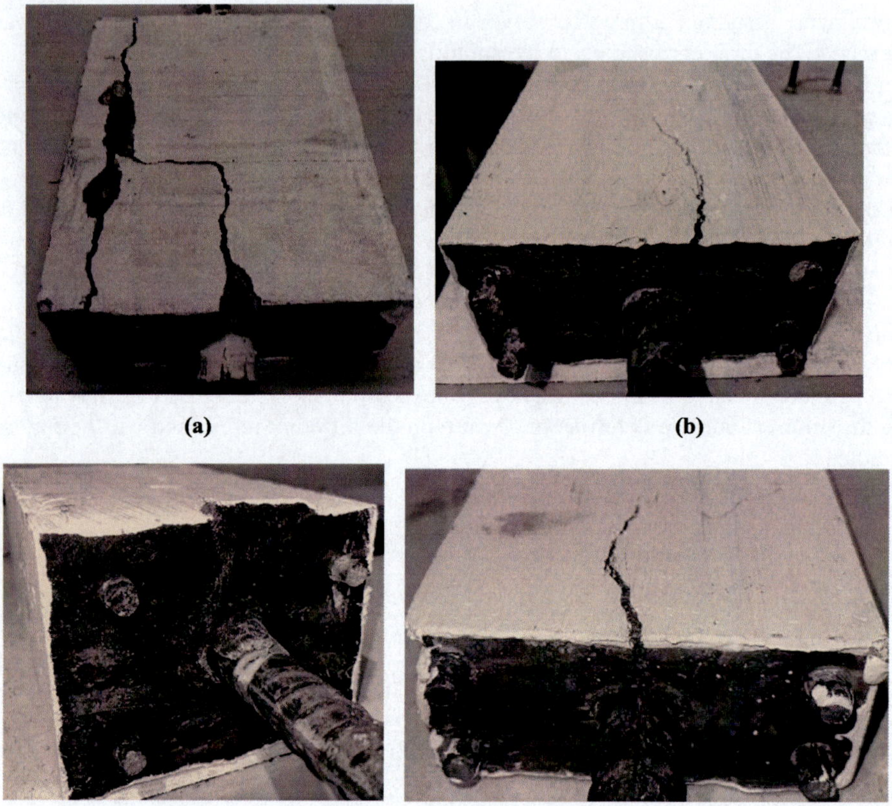
(a) (b) (c) (d)

Fig. 7. Failure modes

4 Conclusions

This study investigates the bond-slip behavior between deformed steel rebars and high-strength Strain Hardening Cementitious Composites (SHCC) or concrete under direct tension test. The effect of different rebar sizes and cover-to-rebar diameter ratios were experimentally evaluated. Based on the obtained results, the following conclusion can be drawn:

(1). Due to the good control of splitting cracks in SHCC, it can bond much better with steel rebar than normal concrete.
(2). With increasing cover-to-rebar diameter ratio the bond strength shows a general increasing trend but may reach a constant value beyond a certain cover thickness.
(3). The bond strength does not show a clear trend with the diameter of rebar within the range of this study.

The findings of this study can hopefully improve the current understanding of the bonding characteristics of SHCC with deformed steel rebars and can facilitate the structural design of reinforced SHCC members.

References

1. Li, V.C., Leung, C.K.Y.: Steady-state and multiple cracking of short random fiber composites. J. Eng. Mech. **118**(11), 2246–2264 (1992)
2. Choi, H., Choi, Y., Choi, C.: Development and testing of precast concrete beam-to-column connections. Eng. Struct. **56**, 1820–1835 (2013)
3. Wei, J., Wu, C., Chen, Y., Leung, C.K.Y.: Shear strengthening of reinforced concrete beams with high strength strain-hardening cementitious composites (HS-SHCC). Mater. Struct. **53**(4), 1–15 (2020). https://doi.org/10.1617/s11527-020-01537-1
4. Chen, Y., Yu, J., Leung, C.K.Y.: Use of high strength strain-hardening cementitious composites for flexural repair of concrete structures with significant steel corrosion. Constr. Build. Mater. **167**, 325–337 (2018)
5. Shi-lang, X.U., Hong-chang, W.: Experimental study on bond-slip between ultra high toughness cementitious composites and steel bar. 工程力学 **25**(11), 53–061 (2008)
6. Xiao, J., Falkner, H.: Bond behaviour between recycled aggregate concrete and steel rebars. Constr. Build. Mater. **21**(2), 395–401 (2007)
7. Chu, S.H., Kwan, A.K.H.: A new method for pull out test of reinforcing bars in plain and fibre reinforced concrete. Eng. Struct. **164**, 82–91 (2018)
8. Desnerck, P., Lees, J.M., Morley, C.T.: Bond behaviour of reinforcing bars in cracked concrete. Constr. Build. Mater. **94**, 126–136 (2015)
9. Haskett, M., Oehlers, D.J., Ali, M.M.: Local and global bond characteristics of steel reinforcing bars. Eng. Struct. **30**(2), 376–383 (2008)
10. Mousavi, S.S., Dehestani, M., Mousavi, K.K.: Bond strength and development length of steel bar in unconfined self-consolidating concrete. Eng. Struct. **131**, 587–598 (2017)
11. Chen, Y., Yu, J., Younas, H., Leung, C.K.: Experimental and numerical investigation on bond between steel rebar and high-strength strain-hardening cementitious composite (SHCC) under direct tension. Cement Concr. Compos. **112**, 103666 (2020)
12. Cheung, A.K., Leung, C.K.: Effective joining of pre-cast concrete slabs with self-compacting HSFRCC. J. Adv. Concr. Technol. **9**(1), 41–49 (2011)
13. Jin, Q., Leung, C.K., Yu, C.: Effective joining method for pseudo-ductile permanent formwork. Mater. Struct. **46**(3), 345–360 (2013)
14. Lee, S.W., Kang, S., Tan, K.H., Yang, E.: Experimental and analytical investigation on bond-slip behaviour of deformed bars embedded in engineered cementitious composites. Constr. Build. Mater. **127**, 494–503 (2016)
15. Deng, M., Pan, J., Sun, H.: Bond behavior of steel bar embedded in Engineered Cementitious Composites under pullout load. Constr. Build. Mater. **168**, 705–714 (2018)
16. Cai, J., Pan, J., Tan, J., Li, X.: Bond behaviours of deformed steel rebars in engineered cementitious composites (ECC) and concrete. Constr. Build. Mater. **252**, 119082 (2020)
17. Rokugo, K., Yokota, H., Sakata, N., Kanda, T.: Overview of "Recommendations for design and construction of high performance fiber reinforced cement composite with multiple fine cracks" published by JSCE. Concr. J. **45**(3), 3–9 (2007)

Crack Width Evaluation of DFRCC Members Reinforced with Braided AFRP Bar

Shugo Takasago[1(✉)], Toshiyuki Kanakubo[2], and Hiroya Kobayashi[1]

[1] Degree Program in Engineering Mechanics and Energy, University of Tsukuba, Tsukuba, Japan
S2220856@s.tsukuba.ac.jp

[2] Division of Engineering Mechanics and Energy, University of Tsukuba, Tsukuba, Japan
kanakubo@kz.tsukuba.ac.jp

Abstract. This study evaluates the crack width in FRP-reinforced DFRCC members and confirms the adaptability of the proposed formula. To achieve these goals, pullout tests followed by uniaxial tension tests were performed on braided aramid FRP (AFRP)-reinforced DFRCC specimens using PVA fibers. The experimental parameters included the cross-sectional area (100 × 100 mm^2, 120 × 120 mm^2, 140 × 140 mm^2) and the fiber volume fraction (0%, 1%, 2%) added to every specimen. First, the pullout test results show that the maximum bond stress increases as the fiber volume fraction increases. The trilinear models are employed for the bond stress–slip relationships to formulate the theoretical calculation of crack width. Finally, the uniaxial tension test results show that crack width increases with increasing fiber volume fraction and decreasing the cross-section of the specimens. The theoretical calculations are compared with the crack width obtained from the uniaxial tension test. In conclusion, most specimens' theoretical curves showed good adaptability to evaluate crack width from the experiment.

Keywords: DFRCC · AFRP · Bond constitutive law · Crack width · Bridging law

1 Introduction

Fiber-reinforced polymer (FRP) bars are corrosion-resistant and exhibit better elastic behavior than steel reinforcement. These properties can improve the repairability and durability of concrete structures when FRP bars replace steel reinforcements [1]. However, FRP bars with low elastic modulus could cause a more considerable deformation of concrete members reinforced with FRP bars than steel rebar. The deformation increases due to wider induced crack width in the concrete members. Therefore, ductile fiber-reinforced cementitious composite (DFRCC) can be adopted to control crack width. DFRCC is cementitious material reinforced with short discrete fibers showing ductile behavior of composite, especially on the tensile side. The main advantage of DFRCC remains its ability to control crack width by the bridging effect of fibers across cracks. Easy maintenance and high durability are expected in structures where DFRCC are

M. Kunieda et al. (Eds.): SHCC 2022, RILEM Bookseries 39, pp. 156–166, 2023.
https://doi.org/10.1007/978-3-031-15805-6_17

reinforced with FRP bars. It is essential to quantitatively evaluate crack width for practical application, considering both fiber bridging effect and bond interaction between DFRCC and reinforcement. The previous study [2] proposed a crack width evaluation formula to predict crack width in steel-reinforced DFRCC members. It is based on steel strain expressed by a linear relationship between the bond stress-slip relationship between the rebars and DFRCC. This formula, which includes the elastic modulus of the reinforcement, can also be adopted for FRP-reinforced DFRCC members.

This study evaluates the crack width in FRP-reinforced DFRCC members and confirms the adaptability with the existing formula. To achieve this goal, the pullout bond test was conducted to model the bond constitutive law between FRP bar and DFRCC prism specimens. Furthermore, the uniaxial tension test was conducted for FRP-reinforced DFRCC prism specimens with slits to experimentally measure the crack width. Then the crack width prediction formula expressed by the bond constitutive law model is validated.

2 Pullout Bond Test

2.1 Outline of Pullout Bond Test

Specimens. Table 1 shows the list of specimens, and Fig. 1 shows an example of specimen details. Three types of rectangular DFRCC blocks with a height of 100 mm were designed. The cross-section was 100 × 100 mm^2, 120 × 120 mm^2, and 140 × 140 mm^2. One braided AFRP bar was arranged in the center of the DFRCC block. A steel coupler was attached to the reinforcement end to fix the testing machine's chuck. The bond length was 54mm, approximately four times bar diameter. The experimental parameters included the cross-section of the specimens and the fiber volume fraction of DFRCC (0%, 1%, 2%). PVA fibers were used for DFRCC. Three specimens were manufactured for each parameter, and 27 specimens were tested.

Table 1. Specimen list

Type	Common factor	Cross-sectional size	Volume fraction	Number of Specimens
MT-A	Height: 100 mm Bond length: 54 mm (=4*d*) Reinforcement: Braided AFRP bar (Diameter 13.58 mm) Fiber type of DFRCC: PVA	100 mm × 100 mm (A series)	–	3
PVA1%-A			1.0%	3
PVA2%-A			2.0%	3
MT-B		120 mm × 120 mm (B series)	–	3
PVA1%-B			1.0%	3
PVA2%-B			2.0%	3
MT-C		140 mm × 140 mm (C series)	–	3
PVA1%-C			1.0%	3
PVA2%-C			2.0%	3

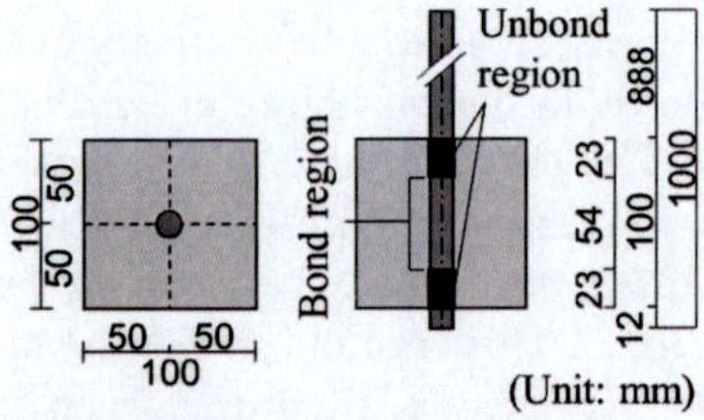

Fig. 1. Dimension of specimens

Used Materials. Figure 2 shows the surface shape of the braided AFRP bar. The mechanical properties of the braided AFRP bar are shown in Table 2. In Fig. 3, the PVA fiber used in DFRCC is shown. Table 3 shows the mechanical properties of the PVA fibers. Table 4 shows the mixture proportion of DFRCC and compression properties of ϕ100 mm × 200 mm cylinder test pieces by compression test. The fiber volume fraction of PVA fibers was set to 0% (mortar), 1%, and 2%.

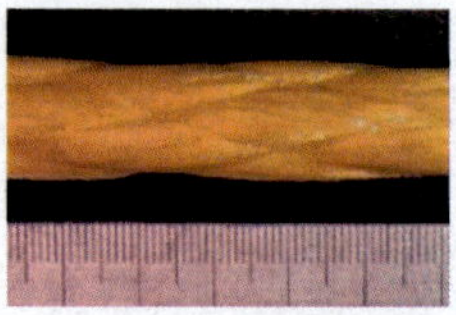

Fig. 2. Braided AFRP bar

Fig. 3. PVA fiber

Table 2. Mechanical properties of reinforcement

Reinforcement	Diameter (mm)	Tensile strength (MPa)	Elastic modulus (GPa)
Braided AFRP bar	13.58	1261	66.0

Table 3. Mechanical properties of fiber

Fiber	Length (mm)	Diameter (mm)	Tensile strength (MPa)	Elastic modulus (GPa)
PVA	12	0.10	1200	28

Table 4. Mixture proportion and mechanical properties of DFRCC

Type	Unit weight (kg/m^3)				PVA fiber (kg)	Compressive strength (MPa)	Elastic modulus (GPa)
	W	C	S	F			
MT	380	678	484	291	0	48.8	17.5
PVA 1%					13	46.2	17.0
PVA 2%					26	47.1	16.4

W: Water, C: High early strength Portland cement,
F: Fly ash of Type II of Japanese Industrial Standard (JIS A 6202),
S: Sand of Size under 0.2 mm

Loading Method. Figure 4 shows the loading method. The monotonic pullout load was applied until the reinforcement slipped out from the block under the controlled displacement. Teflon sheets were placed between the specimen and the reaction plate to facilitate lateral displacement of the block. The LVDT was set to measure slip at the free end. Measurement items were the pullout load and the reinforcement slip at the free end.

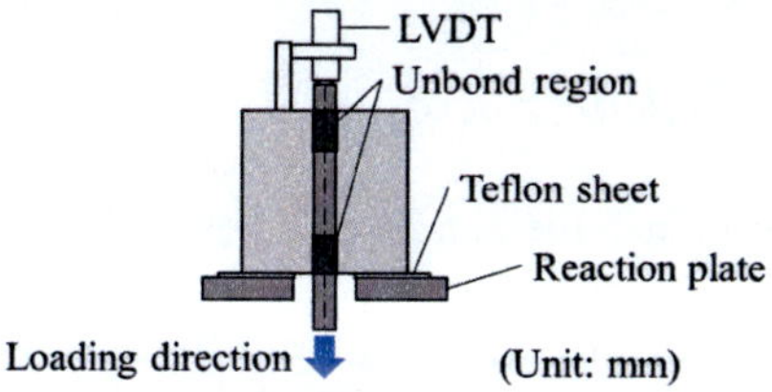

Fig. 4. Loading method

2.2 Experiment Results

Figure 5 shows the bond stress–loaded end slip relationship obtained from the pullout bond test. The loaded end slip was calculated by neglecting the deformation of the DFRCC. It is the sum of reinforcement elongation and free-end slip under the assumption that the stress of the bond is evenly distributed among the bond region. The bond stress was calculated by dividing the measured load by the surface area of the reinforcement in the bond length ($\pi \times$ diameter $\times$ bond length).

The results show a rapid decrease in bond stress with a widening crack after the maximum bond stress in MT specimens. In PVA specimens, the bond stress increases even after cracking. Furthermore, the maximum bond stress in PVA specimens was higher than that of MT specimens. In addition, PVA specimens show more ductile behavior after the maximum bond stress due to the fiber bridging effect.

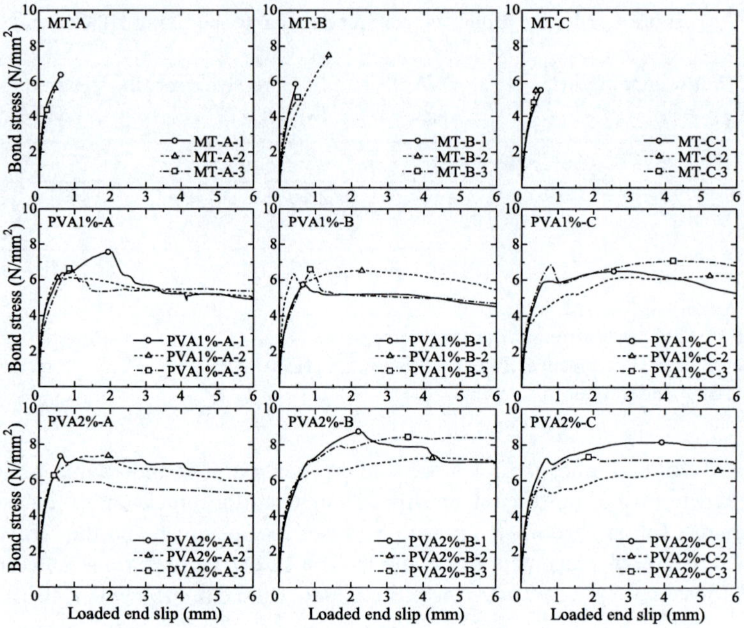

Fig. 5. Bond stress–loaded end slip relationship

3 Crack Width Prediction Formula

Sunaga et al. [2] have proposed a theoretical calculation formula to predict crack width in steel-reinforced DFRCC, as shown in Eq. (1). The formula is obtained by solving the force equilibrium and compatibility conditions between DFRCC and reinforcing bar considering bond stress – slip relationship, fiber bridging law (bridging stress – crack width relationship), and condition of crack occurrence. For the bridging law, a trilinear model was used, which modeled the calculation result using the same materials [3].

$$\varepsilon_{s(LOAD)} = \frac{\varphi_s}{A_c\{\sigma_{cr} - \sigma_{br}(w_{cr})\}} \int_0^{s_l} \tau_x ds_x + \frac{1+np}{2npE_c}\{\sigma_{cr} + \sigma_{br}(w_{cr})\} \tag{1}$$

where, $\varepsilon_{s(LOAD)}$ is the reinforcement strain at the loaded end. φ_s is the perimeter of reinforcement. σ_{cr} is the cracking strength of DFRCC. w_{cr} is the crack width. $\sigma_{br}(w_{cr})$ is the fiber bridging stress at crack [3]. n is the ratio of elastic modulus ($= E_s/E_c$). p is the reinforcement ratio ($= A_s/A_c$). E_s is the elastic modulus of reinforcement. E_c is the elastic modulus of DFRCC. A_s is the cross-sectional area of reinforcement. A_c is the cross-sectional area of DFRCC. s_l is the loaded end slip.

Equation (1) considers the elastic modulus of the reinforcing bars. Thus, it can also be adopted for FRP-reinforced DFRCC members. The integral of the right-hand side of Eq. (1) is obtained from the bond stress-slip relationship (bond constitutive law). The relationship between the reinforcement strain at the loaded end and crack width can be obtained by modeling the bond constitutive law and solving for this integral. The trilinear model is used as the bond constitutive law in this study. Figure 6 shows the trilinear model. Each characteristic value was determined as follows.

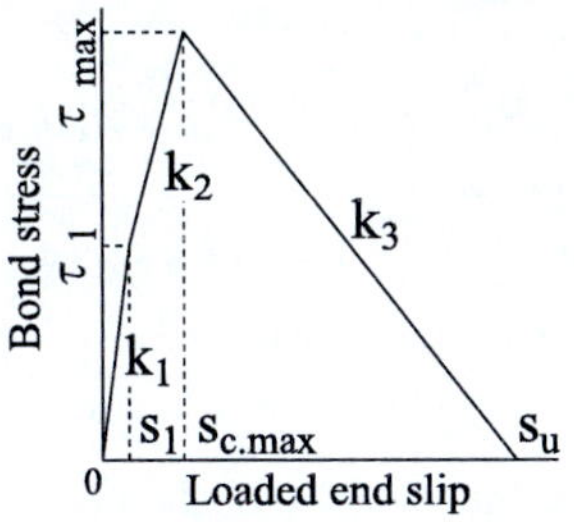

Fig. 6. Trilinear model

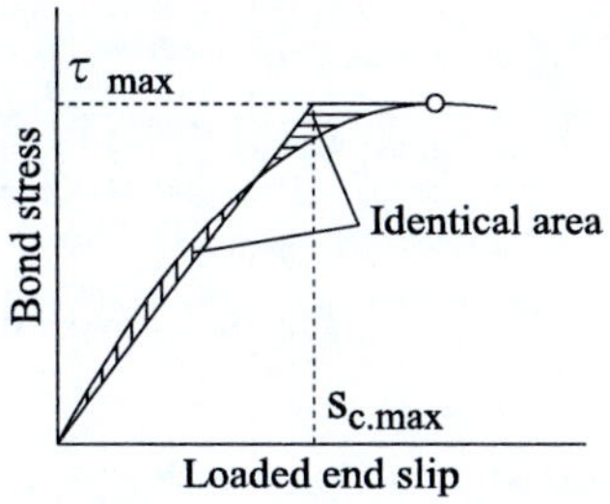

Fig. 7. Complementary energy

The maximum bond stress τ_{max} was the average of the maximum bond stresses among the identical three specimens. The slip at maximum bond stress is the corrected loaded end slip $s_{c.max}$, which was determined based on complementary energy equivalence (Fig. 7). The initial slope k_1 was empirically set to 1/5 of the maximum bond stress. τ_1 was set to 1/2 of the maximum bond stress. The softening slope k_3 was determined so that the bond fracture energy after $s_{c.max}$ was equivalent. In the case of MT specimens, since experimental results of the softening branch could not be obtained, the ultimate slip s_u was set to 1.5 mm. Table 5 shows the list of characteristic values of the trilinear model. Figure 8 shows an example of the comparison of the experimental results with the trilinear model. The trilinear model shows good adaptability with the experimental results within the range of up to a few millimeters of slip at the loaded end.

Table 5. List of characteristic values of the trilinear model

Type	τ_{max} (MPa)	$s_{c.max}$ (mm)	τ_1 (MPa)	s_1 (mm)	s_u (mm)	k_1 (N/mm^3)	k_2 (N/mm^3)	k_3 (N/mm^3)
MT-A	5.46	0.25	2.73	0.035	1.50	77.34	12.91	−4.35
PVA1%-A	6.77	0.47	3.39	0.047	20.95	71.35	7.99	−0.33
PVA2%-A	6.99	0.34	3.50	0.041	23.69	84.85	11.80	−0.30
MT-B	6.15	0.52	3.08	0.083	1.50	36.91	7.02	−6.29
PVA1%-B	6.29	0.43	3.14	0.054	20.38	58.28	8.46	−0.32
PVA2%-B	8.15	0.91	4.07	0.067	25.00	60.79	4.81	−0.34
MT-C	5.26	0.28	2.63	0.052	1.50	50.07	11.50	−4.31
PVA1%-C	6.61	0.94	3.30	0.086	24.03	38.56	3.88	−0.29
PVA2%-C	7.36	0.82	3.68	0.081	26.68	45.49	5.00	−0.28

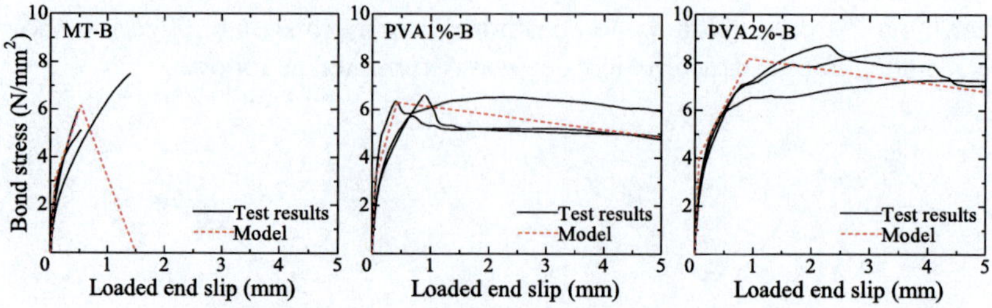

Fig. 8. Trilinear model of the bond constitutive law

In uniaxial tension specimens, it can be assumed that the slip at the crack position gives half of the crack width because of the symmetry. The right-hand side of Eq. (1) is a function of the crack width. Therefore, Eq. (2), expressing the relationship between the reinforcement strain at the loaded end and crack width, was obtained using the trilinear model of bond constitutive law.

$$(w_{cr} < 2 \cdot s_1)$$

$$\varepsilon_{s(LOAD)} = \frac{k_1 \varphi_s}{8A_c\{\sigma_{cr} - \sigma_{br}(w_{cr})\}} w_{cr}^2 + \frac{1+np}{2npE_c}\{\sigma_{cr} + \sigma_{br}(w_{cr})\}$$

$$(2 \cdot s_1 < w_{cr} < 2 \cdot s_{c.max})$$

$$\varepsilon_{s(LOAD)} = \frac{\varphi_s}{8A_c\{\sigma_{cr} - \sigma_{br}(w_{cr})\}}\left\{k_2 w_{cr}^2 + 4(\tau_{max} - k_2 s_{c.max}) w_{cr} - 4 s_1(\tau_{max} - k_2 s_{c.max})\right\} + \frac{1+np}{2npE_c}\{\sigma_{cr} + \sigma_{br}(w_{cr})\} \quad (2)$$

$$(w_{cr} > 2 \cdot s_{c.max})$$

$$\varepsilon_{s(LOAD)} = \frac{\varphi_s}{8A_c\{\sigma_{cr} - \sigma_{br}(w_{cr})\}}\left\{k_3 w_{cr}^2 - 4k_3 s_u w_{cr} + 4(k_3 s_u s_{c.max} + \tau_1 s_{c.max} - \tau_{max} s_1)\right\} + \frac{1+np}{2npE_c}\{\sigma_{cr} + \sigma_{br}(w_{cr})\}$$

4 Uniaxial Tension Test

4.1 Outline of Uniaxial Tension Test

Figure 9 shows an example of specimen details with the setup of displacement transducers. Three types of rectangular DFRCC blocks with a length of 600 mm were designed. The cross-section was set to $100 \times 100\ \text{mm}^2$, $120 \times 120\ \text{mm}^2$, and $140 \times 140\ \text{mm}^2$. One braided AFRP bar was arranged in the center of the DFRCC block. Steel couplers were attached to the reinforcement ends to fix the chuck. Slits were set to control the crack position by cutting the hardened DFRCC. The depth of the slits was set by 60% of the entire section. The experimental parameters were the cross-section of the specimens and the fiber volume fraction of DFRCC (0%, 1%, 2%). PVA fibers were used for DFRCC. Three specimens were manufactured for each parameter, and 27 specimens were tested. The materials, including DFRCC and reinforcement, are similar to those used in the pull-out bond test. The specimen IDs are same as those of the pullout bond test specimens. Uniaxial monotonic tension loading was conducted using a universal testing machine. When the tensile load reached approximately 105 kN (725 MPa in reinforcement stress), the loading was completed, and the specimens were unloaded. Measurement items were tensile load, crack width measured by π-type displacement transducers arranged at each slit position, and total deformation.

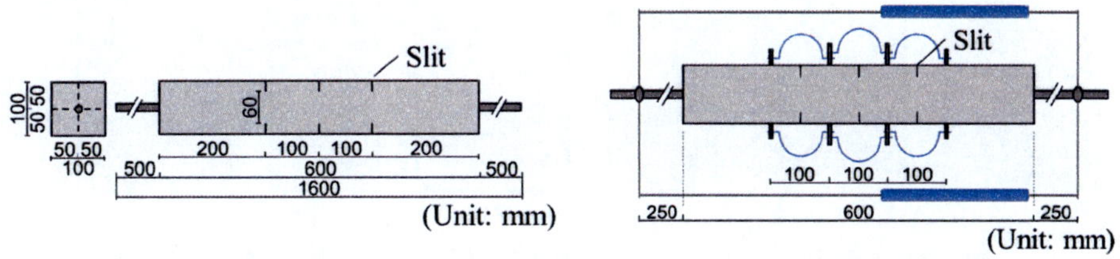

Fig. 9. Detail of the specimens and setup of displacement transducers

4.2 Experiment Results

Figure 10 shows an example of cracks with a tensile load of approximately 105 kN. In MT specimens, crack width increased with the increase of the tensile load after the first crack generated at the slit position. A single crack was only observed. In DFRCC specimens, multiple cracks were observed after the first crack generated at the slit position. The number of cracks increases with increasing fiber volume fraction and decreasing the cross-section of the specimens. Moreover, cracks in all specimens were closed when the load was removed.

4.3 Adaptability of Calculation Formulas with Experimental Results

Figure 11 shows the reinforcement strain at the loaded end and crack width relationship obtained from the uniaxial tension test. The reinforcement strain at the loaded end was determined by dividing the tensile load by the elastic modulus and the cross-sectional area of the reinforcement (Table 2). Experimental crack width values were used until the second crack was observed in the target region (100 mm) (Fig. 12). The experimental values for each specimen were linearly approximated by the least-squares method, and the average line was drawn by calculating the average value of an intercept and a slope of the approximated results.

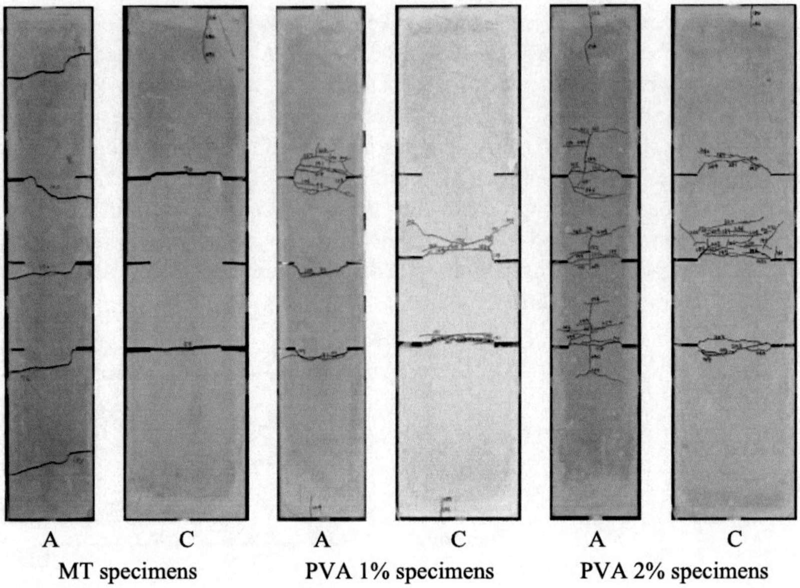

Fig. 10. Example of crack

Figure 11 also shows the reinforcement strain at the loaded end and crack width relationship obtained from the crack width prediction formula, Eq. (2). The tensile strength of DFRCC was the average strength at cracking from the uniaxial tension test (2.29 MPa). It is obtained by dividing the tensile load at cracking by the cross-sectional area at the slit position. For the fiber bridging law, a DFRCC bridging model [3] obtained with the same used materials as this study was employed. The reinforcement strain at the loaded end and crack width relationship obtained from the crack width prediction formula gives a possible maximum crack width at an arbitrary strain of reinforcement [2]. Thus, the adaptability between the crack width prediction formula and the experimental results can be considered reasonable. In specimens with 2% PVA, the theoretical reinforcement strain decreases at crack width equal to 0.2 mm due to the softening of fiber bridging law in this region. This phenomenon is not observed in the experimental results because of the difficulties of keeping the equilibrium in the softening branch. In addition, the slope of the average line of experimental results becomes more significant as fiber volume fraction increases. The fiber bridging effect brings the bond stress loss and crack width opening control at the same reinforcement strain. Also, the slope of the average line becomes smaller with increasing the specimens' cross-section due to the delay of cracking in specimens with larger cross-sections. In PVA2%-C specimens, the increase in the average line slope is likely due to cracking at a location where the bridging law did not reach the softening branch.

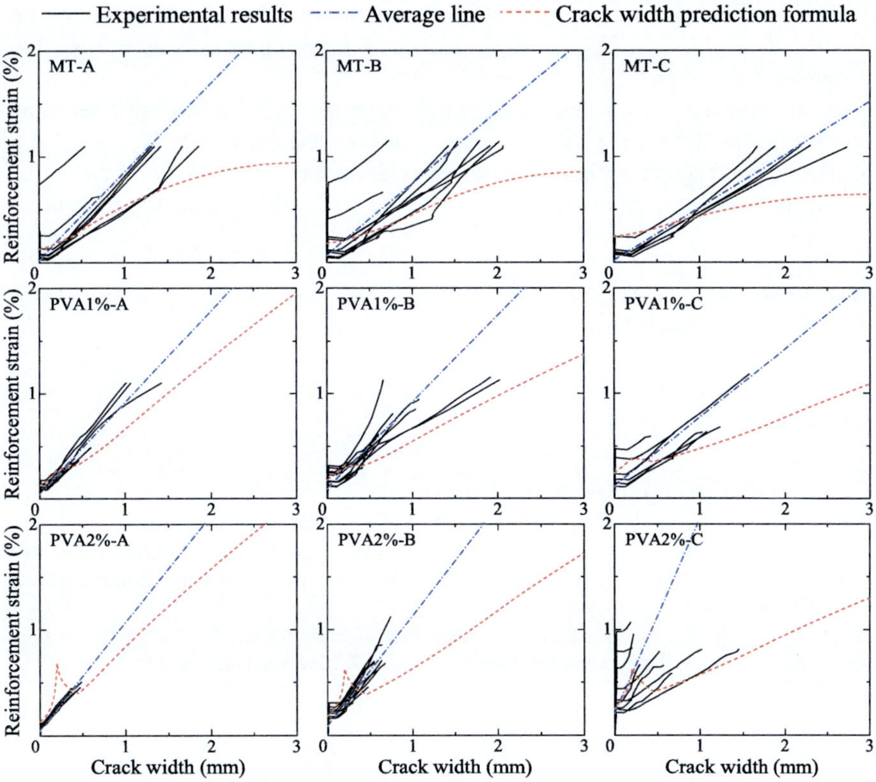

Fig. 11. Reinforcement strain at the loaded end and crack width relationship

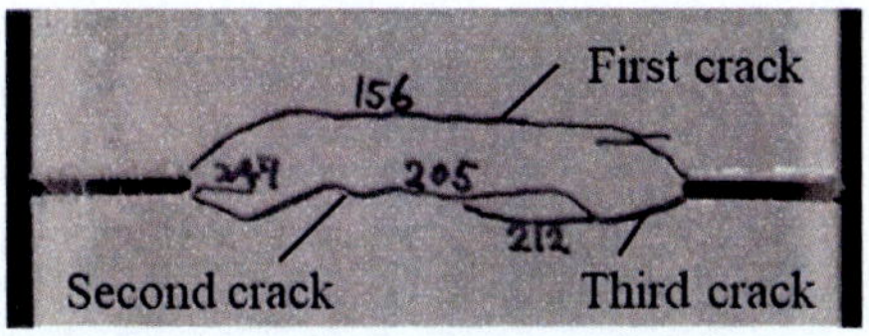

Fig. 12. Examples of cracking process

5 Conclusion

This study conducted a pullout bond test and uniaxial tension test to evaluate the crack width of aramid FRP reinforced DFRCC members. The main experimental parameters included the cross-section of the specimens and the fiber volume fraction of DFRCC. The main conclusions of this research are summarized below:

(1) In the pullout test, the decrease in bond stress becomes smaller as the fiber volume fraction of DFRCC increases after the maximum bond stress is reached.

(2) The reinforcement strain-crack width formula was proposed by modeling the bond stress-loaded end slip relationship, obtained from the pullout load test by the trilinear model.
(3) In the uniaxial tension test, crack width decreases with increasing fiber volume fraction and decreasing the cross-section of the specimens.
(4) The reinforcement strain-crack width relationship obtained from the uniaxial tension test shows good adaptability with the crack width prediction formula.

Acknowledgments. The AFRP bars and PVA fibers were provided from Fibex Co., Ltd. and Kuraray Co., Ltd., respectively.

References

1. American Concrete Institute: Guide for the Design and Construction of Structural Concrete Reinforced with Fiber-Reinforced Polymer (FRP) Bars, ACI 440.1R-15 (2015)
2. Sunaga, D., Namiki, K., Kanakubo, T.: Crack width evaluation of fiber-reinforced cementitious composite considering interaction between deformed steel rebar. Construct. Build. Mater. **261**, 119968 (2020)
3. Ozu, Y., Miyaguchi, M., Kanakubo, T.: Modeling of bridging law for PVA fiber-reinforced cementitious composite considering fiber orientation. J. Civil Eng. Archit. **12**, 651–661 (2018)

Cracking Behaviour of Strain-Hardening Cementitious Composites (SHCC) Under Practical Creep Conditions

K. A. Shan D. Ratnayake(✉), Ka Wai Li, and Christopher K. Y. Leung

Hong Kong University of Science and Technology, Clear Water Bay, Hong Kong, China
kasdr@connect.ust.hk

Abstract. Strain-Hardening Cementitious Composites (SHCC) have become increasingly popular as a material which could be adopted for enhancing the durability of structures. Studies have shown that the cracks formed during short-term tensile loading in SHCCs can be controlled in the range of 50–100 μm as compared to conventional reinforced concrete, where cracks are designed to be around 200–300 μm at serviceability state. Thereby, applying SHCC as an additional protective layer (e.g. permanent formwork) on the concrete tensile face can effectively control the ingress of water and harmful chemicals due to the fine crack size, thus improving the durability of the structure. Although the crack widths are small under short-term loading, imposing a constant sustained load on SHCC over a longer period has shown to increase the crack size beyond acceptable levels (to over 200 μm). In real structures, however, the deformation is limited by design as well as load sharing between structural elements, therefore the crack patterns in such cases will follow a different propagation mechanism to that under constant loading tests. Specifically, for a reinforced concrete member with a SHCC layer on the surface, any tensile creep deformation in the SHCC will redistribute load to the steel reinforcement and/or concrete, whereby the stress on SHCC gradually reduces with deformation. This study aims to investigate the creep of SHCC under realistic conditions by the experimental simulation of different cases. A novel testing set-up is developed to impose a constant moment that is shared between reinforcement and an SHCC layer, so the ensuing crack formation and opening in SHCC under restrained creep conditions is studied. It was observed that the creep strain increase was predominantly due to widening of pre-existing cracks while the number of cracks stays the same. The crack widths increased by about 10 μm on average (20% increase) which is far less than the value reported for constant stress test and stabilised after 1000 h. Such a marginal increase would have a very low impact on the durability performance reported in short-term tests.

Keywords: Strain hardening cementitious composite · Creep · Cracking · Sustained loading

M. Kunieda et al. (Eds.): SHCC 2022, RILEM Bookseries 39, pp. 167–177, 2023.
https://doi.org/10.1007/978-3-031-15805-6_18

1 Introduction

The behaviour of strain hardening cementitious composites (SHCC) has been studied extensively to understand the mechanics [1–4] and demonstrate numerous advantages of its intrinsic tight crack control to below 100 microns [5–7]. The superior durability of SHCC as compared to normal concrete under certain scenarios has been one of the key factors in its rise in popularity.

In cases where SHCC is used in combination with concrete, the tensile properties of SHCC can be exploited to curtail undesirable effects arising from the weak tensile performance of conventional concrete. For example, formation of cracks larger than 100 microns in concrete can make it highly susceptible to water/chemical ingress [8]. Maalej and Li [9] were able to significantly reduce flow of aggressive substances by using an SHCC layer within the tension zone of a reinforced concrete flexural member.

While most studies have been focused on the short-term loading behaviour, the long-term behaviour showcased literature indicate that the durability figures exhibited under short term loading tests may have overestimated the performance when SHCC is used in a real structure, where it may be subjected to stresses or strains sustained over very long periods of time. Experiments carried out by Boshoff et al. [10] on sustained loading of SHCC specimens which were previously loaded to 1% strain showed that an existing crack, among others, widened significantly from 78 μm to 244 μm in 8 months under the constant stress. In addition, new cracks were also generated when the sustained load was sufficiently high.

Considering the design life of real structures being in the order of decades, the properties exhibited by SHCC in short term tests may not be representative over such a period. In the conventional constant stress creep test, the deformation is e allowed to increase without limit, which is different to an actual structure, where the deformation is limited by serviceability design criteria. In a scenario where an SHCC layer is used on the tension side of a reinforced concrete member (with the objective of enhancing durability), SHCC will be a load sharing partner in the system. The stress on SHCC is likely to decrease with time as SHCC creeps faster than concrete and steel, thereby resulting in a force redistribution to the other elements. Hence the cracking behaviour observed in constant stress creep tests may be different from that in practical cases. In this study, the creep behaviour of a SHCC layer on the surface of a structural member is investigated through a novel experimental setup.

2 Experimental Setup

2.1 Materials and Preparation

The cement used in the experiment was CEM I Class 52.5N Hong Kong Green Island ordinary Portland cement which contains at least 95% ground clinker. Silica fume used was Elkem™ 920U. Commercially available river sand sieved through 5 mm sieve and silica sand (Mesh 80–120) were used as fine aggregate for concrete and SHCC respectively. 10 mm crushed aggregate was used as coarse aggregate. A high range water reducing superplasticiser, PCA-I from Subote™, was used to control the workability of the mixes. Quantaflex™ Polyethylene (PE) fibres were used to produce SHCC. Specimens were cured at a temperature of 23 ± 2 °C and relative humidity of 95 ± 5%.

Creep tests focus on the behaviour of a material loaded for an extended duration. Therefore, it would be ideal for properties of the test specimens to remain the same over the duration of the test. Specifically, a sustained loading being imposed on an early age specimen would give a different creep response compared to a specimen of the same material at a late age due to, but not limited to, the effect from continued hydration. It is therefore preferable to test specimens whose mechanical and chemical properties have minimal change after starting a creep test. To begin testing specimens for creep as early as possible, SHCCs having fly ash and ground granulated blast-furnace slag in their composition were avoided. A high strength SHCC developed previously by Chen et al. [11] was used for this study. The mix proportions by mass of the SHCC and concrete is shown are Table 1.

Table 1. Mix proportions by mass used for concrete and SHCC

Mix	Cement	Silica fume	Fine aggregate	Coarse aggregate	Water	Superplasticiser	Fibre % (by volume)
Concrete	1	–	1.88	1.47	0.44	0.008	–
SHCC	0.8	0.2	0.5	–	0.18	0.028	2.2

2.2 Quasistatic Characterisation Tests on SHCC

The characteristic tensile properties were determined by using 250 kN servo-hydraulic MTS 810. 13 mm and 20 mm thick dog bone-shaped samples were selected since the 15 mm SHCC layer thickness (used for the creep test) falls between these two values. Two linear variable differential transformers (LVDT) were used to capture the deformation in the 80 mm gauge length. The corresponding strain value was taken by averaging two LVDT data. The specimen was gripped at the ends and a strain rate of 6.25×10^{-5}/s was applied until sample failure. The compressive strength was determined using 40 mm and 100 mm cubes for SHCC and concrete respectively, tested at a loading rate of 0.625 MPa/s.

2.3 Restrained Creep Test Setup

The design of this experiment was based on the need to characterise the creep behaviour of SHCC in a scenario which is more likely to occur in a practice. An example of such an instance is the tension zone of a reinforced concrete flexural member with a surface layer of SHCC for durability enhancement. In section analysis, the concrete in this zone is assumed to be in cracked state therefore no tensile contribution is considered. Therefore, a simulation of a constant tensile load being shared exclusively between SHCC and the steel reinforcement (not within the SHCC) as shown in Fig. 1(a) would be analogous to this type of system. A constant moment imposed about the hinge would ensure that the load shed by SHCC over time due to creep is picked up by the reinforcement bar.

This arrangement was realised by employing a frame made of two concrete half beams as shown in Fig. 1(b). The full beam measured 1.7 m × 0.1 m × 0.2 m and a discontinuity (150 mm over the top 70 mm, 12 mm for the remaining height) was introduced at the centre using PVC plates. A 25 mm diameter steel pin was employed as the hinge which allow the two halves of the beam to rotate in the vertical plane. The pin was surrounded by a steel block on each side having a groove of the same diameter (to facilitate rotation). The steel blocks were fixed firmly to the concrete member with horizontal anchors. The concrete beam is reinforced by two 16 mm Ø high yield steel bars for tension and two 10 mm Ø high yield steel bars for compression. Stirrups of 6 mm Ø mild steel at 125 mm spacing provide shear reinforcement. This setup will act as a frame for testing the SHCC in a restrained condition.

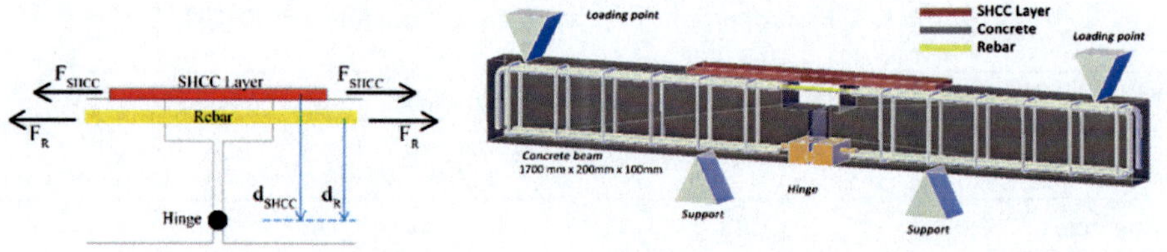

(a) (b)

Fig. 1. (a) Schematic of an analogous tension load sharing system between rebar and SHCC and (b) fixture used to create a load sharing system exclusively between rebar and SHCC layer

A single 500 mm rebar (with 2 strain gauges attached and pre-calibrated in a universal testing machine) bridging the two halves was placed at 175 mm above the bottom surface prior to casting the beam with medium strength concrete. The bridging rebar is bent 90° at the ends with a 100 mm anchorage into the concrete. The surface of the concrete at the edges above the hinge was treated with a waterjet to expose aggregates just after casting to increase the bond with the SHCC layer to be placed later. The beam was cured until 14 days after casting.

A 15 mm layer of SHCC was then placed on the central 500 mm of the top surface, using a temporary formwork to support the bottom of the SHCC in the middle 150 mm gap. The width was reduced from 100 mm to 50 mm in the central 80 mm region (resulting in a dog bone shape). Two lifting hooks fixed in the concrete ~180 mm from the centre on either side also serves to anchor the SHCC onto the concrete. The SHCC and concrete was then cured until 14 days after casting SHCC.

To impose a uniform moment throughout the midspan, a 4-point bending arrangement was employed, with a middle span of 600 mm and loading points 100 mm from each end of the beam. To maintain a constant load at each end, hydraulic cylinders fitted with a valve for maintaining pressure was used. The cylinders were fixed to a load cell and rested on a 100 mm strip of steel pasted onto the top surface of the beam at the loading point. Manual hydraulic pumps were used for stroke control of the cylinder. The steel pin was cleared of debris and lubricated with grease prior to loading and the surface of

SHCC was painted white to better distinguish cracks. A 25 mm LVDT and its target was anchored on the SHCC 75 mm from the centre on each side. Since it was expected that the cracks would be concentrated in the middle 80 mm gauge length of the SHCC with reduced area, the LVDT would predominantly capture the deformation in the gauge length in addition to the elastic deformation in the outer 35 mm on each side. Since the elastic strain is negligible compared to cracking strain, it was not considered.

Loads at both ends must be equal for a uniform moment to be induced within the midspan, hence loading was increased gradually keeping the load at both ends equal as much as possible. The final sustained constant moment was selected such that the rebar was safely below its yielding point, but also had sufficient strain to induce cracking in the SHCC layer. Hence, two load cases, a constant 10 kN (LC1) and 8 kN (LC2) were chosen for the system, which translated to 363 MPa and 290 MPa in the rebar in the event of complete SHCC relaxation (stress drops to zero in SHCC layer). Two specimens (S1 and S2) were used for each load level. During the quasistatic loading phase to reach the predetermined load level, a Canon EOS Mark 6D II camera was used to take photographs of SHCC along the gauge length at 20 s intervals to capture the crack pattern formation. The image at the end of the quasistatic phase was then compared with one taken at the end of the restrained creep test to check for formation of new cracks. A USB microscope having a resolution of ~0.5 μm/pixel was used to capture a section of each crack at different times, and this was used to monitor the change of crack width over time. A line was drawn across the surface cracks to mark and identify the monitoring region of each. When the load in hydraulic cylinders dropped due to the creep deformation of SHCC, the load was increased by using the manual pump. Hence, the load was largely maintained within 0.5 kN of the target load for the duration of the test. The test was conducted at 23 ± 2 °C and relative humidity 60% ± 10% and lasted 61–77 days.

The idealised mechanics of stretching of a line element (representing the SHCC specimen) connected between the two half beams at the centre is shown in Fig. 2. Geometrically, the connection points will move in an arc. However, as the deformations will be relatively small, the bending effects can be considered insignificant.

Fig. 2. Elongation of an element attached to the concrete beam between two nodes AB at the centre

As this is a symmetrical arrangement, a single side will be considered for simplicity.

Due to strain increase in the element, rotation about the pivot increases the angle α by a small angle of $\Delta\theta$.

$$\Delta\theta = \frac{\Delta L_{curve}}{d_{curve}} = \frac{\left(\frac{\Delta L}{cos\alpha}\right)}{\left(\frac{d}{cos\alpha}\right)} = \frac{\Delta L}{d} = \frac{\varepsilon \times L_i}{2 \times d} \tag{1}$$

Considering two elements connected to the nodes (SHCC layer and the steel bar), the angle $\Delta\theta$ will be common to both if they are rigidly fixed. Therefore, the relationships will be identical for the steel bar, with changes to α, d and L_i.

Thereby, considering compatibility,

$$\Delta\theta = \frac{\varepsilon_{rebar} \times L_{i,rebar}}{2 \times d_{rebar}} = \frac{\varepsilon_{SHCC} \times L_{i,SHCC}}{2 \times d_{SHCC}} \tag{2}$$

$$\text{Considering Moment balance, } M = d_{rebar} \times F_{rebar} + d_{SHCC} \times F_{SHCC} \tag{3}$$

From material properties and substituting from (2),

$$\sigma_{SHCC} = \left[-\left(\frac{d_{rebar}}{d_{SHCC}}\right)^2 \times \left(\frac{L_{i,SHCC}}{L_{i,rebar}}\right) \times \left(\frac{A_{rebar}}{A_{SHCC}}\right) \times E_{rebar}\right] \times \varepsilon_{SHCC} + \frac{M}{d_{SHCC} \times A_{SHCC}} \tag{4}$$

In the fixture, during the creep test, M is held at a constant value, thereby Eq. (4) takes the form of y = mx + c, where E and A are the elastic modulus and cross-sectional area of the element respectively.

The stress-strain relationship derived for the partially restrained system (adjusted for the 80 mm central gauge length) is illustrated in Fig. 3. In contrast to a constant stress or strain test, the system controls dynamically both the stress and strain which will change with time in the direction of the arrow indicated in each case. In this graph, the constant stress and strain creep tests are drawn assuming that SHCC will be under a tensile stress of 10 MPa upon quasistatically reaching the respective target load.

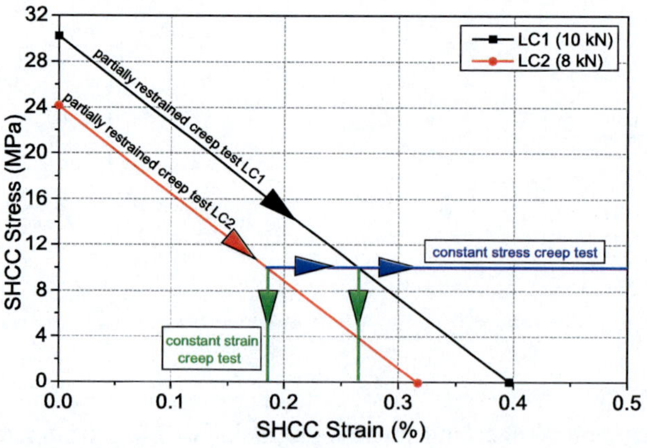

Fig. 3. Stress-strain relationship of the partially restrained system

3 Results and Discussion

From the quasistatic characterisation tests, the representative tensile stress-strain curves of SHCC at different ages is shown in Fig. 4. The mechanical properties of the SHCC and concrete are shown in Table 2.

Table 2. Mechanical properties of SHCC and concrete

Mix	Age	Compressive strength (MPa)	Tensile strength (MPa)	Tensile strain capacity (%)
SHCC (13 mm)	14	124.7 (7.5)	10.1 (1.0)	2.52 (0.50)
SHCC (20 mm)	14		9.4 (0.4)	2.38 (0.83)
SHCC (13 mm)	28	132.6 (8.8)	10.1 (1.3)	1.70 (1.06)
SHCC (20 mm)	28		13.1 (0.4)	1.63 (0.67)
SHCC (13 mm)	90	146.1 (9.6)	11.3 (1.1)	1.54 (0.02)
SHCC (20 mm)	90		11.4 (0.2)	0.92 (0.45)
Concrete	7	62.3 (2.5)	–	–
Concrete	28	74.6 (4.6)	–	–

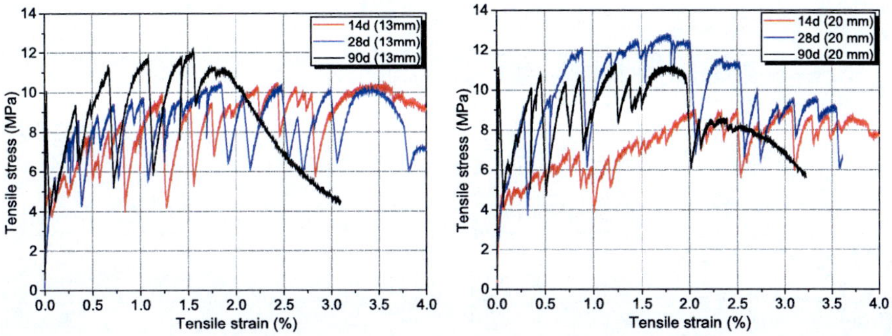

(a) (b)

Fig. 4. Representative tensile stress-strain curves of (a) 13 mm and (b) 20 mm SHCC at 14, 28 and 90 days

The compressive and tensile strength have not changed significantly after 14 days, but the tensile strain capacity of the SHCC has dropped over a 90-day period from ~2.4% to ~1%. However, the strain capacity exceeds the maximum strain expected from the stress-strain curve during the creep test (as shown in Fig. 3).

A comparison of the cracking pattern after reaching the target is shown in Fig. 5. From visual inspection, no new cracks were observed by the end of the test and only the pre-existing cracks had widened. The variation of width of each of the cracks observed within the gauge length is shown in Fig. 6 and summarised in Table 3. The tensile stress in the SHCC layer when the test ended was between 1–2 MPa.

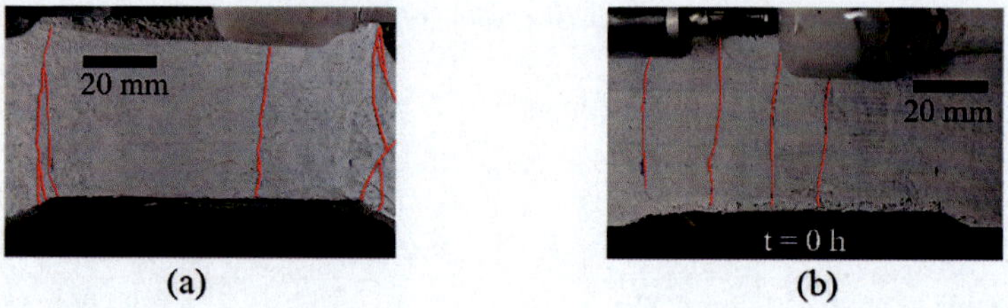

Fig. 5. The surface crack pattern on (a) LC1 S2 and (b) LC2 S2 (traced in red for clarity)

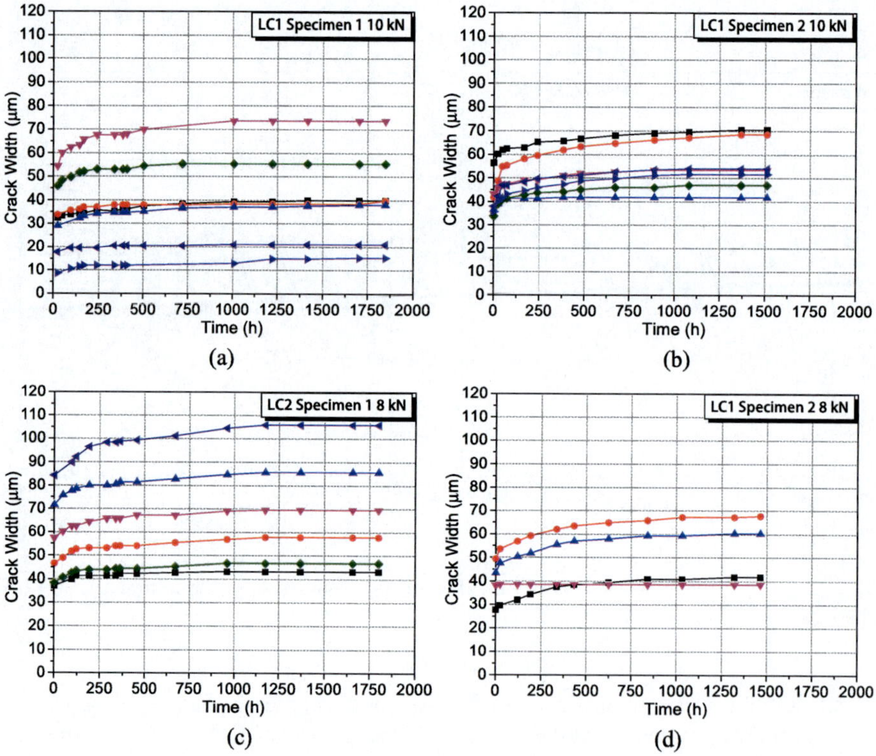

Fig. 6. Graphs of crack width with time for (a) LC1 S1 (b) LC1 S2 (c) LC2 S1 (d) LC2 S2

The crack widths were less than 100 μm after the quasistatic loading phase in all cases. Most of the change in crack widths (8 μm) occurred over the first 500–750 h and only a marginal average increase of 2 μm was registered from 750–1500 h. A similar trend was observed in the strain data from the LVDTs. No significant difference of the average increase and percentage increase in crack widths is observed across the two load cases. Figure 7 shows a photograph a crack in each specimen at the start and end of the test.

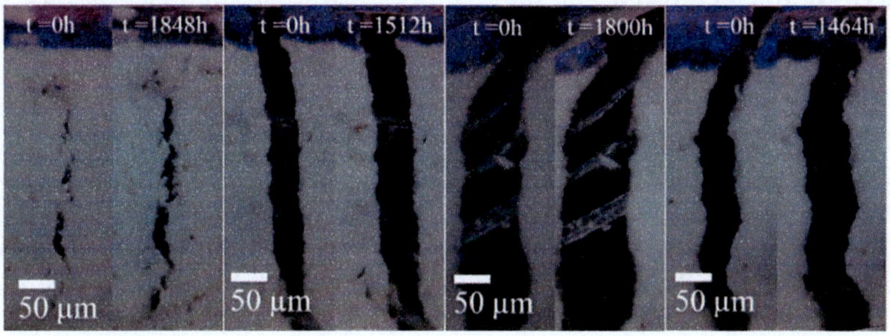

Fig. 7. Initial and final crack image in (a) LC1 S1 (b) LC1 S2 (c) LC2 S1 (d) LC2 S2

Table 3. Average and percentage increase in crack widths with time

Time (h)	Time (days)	LC1 S1	LC1 S2	LC2 S1	LC2 S2
		Average crack width (μm)			
24	1	31.8 (15.5)	45.1 (7.6)	50.5 (25.5)	42.7 (10.7)
72	3	34.1 (17.2)	48.3 (7.9)	52.7 (26.2)	43.7 (11.0)
168	7	36.4 (18.2)	49.7 (8.1)	55.8 (27.3)	45.8 (11.6)
336	14	37.4 (18.7)	51.5 (8.8)	56.9 (28.3)	48.8 (12.4)
672	28	38.9 (19.8)	53.7 (9.6)	58.5 (29.1)	50.7 (13.4)
1344	56	40.2 (19.9)	55.3 (10.6)	60.8 (30)	52.3 (14.1)
		Percentage increase in average crack width (%)			
72	3	6.7	6.8	4.1	2.4
168	7	12.5	9.3	9.4	6.8
336	14	14.9	12.6	11.3	12.5
672	28	18.3	16.1	13.6	15.9
1344	56	20.8	18.5	16.9	18.5

4 Conclusions and Limitations

A novel test method with a load sharing system was used as a pragmatic approach of simulating long-term loading on SHCC. A constant moment of 3.6 kNm and 4.5 kNm were imposed in the midspan for the two load cases tested.

1) In the cases tested, the crack widening was not as pronounced as those observed in previous studies where constant stress systems were used for loading SHCC. All cracks were below 100 μm at the start of the creep test and increased only by 9–10 μm on average in all load cases. This will have a very low adverse effect on the durability properties established under short-term tests.
2) The increase of average crack width did not differ significantly between the loading cases tested. This may be due to SHCC being within a relatively low tensile strain for the duration of the testing and its steady-state properties do not vary significantly at low strain levels.
3) A 16–20% increase of crack width was observed across all load cases but reaches a plateau after ~1000 h due to the continuous stress decrease. Although the crack sizes were monitored only for 1500–2000 h, it is most likely that the crack stops widening after the stress drops to very low levels.
4) In contrast to previous studies which applied a constant stress on SHCC, new cracks were not formed during this test in any specimen. This can be attributed to the SHCC stress dropping to around 1–2 MPa towards the end of the test, which is ~60% lower than the matrix cracking strength.

The stress of the SHCC at the end of the quasistatic loading (to reach a certain target) can vary due to the fluctuation of stress during crack formation. Hence the initial stress will not be constant across specimens for the same load case, thereby the speed of relaxation may differ. Due to the circular movement of the two half-beams about the hinge, this setup will introduce bending effects at high deformations, hence the final strain of the SHCC needs to be controlled to an acceptable range by using a suitable load sharing element. No parallel measurement of shrinkage properties was performed; hence the time dependant strain changes implicitly include shrinkage strains as well.

References

1. Bentur, A., Mindess, S.: Fibre Reinforced Cementitious Composites, 2nd edn. CRC Press, Boca Raton (2007)
2. Li, V.C., Leung, C.K.: Steady-state and multiple cracking of short random fiber composites. J. Eng. Mech. **118**(11), 2246–2264 (1992)
3. Leung, C.K.: Design criteria for pseudo-ductile fiber reinforced cementitious composites. In: Proceedings of the 10th ASCE Engineering Mechanics Conference (1995)
4. Kanda, T., Li, V.C.: Practical design criteria for saturated pseudo strain hardening behavior in ECC. J. Adv. Concr. Technol. **4**(1), 59–72 (2006)
5. Li, V., Wang, S., Wu, C.: Tensile strain-hardening behavior of polyvinyl alcohol engineered cementitious composite (PVA-ECC). Mater. J. **98**(6), 483–492 (2001)
6. Lepech, M.D., Li, V.C.: Application of ECC for bridge deck link slabs. Mater. Struct. **42**(9), 1185–1195 (2009)

7. Li, V.C.: Engineered cementitious composites (ECC)–tailored composites through micromechanical modelling. In: Fibre Reinforced Concrete: Present and the Future, pp. 1–38 (1997)
8. Wang, K., et al.: Permeability study of cracked concrete. Mater. Struct. **32**, 370–376 (1999)
9. Maalej, M., Li, V.C.: Introduction of strain-hardening engineered cementitious composites in design of reinforced concrete flexural members for improved durability. ACI Struct. J. **92**(2), 167–176 (1995)
10. Boshoff, W.P., Mechtcherine, V., van Zijl, G.P.A.G.: Characterising the time-dependant behaviour on the single fibre level of SHCC: Part 1: mechanism of fibre pull-out creep. Cem. Concr. Res. **39**(9), 779–786 (2009)
11. Chen, Y., Yu, J., Leung, C.K.Y.: Use of high strength Strain-Hardening Cementitious Composites for flexural repair of concrete structures with significant steel corrosion. Constr. Build. Mater. **167**, 325–337 (2018)

Influence of Loading Frequency and Force Level on the Cyclic Performance of Strain-Hardening Cement-Based Composites (SHCC)

Dominik Junger(✉) and Viktor Mechtcherine

Institute of Construction Materials, Technische Universität Dresden, 01062 Dresden, Germany
dominik.junger@tu-dresden.de

Abstract. Structures are exposed to a variety of quasi-static and dynamic/cyclic loads. For a safe, material-minimized structural design, a comprehensive knowledge of the material behavior under various loading conditions is required. Previous studies showed that Strain-Hardening Cement-based Composites (SHCC), in literature also often called Engineered Cementitious Composites (ECC) are a promising class of materials that exhibits an outstanding mechanical resistance under both quasi-static and cyclic loading regimes. However, a profound understanding of the mechanisms leading to the specific behaviors under cyclic loads is missing.

The article at hand presents experimental results from cyclic tension-swelling and alternating tension-compression tests performed on uniaxially loaded, notched dogbone-shaped specimens made of high-strength SHCC with a polyethylene fiber content of 2% by volume. The samples were exposed to harmonic loads with different frequencies, i.e., 1 Hz and 20 Hz for a certain number of load cycles. The chosen stress level in the tension-swelling tests corresponded to 80% of the first crack strength while for the alternating cyclic loading tests 25% of the compressive strength and 80% of the first crack strength were defined as reversal points. In addition, morphological analysis of the fracture surfaces and crack patterns were carried out by means of microscopy in order to determine the degradation condition of each phase, i.e., polyethylene fiber and matrix. Finally, the results were discussed referring to the physical phenomena causing the observed behavior.

Keywords: SHCC · ECC · Cyclic loading · Fiber degradation · Failure mechanisms

1 Introduction

Strain-Hardening Cement-based Composites (SHCC) are a special group of highly ductile fiber-reinforced concrete-like materials exhibiting a high tensile strain capacity caused by a multiple cracking behavior [1]. Due to this behavior, SHCC is a promising material for structures exposed to dynamic or cyclic loads, such as impact and traffic

M. Kunieda et al. (Eds.): SHCC 2022, RILEM Bookseries 39, pp. 178–187, 2023.
https://doi.org/10.1007/978-3-031-15805-6_19

loading. By now, the quasi-static behavior of SHCC and also partially its dynamic performance [2] are reasonably well investigated and understood whereas there is still a lack of knowledge concerning the cyclic behavior. In the past, several studies were performed on SHCC made of a normal strength matrix and polyvinyl alcohol (PVA) fibers subjected to this type of loading [3, 4]. Multiple degradation mechanisms of the fiber such as abrasion, defibrillation, buckling, squashing or fatigue could be found by the authors by applying different loading scenarios, i.e., pure tension and reversed cyclic loading. Beside these degradation phenomena a decrease in the mechanical properties such as strain capacity and tensile strength was observed with increasing level of maximum compressive stress and decreasing strain increment per cycle [3].

As polyethylene (PE) fibers exhibit a superior mechanical behavior compared to PVA fibers, i.e., higher tensile strength and modulus of elasticity under quasi-static loading [5]. This makes them a promising alternative to PVA fiber also with respect to the cyclic performance. Hence this study should provide first results of the mechanical behavior of PE-SHCC under pure tension and alternating cyclic loading conditions on the macroscopic level of consideration. In case of the pure cyclic tension also the influence of the loading frequency was investigated.

2 Experimental Setup and Testing Program

2.1 Material

The material used in the study was a high-strength PE-SHCC, developed at the Institute of Construction Materials, TU Dresden [5]. Table 1 gives its composition. The mixture consists of Portland cement type CEM I 52.5 R-SR3/NA and micro silica as the binder material. Due to the hydrophobic properties of the PE fibers the micro silica is needed to densify/strengthen the interfacial transition zone (ITZ) and so to enhance the bond between fibers and cementitious matrix. To ensure a proper fiber distribution only very fine quartz sand with particle sizes between 0.06 to 0.2 mm was used. Besides water, a polycarboxylate ether-based superplasticizer BASF Glenium ACE460 was added to achieve appropriate workability.

Table 1. Composition of SHCC under investigation (in kg/m^3)

CEM I 52.5R-SR3/NA	Micro silica ELKEM 971	Quartz sand 0.06–0.2 mm	Superplasticizer BASF Glenium ACE 460	Water	PE fibers 2.0 Vol.-%
1460	292	145	35	315	20

The fiber under investigation is the commercially available PE fiber SK78 from DSM Dyneema® with a length of 12 mm and a diameter of 20 μm, according to the manufacturer. For the experiments a fiber content of 2.0% by volume was used.

2.2 Manufacturing and Preparation

The production of the dumbbell-shaped SHCC specimens was divided into several steps. In the first step, all dry components were mixed for 90 s to ensure a sufficient homogeneity. Subsequently, approximately 80% of the water and superplasticizer was added and the components are mixed for another 60 s. Then, the fibers were gradually given into the vessel and the mixture is stirred until consistency becomes more plastic. Finally, the remaining liquid components were added and the material is mixed for another 60 s at a higher speed.

The fresh SHCC was then filled layer by layer into the dumbbell-shaped steel molds. Due to the filling process the fibers oriented slightly towards the load direction. At the age of 24 h the specimens were demolded and stored at 20 °C and 65% RH for another 55 days.

Since SHCC specimens were produced with some excess of material, the filling side had to be flattened with a grinding machine prior to testing in order to achieve the proper sample thickness. To analyze the crack formation by means of digital image correlation (DIC) a black and white speckle pattern was sprayed onto the sample's surface.

2.3 Testing Program

The experiments were performed on notched dumbbell-shaped SHCC specimens with a notch depth of 10 mm. The notch was used to minimize the number of cracks and to force the crack development in a certain area, so that the degradation of the material was concentrated predominantly one crack plain within the gauge length and not outside of the measuring range. Due to the multiple cracking behavior of SHCC, several cracks nevertheless developed.

The deformation measurement was carried out by two linear variable differential transformers (LVDT) at a gauge length of 100 mm. The testing setup can be taken from Fig. 1b. The load-controlled cyclic uniaxial tension and tension-compression tests were executed at a loading frequency of 1 Hz and 20 Hz (only for pure tension regime). The upper reversal point was set to 80% of the first crack force F_{te} whereas the lower reversal point was 10% of F_{te} or 25% of the maximum compressive force F_c. Prior to the cyclic loading stage, the samples were pre-damaged until a displacement of 0.35 mm of one pre-defined LVDT was reached at a deformation rate of 0.1 mm/s. Due to the uneven crack opening on both sides of the specimen, the displacement on the "non-controlled" side could be much larger than 350 μm. Then, 150,000 loading cycles were performed before the samples were pulled until failure under quasi-static loading conditions at a displacement rate of 0.1 mm/s. Figure 1a displays a schematic overview of the experimental procedure.

The overview of the experimental program and IDs of the series of specimens for the pure cyclic tension and alternating tension-compression tests are given in Table 2.

After a predefined number of loading cycles, images of the specimens were taken by a DSLR camera NIKON Z 6II directly connected to the control of the testing machine.

The morphological analyses were carried out using an environmental scanning electron microscope QUANTA FEG 250.

Table 2. Overview of the experimental program

ID	Frequency [Hz]	Upper force level [kN]	Lower force level [kN]
S_P2H_1	Quasi-static loading regime		
Z2_P2H_1	1	2.7	0.3
Z2_P2H_s	20	2.7	0.3
W1_P2H_1	1	2.7	−16

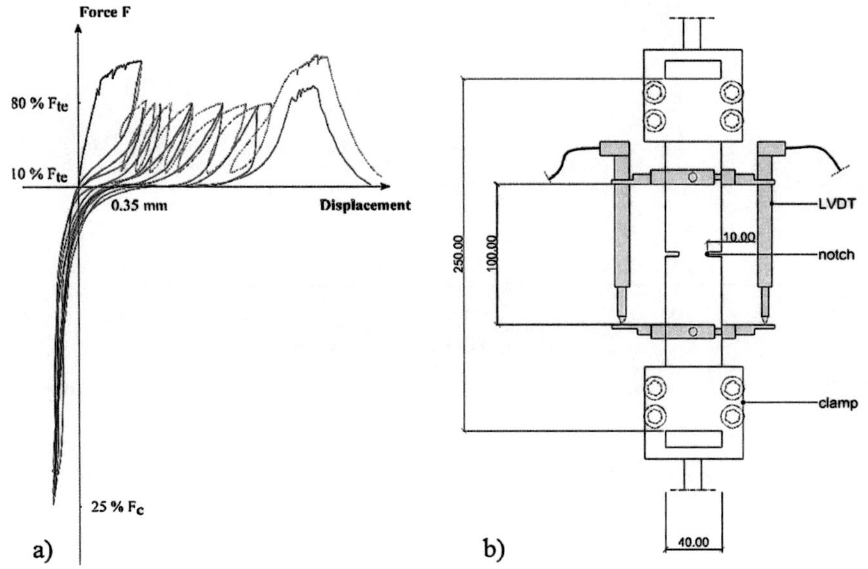

Fig. 1. Cyclic loading experiments. a) schematic representation of the experimental procedure, b) testing setup for the cyclic experiments

3 Test Results

Prior to performing the cyclic tests, uniaxial tension tests on dumbbell-shaped specimens and uniaxial compression tests on cubes ($10 \times 10 \times 10\,\text{cm}^3$) were conducted to determine the compressive strength, the tensile strength and first-crack stress. The material yielded values of −133 MPa, 8.3 MPa and 6.9 MPa, respectively. As a result, the upper force limits for the cyclic loading tests were set to −16 kN and 2.7 kN in the alternating tests and 0.3 kN and 2.7 kN in the swelling tests (pure tension).

Due to structural inertia or misalignment a few specimens could not bear the defined number of loading cycles in pure tension loading regime. These samples are not taken into consideration.

Figure 2 shows the stress–cumulative crack-opening relation for representative specimens of all test series. Cumulative crack-opening describes the deformation of the sample in the area of the notch, i.e. approximately 5 mm above and below the notch, as the sum of the single crack widths in this range.

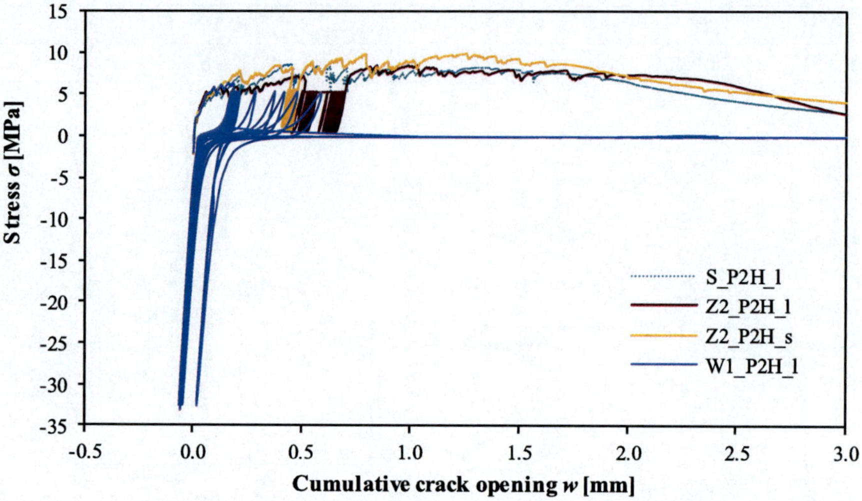

Fig. 2. Stress–crack-opening relationship for pure tension and alternating cyclic loading tests

In terms of better clarity, for cycle number 1 to 1,000 every 100th cycle and for cycles 1,000 to 150,000 every 10,000th cycle is displayed in the diagram in the case of tension-swelling tests. For the specimens tested under an alternating loading regime every 500th cycle is depicted. As can be seen in Fig. 2, the quasi-static pull-out stage after the cyclic pure tension loading leads to a strain-hardening behavior of the specimens with a more or less pronounced increase in tensile stress until their final failure indicating that, independent from the loading frequency, the load-bearing capacity is not exceeded due to the repeated loading. While the samples of the Z2_P2H_l series could achieve strengths of around 8 MPa, the specimens of the Z2_P2H_s series could bear tensile stresses of over 9 MPa. The average strengths can be taken from Table 3.

Table 3. Mechanical properties (standard deviation given in parenthesis)

ID	Ultimate tensile strength σ_{tu} [MPa]	Change in notch stiffness E_n [%]	Plastic strain ε_{pl} [%]
Z2_P2H_l	7.98 (–)	−25.4	2.85 (–)
Z2_P2H_s	9.23 (0.82)	−4.09	0.28 (0.06)
W1_P2H_l	– (–)	– (–)	– (–)

When applying a tension-compression loading regime all of the samples failed in the cyclic loading phase after only a few thousand loading cycles. Hence, no ultimate tensile strength could be measured separately. Rather, the upper tensile stress can be regarded as cyclic tensile strength.

In order to assess the state of degradation of the SHCC samples caused by the repeated loading, the stiffness of the material was calculated following Eq. (1).

$$E = \Delta\sigma/\Delta\varepsilon = (\sigma_{max} - \sigma_{min})/(\varepsilon_{max} - \varepsilon_{min}) \quad (1)$$

Here, the maximum and minimum values of the stress and strain of each single cycle were used. For this purpose, the total displacement of the gauge length is related to the notched area as described above in order to estimate the strain. To evaluate the impact of the loading regime, the change in the material's stiffness in the notched area was determined by subtracting the values of the initial and final loading cycle. The results are also shown in Table 3. Representative stiffness curves are exhibited in Fig. 3.

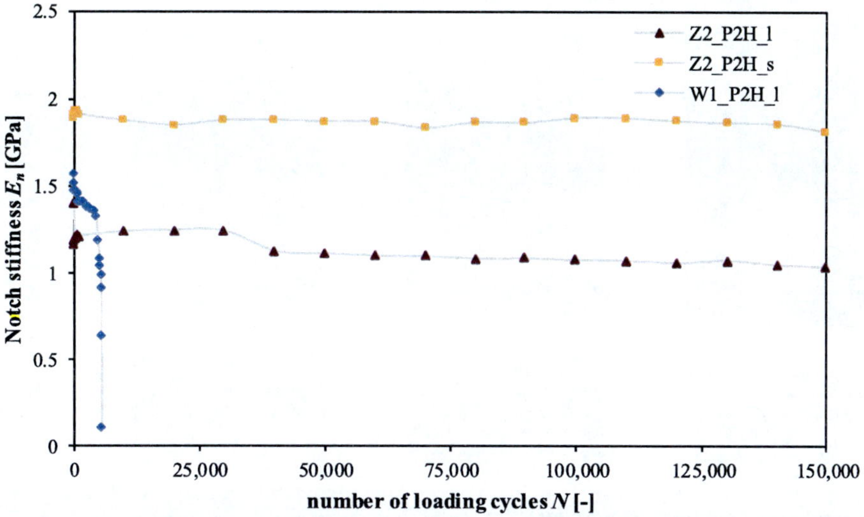

Fig. 3. Development of the notch stiffness E_n over the increasing number of loading cycles

As can be seen from the graphs, the stiffness of the samples tested under tensile swelling loading conditions changes only slightly with increasing numbers of loading cycles. While the low testing frequency leads to a decrease of about 25%, a high testing frequency of 20 Hz causes a much lower reduction of just 4%. In contrast, when applying an alternating loading regime, the stiffness decreases significantly until the samples fail prematurely before the end of the cyclic phase is reached due to a more severe degradation of the composite.

Besides the notch stiffness, also the plastic strain is affected by the applied loading frequency. The plastic strain can be calculated following Eq. (2):

$$\varepsilon_{pl} = \varepsilon_i - \varepsilon_0 \quad (2)$$

Here, ε_i is the total strain in the individual loading cycle. The plastic strain continuously increases over the test duration with varying slope and final strain values; see Table 3. The lower strain rate causes a steeper increase in plastic strain with ultimate strains of 2.85% on average. The higher strain rate leads to lower final strains and a lower increase in plastic strain per cycle. This observation stands in line with previous findings on the influence of the strain rate on the mechanical behavior of SHCC under quasi-static loading conditions [6, 7]. A higher strain rate induces a stronger bond between the fibers and the cementitious matrix leading to a less pronounced increase in strain from one loading cycle to the next, since the fibers are pulled out less compared to a lower frequency. Due to the smaller deformations at higher frequencies the stiffness shows only a slight decrease. The smaller strain coincides with higher ultimate tensile strength as shown above; see Table 3. Since the cracks are completely closed under reversed cyclic loading, no plastic strain is given.

A comparison of the deformations at the end of the cyclic loading stage is depicted in Fig. 4a, b. Note that the majority of cracks developed in the pre-damaging phase.

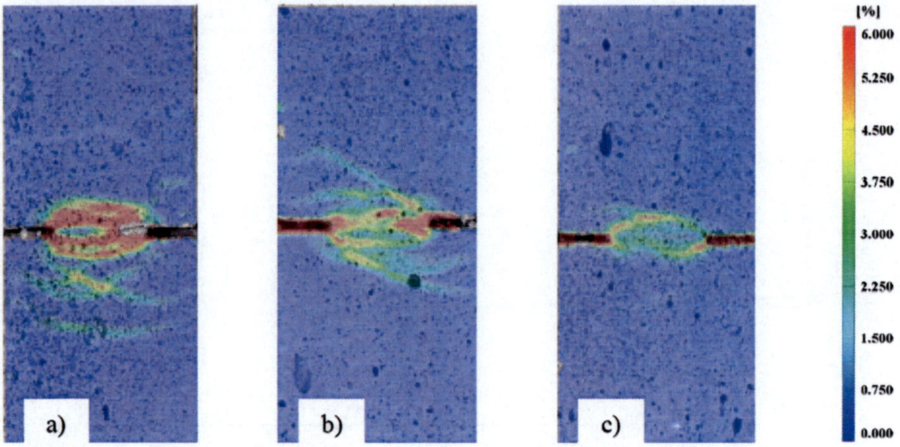

Fig. 4. Photogrammetric analyses of the representative repeatedly loaded SHCC samples: a) Z2_P2H_l, b) Z2_P2H_s, c) W1_P2H_l

As can be seen from the photogrammetric analyses lower strains were measured at higher loading frequencies. While at a frequency of 1 Hz strains of 6% and more were attained, the majority of specimens tested at a frequency of 20 Hz showed strains of maximum 5% which were in line with the measurements by the LVDTs. Previous studies showed that the crack width has a significant influence on the deterioration of the polymer fibers [8] since the free length of the fiber gets larger with increasing crack opening causing the probability of a more pronounced buckling and squeezing when closing the crack. Note that despite the pure tension loading regime, local compressive stresses may occur between the crack flanks, due to loose matrix particles and other derangements. Furthermore, the images reveal that the cracks develop also in unnotched areas of the sample. While the origin of the cracks in the Z2_P2H_s series is located at the notch, a few cracks in the Z2_P2H_l series also develop outside the notched region.

When the loading regime shifts from pure tension to alternating tension-compression loading conditions, the samples show only very few cracks; see Fig. 2 and Fig. 4c. It should be mentioned that the photogrammetric presentation of the W1_P2H_l series indicate the correct number of cracks; however, in contrast to the other testing series the individual photos were not taken at the maximum force. Thus, the strains are not correctly reproduced as the cracks are not completely opened. From the stress-strain relationship it can be seen, that the displacement is much larger than in the pure cyclic tension loading regime indicating a wider opening of the individual cracks. These larger crack widths lead to a more severe degradation of the fibers as stated before causing a rapid decrease in the stiffness values and finally in a premature failure.

These assumptions can be supported by the morphological analyses of the crack surfaces; see Fig. 5. The images reveal that in the testing series Z2_P2H_l and Z2_P2H_s most of the fibers are simply pulled out of the matrix in the post-cyclic loading stage without critical signs of deterioration except some peeled-of fibrils. Nevertheless, there seems to be areas located near to the notch that are exposed to higher loads as a higher degree of degradation could be observed. Instead of being pulled out some fibers tend to rupture; cf. Fig. 5a, b.

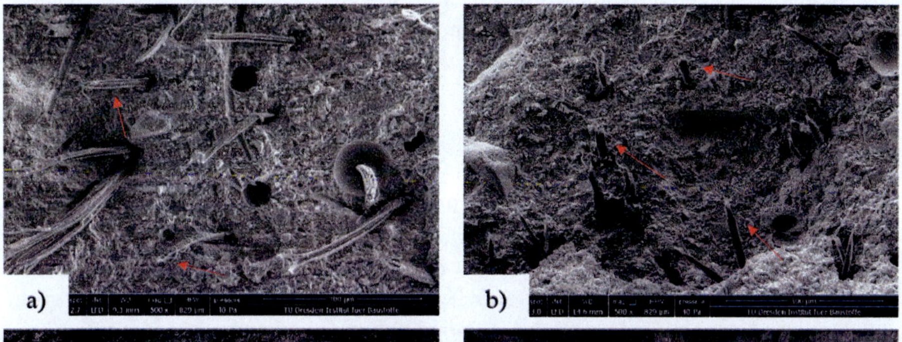

Fig. 5. ESEM images of the crack surfaces of the tested samples: a) Z2_P2H_l, b) Z2_P2H_s, c) W1_P2H_l; d) mashed polyethylene fibers in reversed cyclic loading

A closer look at the fiber surfaces reveals buckling or bumping of the fiber due to repeated push-in of the fiber in the unloading stage of the cycle or a partly abrasion of the fiber surface material caused by the gradual pull-out in the tension phase. Despite occasional micro cracks and spalled matrix particles caused by the pull-out of inclined fibers the cementitious matrix seems to be in good condition.

In the reversed cyclic loading series two degradation processes can be determined that stand in line with the findings from the mechanical tests. As can be seen in Fig. 5c, the fibers are heavily grinded between the crack faces causing a total defibrillation and finally the rupture of the polymer fibers. There are just remains of the fibers visible at the edge of the crack that are exemplary pointed out above. In addition to the fiber degradation the matrix shows a high extend of loose particles caused by the forceful pressing of the crack faces. These phenomena could be detected in previous studies too [3]. Beside the heavy crushing of the fibers another observation could be made. The ESEM images show also large regions where the fibers seem to be completely pulled out of the matrix and then buckled and squashed between the crack faces. Hence, the fibers show big bumps and are partly mashed up to a single polyethylene lump; see Fig. 5d.

It is assumed that this phenomenon is related to the above mentioned completely deteriorated regions of the sample. The high deterioration results in higher stresses to be carried by the remaining fibers. To reach the defined upper force level the fibers are gradually pulled out until it is impossible to transfer the forces anymore. So the fibers are completely pulled out by the testing machine as can be seen in Fig. 2 as the machine control tries to reach the defined force limit. Subsequently, the fibers are crushed until the test is stopped.

4 Conclusions

The paper at hand focuses on the cyclic behavior of strain-hardening cement-based composites (SHCC) under pure tension and alternating tension-compression loading regimes. In case of tension-swelling tests the influence of the frequency was also studied. To investigate the cyclic performance of SHCC, after pre-damaging the samples a cyclic loading phase with 150,000 cycles was applied before the specimens were pulled apart under quasi-static loading conditions.

When applying a pure tension cyclic regime, the samples could bear the defined number of loading cycles with less critical signs of deterioration. A few fibers failed in the cyclic phase caused by ongoing abrasion or buckling whereas the majority of fibers were pulled out in the quasi-static loading stage.

The tests showed that the application of compressive forces in the cyclic tests leads to a more severe deterioration of the material compared to the pure tension regime caused by the repeated crushing of the fibers between the initial crack faces in the compressive phase of the loading cycle. This increasing fiber degradation, i.e., fiber rupture and buckling, leads to higher strains than in pure cyclic tension regime and a premature failure.

When increasing the loading frequency, the strain of the samples decreases due to a higher bond between the fibers and the surrounding cementitious matrix. Smaller strain leads to a smaller decrease in stiffness resulting in higher ultimate tensile strengths after the cyclic loading stage. Nevertheless, the degradation phenomena are very similar. In the region next to the tip of the notch some fibers are more likely to rupture while the majority of the fibers tend to be pulled out in the quasi-static pull-out phase.

The experiments and analyses showed that the cyclic behavior and degradation of PE-SHCC strongly depends on the loading history. Based on the test data the composite shows apparently a higher resistance against the distinct loading regimes in comparison to PVA-SHCC studied in previous research. Nevertheless, the observations have to be proofed by further tests on the microscopic and macroscopic levels of observation.

Acknowledgements. The authors would like to express their gratitude to the German Research Foundation (DFG) for the financial support within the scope of Priority Program 2020 "Cyclic deterioration of High-Performance Concrete in an experimental-virtual lab", project number 352324592.

References

1. Li, V.: From micromechanics to structural engineering-the design of cementitious composites for civil engineering applications. J. Struct. Mech. Earthquake Eng. **10**(2), 37–48 (1993)
2. Curosu, I., Mechtcherine, V., Forni, D., et al.: Performance of various strain-hardening cement-based composites (SHCC) subject to uniaxial impact tensile loading. Cem. Concr. Res. **102**, 16–28 (2017)
3. Müller, S., Mechtcherine, V.: Fatigue behaviour of strain-hardening cement-based composites (SHCC). Cem. Concr. Res. **92**, 75–83 (2017)
4. Ranjbarian, M., Mechtcherine, V.: Cyclic Damage to PVA microfibre embedded in cementitious matrix in alternating tension-compression regime. In: FRC2018: Fibre Reinforced Concrete: from Design to Structural Applications, pp. 321–329, Desenzano, Lake Garda, Italy (2018)
5. Curosu, I., Liebscher, M., Mechtcherine, V., et al.: Tensile behavior of high-strength strain-hardening cement-based composites (HS-SHCC) made with high-performance polyethylene, aramid and PBO fibers. Cem. Concr. Res. **98**, 71–81 (2017)
6. Mechtcherine, V., Silva, F. de A., Butler, M., et al.: Behaviour of strain-hardening cement-based composites under high strain rates. ACT **9**(1), 51–62 (2011)
7. Boshoff, W.P., Mechtcherine, V., van Zijl, G.P.A.G.: Characterising the time-dependant behaviour on the single fibre level of SHCC: Part 2: the rate effects on fibre pull-out tests. Cem. Concr. Res. **39**(9), 787–797 (2009)
8. Ranjbarian, M., Ma, X., Mechtcherine, V.: Influence of crack width in alternating tension-compression regimes on crack-bridging behaviour and degradation of PVA microfibres embedded in cement-based matrix. Materials **13**(18), 4189 (2020)

A Novel Deep Learning Model for End-to-End Characterization of Thin Cracking in SHCCs

Avik Kumar Das[1,2](✉) and Christopher K Y Leung[2]

[1] Shenzhen International Graduate School, Tsinghua University, Shenzhen, China
akdas@connect.ust.hk

[2] Hong Kong University of Science and Technology, Hong Kong, China

Abstract. Strain hardening cementitious composites (SHCCs), can be designed to exhibit small crack widths. As a result, even after the material is cracked it restricts the flow of deleterious material. This property among many other properties makes it useful to design durable infrastructures. Assessment of durability of SHCCs requires measurements of surface cracks. The process of development of an SHCC mix for any application thus involves the characterization of cracks to understand its durability component at different strain values. Conventionally, this can be done by using digital cameras to document the images of SHCC surfaces during testing and manually analyzing these images to compute different crack characteristics such as width. This process is laborious and time-consuming thus, non-scalable if the number of examples is large. In this work, we designed a novel deep learning model for this purpose called Strain Hardening Segmentation Network (SHSnet). SHSnet has a very high accuracy of 85% while requiring ~ 4M parameters which is one order less than other state-of-the-art networks like U-net. Due to the inherent thinness of the cracks in SHCCs, the amount of examples for training is fairly low (<1%). To this end, we proposed a loss function (PLF) to train the network efficiently. With this, it takes at least 10x less time for computing crack parameters. We further show applications of the SHSnet empowered technique for durability assessment and detecting crack evolution. These results suggest that SHSnet has the potential to facilitate the autonomous characterization of SHCC cracks in many scenarios.

Keywords: End-to-end segmentation · Deep learning · Strain hardening cementitious composites · Vision-based durability monitoring

1 Introduction

In Strain hardening cementitious composites (SHCCs) cracks can be controlled to small values [1] which lowers the transport of deleterious material [2–4] and promotes autogenous self-healing [5–7]. The utilization of SHCC had been seriously contemplated to enable durable civil infrastructure to exploit these beneficial properties [8–11]. The experimental studies on the durability measurement for multiple-thin SHCC cracks are summarized in [12]. The results of such studies unanimously agree that surface cracks of SHCCs (or statistics of them) are directly correlated to the transportation of materials [12].

M. Kunieda et al. (Eds.): SHCC 2022, RILEM Bookseries 39, pp. 188–198, 2023.
https://doi.org/10.1007/978-3-031-15805-6_20

1.1 Related Work

Over the years, various techniques with different principles such as stress waves [13–18], electrical properties [7, 19, 20], and vision-based systems [21, 22] were developed to monitor the damage in SHCCs [23]. In our understanding, the vision-based technique is the most widely adopted because of the simple implementation as well as direct and easy result interpretation. Zijl et al. [21] surveyed the utilization of non-contact in-situ vision-based documentation methods of SHCC cracks across various laboratories around the globe. Survey results show that digital image correlation or digital image processing are the two most common technologies utilized for this purpose. The former requires extensive instrumentation which is not always implementable however, the latter required extensive manual labor for processing from digital images collected during the testing [21]. Apart from these digital techniques, the crack scope could also be utilized to measure cracks accurately but involves measurement of surface cracks of SHCCs manually [24]. These methods could produce accurate results but they are laborious, time-consuming, and may not be suitable for analyzing a large number of samples.

2 Proposed Method

2.1 Background

We define hypothesis space(H) is a set of all the hypothesis which governs the segmentation of these cracks. Since cracks in SHCCs can be segmented from images manually by experts, therefore, H must be finite. Rather than waiting to discover all these complex hypotheses an (attractive) alternative way would be to 'learn' these hypotheses. Haussler's Theorem [25] states that the probability that a *consistent learner* cannot learn our Hypothesis Space (H) from randomly drawn D examples such that the number of examples ((m) > 1) with a known concept is extremely small ($\leq |H|e^{-\epsilon m}$). Here, $0 < \epsilon < 1$ is a learning error. It is then, possible to show that (Eq. 1) error of a *learner* is related to the number of training examples and expected probability of confidence (δ).

$$error = \frac{1}{m} * (\ln\left(\frac{1}{\delta}\right) + \ln(H)) \quad (1)$$

It is intuitive, (Eq. 1) for a given complexity of the problem (size of H) with a fixed number of training examples the consistency of a learner determines the error. In our previous study [22] found that deep convolutional neural network (CNN) are most consistent in learning cracks in the images. We use these results as the guiding principle for network design.

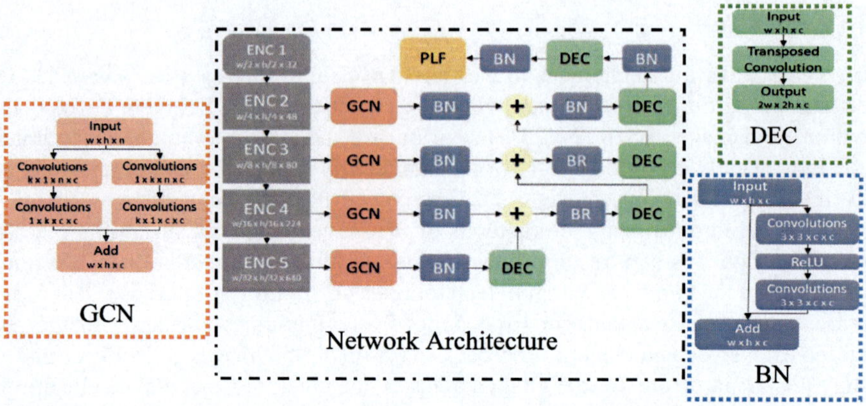

Fig. 1. Architecture of the proposed SHSnet for segmentation of thin-tortuous-dense cracks in SHCCs. Here, w, h, n, c is width, height, no. of channels and no. of classes respectively

2.2 Network Design of "Strain Hardening Segmentation Network" (SHSnet)

For end-to-end pixel-wise segmentation, previous research has established that a special design of CNN called encoder-decoder models is the most suitable solution [26–29]. In such a network, the encoder encodes (ENC) images to low-resolution feature spaces. The role of the decoder (DEC) is to map the low-resolution encoder feature maps to full input resolution feature maps for pixel-wise classification. This will ensure that each pixel in the input layer would have a corresponding pixel in the output layer.

The proposed network-SHSnet is shown in Fig. 1. As will be shown later; in this work, the deep learning model has to deal with diversity in testing conditions, material development, and intractability of cracks due to the formation, and propagation interaction of high-density cracks. A 'deeper' encoder (ENC) is therefore employed but requires large computational resources, for example, U-net has ~ 30 million parameters that should be trained and computed. The computational resource is a bottleneck in material research. We use blocks from EfficientNet [30] which is recently developed by google to adopt a network design that attempts to optimize parameter size to maximize learning [30]. Global convolution networks (GCN)[31] units are then employed here to implement a large kernel that allows a large receptive field to facilitate segmenting the tortuous cracks. To facilitate creating a better delineation of the boundaries of thin cracks a boundary refinement (BN) network (see Fig. 1) is used after GCN. The results at each scale were then up-sampled gradually with a decoder (DEC) as shown in Fig. 1. In comparison to U-net, SHSnet only requires 4 million i.e. one order fewer parameters.

2.3 Proposed Novel Loss Function (PLF)

Development of loss functions that could guide in 'learning' the features correctly is thus, important. Because of the thinness of the crack, the number of pixels of each crack is less than 0.01% percent of the total area of the specimen. Such a dataset is an example of a highly imbalanced dataset. To facilitate learning in this dataset, PLF was implemented (Eq. 2). In PLF, irrespective of the number of pixels; each class (i.e. a crack or background) is treated equally with β to give appropriate weight to false positives and negatives.

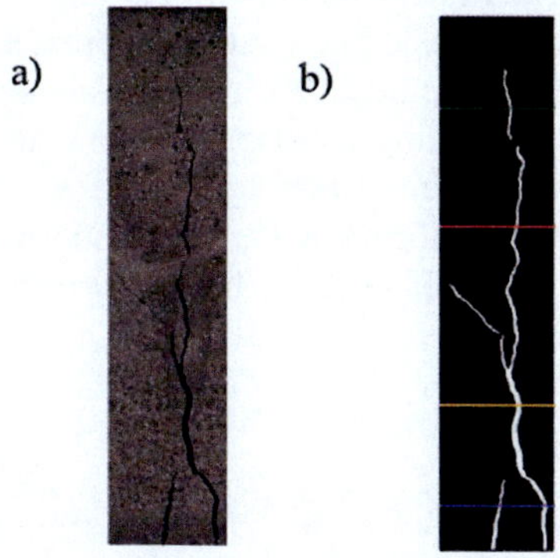

Fig. 2. a) Typical image for segmentation b) Binary image with some examples of scanning lines

$$PLF = 1 - \frac{\sum_{i=1}^{N} \rho_i^c G_i^c}{\sum_{i=1}^{N} \rho_i^c G_i^c + (1-\beta)\sum_{i=1}^{N} \rho_i^c G_i^{\hat{c}} + \beta \sum_{i=1}^{N} \rho_i^{\hat{c}} G_i^c} \tag{2}$$

Here, N is the total number of pixels in the image, G is the ground truth of the image c

2.4 Crack Computation from Segmentation Mask

At first, each captured (RGB) image (Fig. 2a) was converted to a binary image (Fig. 2b where binary values represent different classes) using SHCnet. Then scanning lines parallel to the tensile stress (Fig. 2b) [22] was used to find the change points in the binary images. Here, the change point is the location classes of the pixel changes. Since each crack would lead to 2 change points (i.e. changing from non-crack to crack and vice versa) therefore, the crack number (at that scanning line) is half of the number of such change points. The width of such change is used to represent the crack width. Since each crack is a thin element for such an element crack length is estimated as half of the perimeter of the crack.

3 Data Collection and Training

We have collected ~ 1000 images from various tests of SHCC samples. The total dataset is divided into training, validation, and test sets with the following ratio-0.8 0.1, and 0.1 respectively. During training, the images were cropped randomly into the resolution of 224×224×3 and trained in a batch size of 32. The Adam optimizer with a learning rate-1e-3 was employed in this study to update the weights of the network after each iteration. Training is deemed to be complete when for next the 10 epochs no improvement in network performance is observed. NVIDIA GPU-2080 graphical processing unit (GPU) was employed to train the neural network.

4 Results and Applications

4.1 Training Ability

Typically, a neural network has a large number of 'learnable'. Sometimes, when a novel architecture is designed and applied to a new dataset, there is a possibility of the gradients computed is either vanished or exploded. As a result, the network may stop 'learning'. This can be observed as 'loss' being constant with low accuracy during the training phase. Figure 3 shows the evolution of loss and accuracy during the initial phase with each training iteration. Results confirm that the proposed network architecture is indeed capable of 'learning' from the SHCC dataset. The performance of SHCnet post-training shows 85% accuracy.

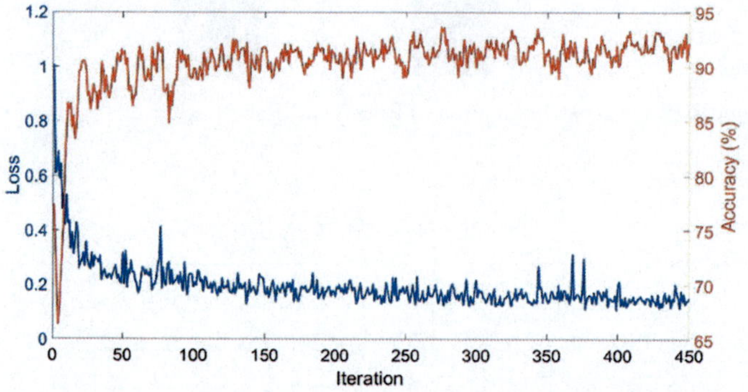

Fig. 3. Evolution of loss and accuracy of SHCnet during first 450 interations of training

4.2 Comparison of the Quality of Crack Parameters

The mechanical properties are determined from stress-strain curves of SHCCs whereas measurement of durability properties requires quantification of cracks. In this section, the study is performed to understand the performance of the proposed network in the estimation of these crack characteristics on the test set. As discussed before for each pixel in the image the proposed network outputs is whether this pixel is a crack or a not-crack. The resultant final image is a binary. These binary images were further analyzed to estimate crack width, crack number and crack length following the procedure in Sect. 2. To evaluate the performance of these computed characteristics, original images were also analyzed manually to compute these crack characteristics. It should be noted the other variables such as maximum crack width, crack map, and crack density are composites of these primary crack features therefore, they are not explored any further. To compare the crack width. Values are those crack widths that are predicted by the proposed deep learning technology and are compared with the width of the same crack computed manually. A total of 56000 such comparisons were made. Figure 4 reports the result, with the y-axis reporting the predicted crack width of cracks with widths on the x-axis. Following these exact steps, we also compare the results for crack number and crack length. These results are reported in Fig. 5 and Fig. 6 respectively. Based on these results it is possible to conclude that in general, these important and primary crack characteristics computed through predicted technology do not deviate too much from those computed manually.

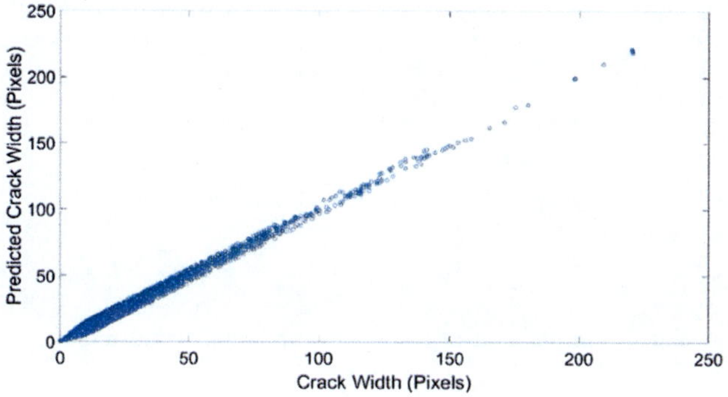

Fig. 4. Comparison of predicted crack width with original ones

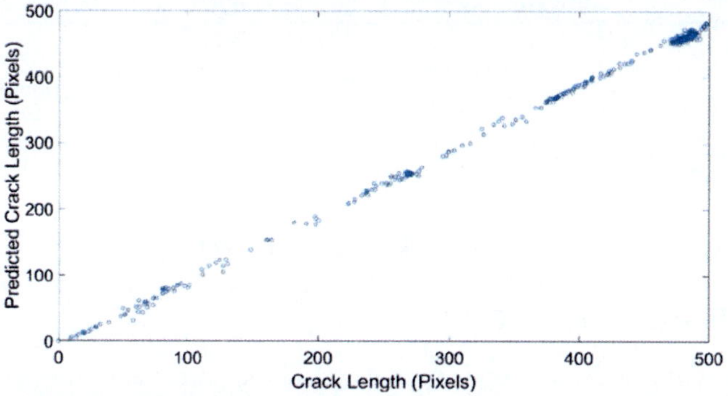

Fig. 5. Comparison of predicted and original crack lengths

4.3 Comparison of Time in Compute Crack Parameters

Research time is scarce. To understand the impact of the proposed technique, we also monitored the total time it takes to densely (for every crack at every location) estimate various parameters a) analyzing the image through the SHSnet, b) analyzing these pictures manually (Manual). Table 1 reports our findings for a typical 20MP image of SHCC crack. It is clearly seen that the proposed method significantly reduces the time complexity for computing various crack parameters. It should be noted that parameters crack width and length takes a very long time because typically SHCCs have a large number of very thin cracks. And because of high density, it may be hard to account for the changing lengths or widths, as a result, it is laborious and time-consuming and requires a large amount of manual effort, unlike our method.

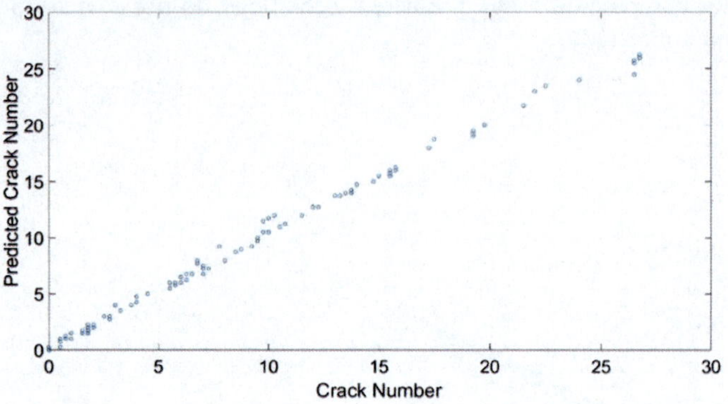

Fig. 6. Comparison of predicted with orginal crack number

Table 1. Measurement of temporal efficiency

Page	Crack Number	Crack Length	Crack Width
With SHSnet	0.3–2 min		
Manual	~6 min	~100 min	~200 min

5 Applications of SHSnet for Studies of SHCC

5.1 Crack Development in SHCC

The main objective of the section is to understand the potential of the *SHSnet* enabled technique to automatically calculate durability during progressive damage. The sample design and experimental process are the same as in our previous work [32, 33]. As shown in Fig. 7 the surface was photographed with a high-resolution 20MP DSLR camera (Canon EOS D6 II). As demonstrated previously, *SHSnet* can accurately identify the cracks in these photographs at different levels of damage. The development of cracks through certain keyframes for the sample is reported in Fig. 8a-e. The results indicate that with the increasing strain the crack density increases as was expected. Moreover, the results show that *SHSnet* is applicable for crack segmentation for SHCCs for a wide range of strain levels and crack density.

5.2 Monitoring of Durability

Generally speaking, the transfer of deleterious materials within the concrete is associated with the deterioration of concrete. Transfer of deleterious materials could take place by

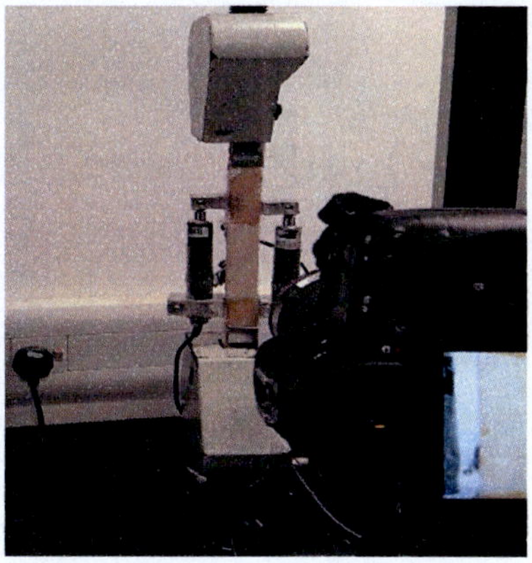

Fig. 7. Tension test setup for SHCC

a) 0.6% b) 3% c)5.4%

Fig. 8. Development of cracks monitored by SHSnet for different strain value; Cracks (marked in red) is overlaid on the original image

diffusion. Intuitively, the diffusion coefficient (d) of the cracked SHCC specimen is a piecewise function of the product of crack width (cw), and length (cl) (Eq. 3).

$$d = \sum K_i cw_i cl_i \tag{3}$$

Here, K is a non-linear multiplier function of crack width.

In a strict mathematical sense, the approximate diffusion coefficient (dc) (Eq. 4) will be greater than the d.

$$d_c = K_{avg} cw_{avg} ncl_{per\ crack} \geq d \tag{4}$$

Here, n is the number of cracks.

It is thus important to measure crack number, crack length (per crack), and crack widths (average) at different strain levels to understand the durability of SHCCs. Figure 9

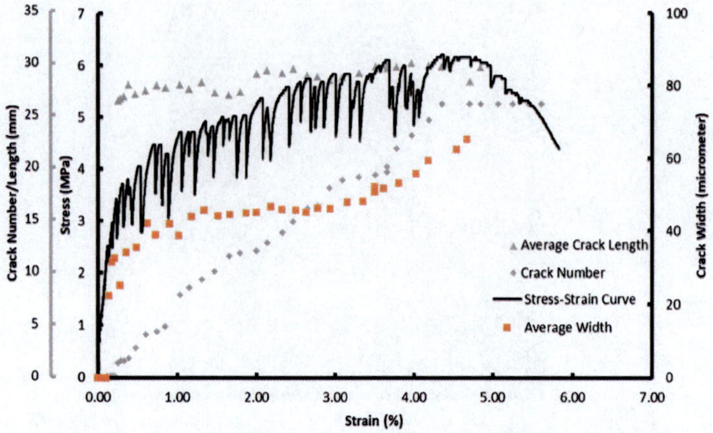

Fig. 9. Evolution of crack morphological parameters with strain

presents the result of the evolution of these cracks' morphological parameters for approximate diffusion coefficient computed with *SHCnet* at different strain values. With increasing stress the number of cracks on the sample increases. Usually, cracks under tensile load propagate to the edges of the sample as a result no significant trend in the crack length was found with applied strain. However, in all cases due to the inherent tortuosity of the cracks, on average the lengths of the cracks can be ~30% higher than the width of the sample. Crack widths (and their statistics average and maximum) show an overall increasing trend with applied strain. This behavior is expected as with increasing strain the fiber bridging stress increases leading to increased crack width.

6 Conclusions

Strain hardening cementitious composites (SHCCs) are being used in many structural applications to create more durable infrastructures. The durability of SHCCs is associated with the characteristics of surface cracks; thus, characterizing the surface crack is an essential part of the material development process. Due to the inherent complexities-high density, thin crack widths, and tortuosity of cracks in SHCCs; characterizing them is essential but is laborious and time-consuming. In this work, we showcased how a 'learning' model could be utilized to minimize such manual effort. In this work, we have assembled a deep learning model - SHSnet to 'learn' to segment SHCC cracks. It was 85% accurate and requires one order fewer parameters than conventional models while saving >95% time in computing crack parameters. It can also facilitate in-situ visualizing visualization of damage progression and automated durability monitoring.

References

1. Li, V.C.: Engineered Cementitious Composites (ECC): Bendable Concrete for Sustainable and Resilient Infrastructure, 1st ed. (2019) https://doi.org/10.1007/978-3-662-58438-5

2. Wang, K., Jansen, D.C., Shah, S.P., Karr, A.F.: Permeability study of cracked concrete. Cement Concr. Res. (1997) https://doi.org/10.1016/S0008-8846(97)00031-8
3. Djerbi, A., Bonnet, S., Khelidj, A., Baroghel-Bouny, V.: Influence of traversing crack on chloride diffusion into concrete. Cement Concr. Res. **38**(6) (2008) https://doi.org/10.1016/j.cemconres.2007.10.007
4. Lepech, M.D., Li, V.C.: Water permeability of engineered cementitious composites. Cement Concr. Compos. **31** (2009) https://doi.org/10.1016/j.cemconcomp.2009.07.002
5. Herbert, E.: Development and Application of Self-healing Engineered Cementitious Composites (ECC) for Durable and Sustainable Infrastructure, PhD (2016)
6. Li, V., Herbert, E.: Robust self-healing concrete for sustainable infrastructure. J. Adv. Concr. Technol. **10**(6) (2012) https://doi.org/10.3151/jact.10.207
7. Das, A.K., Leung, C.K.Y.: A strategy for in situ determination of self-healing state for strain hardening cementitious composites. Cement Concr. Compos. **112**, 103641 (2020). https://doi.org/10.1016/j.cemconcomp.2020.103641
8. Lepech, M.D., Li, V.C.: Design and field demonstration of ECC link slabs for jointless bridge decks (2005) https://www.michigan.gov/documents/mdot/MDOT_Research_Report_RC1471_200102_7.pdf
9. Li, V.C., et al.: Durable link slabs for jointless bridge decks based on strain-hardening cementitious composites (2003)
10. Rokugo, K.: Applications of SHCC in Japan - Tools and Tips for Promoting its Use (2017)
11. Rokugo, K., Kunieda, M., Lim, S.C.: Patching repair with ECC on cracked concrete surface (2005)
12. van Zijl, Gideon P.A.G., Slowik, V.: A Framework for Durability Design with Strain-Hardening Cement-Based Composites (SHCC): State-of-the-Art Report of the RILEM Technical Committee 240-FDS, vol. 22 (2017) https://doi.org/10.1007/978-94-024-1013-6
13. Das, A.K., Leung, C.K.Y.: ICD: A methodology for real time onset detection of overlapped acoustic emission waves. Autom. Constr. **119**, 103341 (2020). https://doi.org/10.1016/j.autcon.2020.103341
14. Das, A.K., Suthar, D., Leung, C.K.Y.: Machine learning based crack mode classification from unlabeled acoustic emission waveform features. Cem. Concr. Res. **121**, 42–57 (2019). https://doi.org/10.1016/j.cemconres.2019.03.001
15. Yang, Y., Lepech, M.D., Yang, E., Li, V.C.: Autogenous healing of engineered cementitious composites under wet-dry cycles. Cement Concr. Res. **39**(5) (2009) https://doi.org/10.1016/j.cemconres.2009.01.013
16. Das, A.K., Leung, C.K.Y.: Fast Tomography: A greedy, heuristic, mesh size–independent methodology for local velocity reconstruction for AE waves in distance decaying environment in semi real-time. Struct. Health Monit. (2021) https://doi.org/10.1177/14759217211036881
17. Das, A.K., Mishra, D.K., Yu, J., Leung, C.K.Y.: Smart Self-Healing and Self-Sensing Cementitious Composites —Recent Developments, Challenges and Prospects (2019) https://doi.org/10.1520/ACEM20190023
18. Das, A.K., Leung, C.K.Y.: A new power-based method to determine the first arrival information of an acoustic emission wave. Struct. Health Monit. (2018) https://doi.org/10.1177/1475921718815058;
19. Hou, T., Lynch, J.P.: Electrical impedance tomographic methods for sensing strain fields and crack damage in cementitious structures . J. Intell. Mater. Syst. Struct. **20**(11)(2009) https://doi.org/10.1177/1045389X08096052
20. Ranade, R., Zhang, J., Lynch, J.P., Li, V.C.: Influence of micro-cracking on the composite resistivity of Engineered Cementitious Composites. Cem. Concr. Res. **58**, 1–12 (2014)
21. Zijl, G.P.A.G., Slowik, V., Filho, R.D.T., Wittmann, F.H., Mihashi, H.: Comparative testing of crack formation in strain-hardening cement-based composites (SHCC). Mater. Struct. **49**(4), 1175–1189 (2015). https://doi.org/10.1617/s11527-015-0567-9

22. Das, A.K., Leung, C.K.Y., Wan, K.T.: Application of deep convolutional neural networks for automated and rapid identification and computation of crack statistics of thin cracks in strain hardening cementitious composites (SHCCs). Cement Concr. Compos. **122**, 104159 (2021). https://doi.org/10.1016/j.cemconcomp.2021.104159
23. Das, A.K.: Development of techniques for monitoring of various aspects of strain hardening cementitious composites (2020) https://doi.org/10.14711/thesis-991012879963603412
24. Schröfl, C., Mechtcherine, V., Kaestner, A., Vontobel, P., Hovind, J., Lehmann, E,: Transport of water through strain-hardening cement-based composite (SHCC) applied on top of cracked reinforced concrete slabs with and without hydrophobization of cracks – Investigation by neutron radiography. Constr. Build. Mater. **76**, 70–86 (2015). https://doi.org/10.1016/j.conbuildmat.2014.11.062
25. Haussler, D.: Decision theoretic generalizations of the PAC model for neural net and other learning applications. Inf. Comput. **100**(1), 78–150 (1992). https://doi.org/10.1016/0890-5401(92)90010-D
26. Lecun, Y., Bengio, Y., Hinton, G.: Deep learning. Nature **521**(7553) (2015) https://doi.org/10.1038/nature14539
27. He, K., Zhang, X., Ren, S., Sun, J.: Deep Residual Learning for Image Recognition, vol. 2016 (2016) https://doi.org/10.1109/CVPR.2016.90
28. Goodfellow, I.: Deep learning (2016)
29. Szegedy, C., et al.: Going deeper with convolutions. In: Proceedings of the IEEE Computer Society Conference on Computer Vision and Pattern Recognition, vol. 07–12 (2015) https://doi.org/10.1109/CVPR.2015.7298594
30. Tan, M., Le, Q.V.: EfficientNet: Rethinking Model Scaling for Convolutional Neural Networks (2019)
31. Peng, C., Zhang, X., Yu, G., Luo, G., Sun, J.: Large Kernel Matters -- Improve Semantic Segmentation by Global Convolutional Network (2017)
32. Das, A.K., Leung, C.K.Y.: Power spectral entropy of acoustic emission signal as a new damage indicator to identify the operating regime of strain hardening cementitious composites. Cement Concr. Compos. **104**, 103409 (2019). https://doi.org/10.1016/j.cemconcomp.2019.103409
33. Das, A.K., Leung, C.K.Y.: A fundamental method for prediction of failure of strain hardening cementitious composites without prior information. Cement Concr. Compos. **114**, 103745 (2020). https://doi.org/10.1016/j.cemconcomp.2020.103745

Parametric Modeling of Flexural Response of Sandwich Composites

Chidchanok Pleesudjai[1] and Barzin Mobasher[2](✉)

[1] Graduate Research Assistant, School of Sustainable Engineering and the Built Env., Arizona State University, Tempe, AZ 85287, USA
[2] School of Sustainable Engineering and the Built Env., Arizona State University, Tempe, AZ 85287, USA
barzin@asu.edu

Abstract. The mechanical response of textile-reinforced aerated concrete sandwich panels was modeled for flexural loading. The core material used in the sandwich composite consisted of plain autoclaved aerated concrete (AAC) and fiber-reinforced aerated concrete (FRAC). The stress skins for the sandwich beams were made out of AR-glass (ARG) textiles embedded in a cementitious matrix. A constitutive material model comprised of a multi-linear tension model for the bottom stress skin and an elastic-perfectly plastic compression model for the top stress skin. The core was modeled using an elastic-perfectly plastic tension and compression model. A detailed parametric study was conducted to determine the effect of the model parameters on the flexural response of the sandwich composite. The model was further applied to simulate the experimental flexural data from the static tests on sandwich composite beams. Flexural strength, stiffness, and energy absorption capacity can be determined for both static loadings. It is observed that textile reinforcement at the tension and compression faces of the beam element results in a ductile behavior using multiple flexural cracking, leading to diagonal tension cracking in the core element. The distributed cracking mechanism in the sandwich composite significantly improves the flexural properties by 5–10 times when compared to the plain aerated concrete specimens which predominantly exhibit single flexural cracks.

Keywords: Aerated concrete · Textile reinforced concrete · Sandwich composite · Impact · Digital image correlation · Back-calculation · Inverse analysis

1 Introduction

Sandwich construction has potential for wall and roof elements since the allows for higher flexural capacity, longer spans, ductile composites, and thermally efficient elements with a low high specific strength. The skin is primarily involved in strength, stiffness, and load carrying capacity, whereas the core, carries the shear stresses as well as provides a thermal barrier and thermal mass [1–3].

M. Kunieda et al. (Eds.): SHCC 2022, RILEM Bookseries 39, pp. 199–208, 2023.
https://doi.org/10.1007/978-3-031-15805-6_21

Aerated concrete (AC) or its autoclaved version (AAC) are manufactured from portland cement, fly ash, quick lime, gypsum, water, and an expansive agent such as aluminum powder [4]. Chemical reactions of calcium hydroxide and aluminum paste generate internal hydrogen gas which expands the volume by producing a porous structure with a dry density of 400–800 kg/m^3 and compressive strength values of 2–6 MPa as common for aerated concrete products [2]. Cellular structures exhibit a considerable amount of residual strength after reaching the peak strength in compression which results in high ductility associated with pore crushing. The flexural and tensile strength are extremely low, therefore the structural applications and limited to block residential construction of one or two-story buildings. To improve ductility, Fiber Reinforced Aerated Concrete (FRAC) are manufactured with 0.5% volume of short polypropylene fibers [5]. Results show that fibers affect the post-cracking mechanical characteristics significantly [6]. The crack bridging by the fibers allows for the use of an elastic–quasi plastic design approach and applications for sandwich composites made with TRC [7].

Cementitious sandwich composites with ductile and brittle components of core and skin elements are used with textile reinforced concrete (TRC) as skin and AAC or FRAC as the core. The use of the tensile strength and stiffness of TRC as the skin element significantly improves the characteristics of the overall system. Flexural tests under quasi-static and low-velocity impact were conducted on sandwich composites with FRAC and AAC as a core element and TRC with alkali-resistant glass (ARG) textile as the skin [6, 7]. Beam specimens with different cross-sections and varying drop heights were tested and modes of failure recorded.

The available analysis techniques include the work by Cuypers & Wastiels [3] who studied the behavior of sandwich panels with TRC faces using a finite element analysis model using ANSYS [8] while Djamai et. al. [9] studied sandwich TRC composites using Abaqus FEM software. Chudoba et al. also presented innovative ways to use TRC materials in structural elements [10]. Multi-parameter constitutive models have been presented to simulate the flexural response of sandwich composites [11]. A tri-linear tension and bi-linear compression model was used to model the behavior of the skin layer, and a bi-linear elastic-perfectly plastic tension and compression model was used for the core properties. This approach enables the development of a predictive tool as the design basis of sandwich composites for structural applications.

2 Skin and Core Material Properties

The basic material properties are defined in the function of two intrinsic parameters, first cracking tensile strain, $\varepsilon_{cr,}$ and Young's modulus, E of the core material. Equations 1 and 2 present a constitutive model of core material in tension and compression as described as a bi-linear segment shown in Fig. 1(a) The normalized tensile strain parameter is defined as the independent variable β , and when $\beta = 1$ hence extreme tensile strain at the bottom of core section experienced $\varepsilon_i = \varepsilon_{cr}$. The terms such as γ, defines E_{cc} stiffness in compression, μ represents the residual tensile strength, and $\gamma\omega$ compressive strength of core material:

$$\textit{Tension};\ \sigma_{cr} = \varepsilon_{cr}E,\ \sigma_{ult} = \mu\sigma_{cr}\ \text{and}\ \varepsilon_i = \beta\varepsilon_{cr} \quad (1)$$

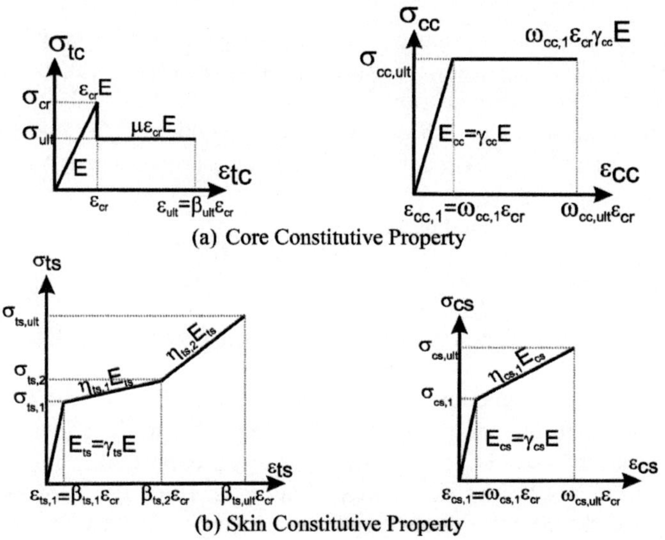

Fig. 1. Constitutive models for (a) skin with multi-linear tension and elastic-perfectly plastic compression (b) core model with elastic-perfectly plastic in tension and compression.

$$\textit{Compression};\ E_{cc} = \gamma_{cc}E,\ \sigma_{cc,ult} = \omega_{cc,1}\gamma_{cc}\sigma_{cr} \text{ or } \varepsilon_{cc,ult} = \omega_{cc,ult}\varepsilon_{cr} \tag{2}$$

A similar approach is adopted for parametric material properties of skin elements. It is assumed that skin and core materials are perfectly bonded allowing for strain compatibility. The stress response of the skin is considered constant through its thickness acting on the area defined as $A_{ts} = A_{cs} = \rho bh$ resulting in force components as $F_{ts} = \sigma_{ts}(\beta)\rho bh$ and $F_{cs} = \sigma_{cs}(\beta)\rho bh$. The normalized compressive and tensile young's modulus of TRC skin can be written as:

$$\gamma_{tc} = 1,\ \gamma_{cc} = \frac{E_{cc}}{E}\ \textit{and}\ \gamma_{ts} = \frac{E_{ts}}{E},\ \gamma_{cs} = \frac{E_{cs}}{E} \tag{3}$$

The normalized stress-strain model for core and skin materials are used as:

$$\begin{aligned} \frac{\sigma_{cc}(\lambda)}{E\varepsilon_{cr}} &= \begin{cases} \gamma\lambda & 0 \le \lambda \le \omega_{cc,1} \\ \gamma\omega_{cc,1} & \omega_{cc,1} < \lambda \le \omega_{cc,ult} \end{cases} \\ \frac{\sigma_{tc}(\beta)}{E\varepsilon_{cr}} &= \begin{cases} \beta & 0 \le \beta \le 1 \\ \mu & 1 < \beta \le \beta_{ult} \end{cases} \end{aligned} \tag{4}$$

λ is the normalized extreme compressive strain corresponding to extreme tensile strain by assuming linear strain distribution and adopting similar triangle formula. Given $\lambda = (\beta * k)/(1 - k)$ when k is neutral axis ratio

$$\begin{aligned} \frac{\sigma_{cs}(\beta)}{E\varepsilon_{cr}} &= \begin{cases} \gamma_{cs}\lambda & 0 \le \lambda \le \omega_{cs,1} \\ \omega_{cs,1}\gamma_{cs} + \eta_{cs,1}\gamma_{cs}(\lambda - \omega_{cs,1}) & \omega_{cs,1} < \lambda \le \omega_{cs,ult} \end{cases} \\ \frac{\sigma_{ts}(\beta)}{E\varepsilon_{cr}} &= \begin{cases} \gamma_{ts}\beta & 0 \le \beta \le \beta_{ts,1} \\ \beta_{ts,1}\gamma_{ts} + \eta_{ts,1}\gamma_{ts}(\beta - \beta_{ts,1}) & \beta_{ts,1} < \beta \le \beta_{ts,2} \\ \beta_{ts,1}\gamma_{ts} + \eta_{ts,1}\gamma_{ts}(\beta_{ts,2} - \beta_{ts,1}) + \eta_{ts,2}\gamma_{ts}(\beta - \beta_{ts,2}) & \beta_{ts,2} < \beta \le \beta_{ts,ult} \end{cases} \end{aligned} \tag{5}$$

The skin stiffness reduction ratio after first cracking is defined as $\eta_{ts,1}$, $\eta_{ts,2}$, &$\eta_{cs,1}$ to account for strain hardening or softening using positive or negative values. The transition points from one strain stage to the other stages are calculated using the first crack strain for all three stages defined with respect to core tensile cracking strain. The same approach was adopted for compression strain with two stages of strain (see Eq. 4).

$$\beta_{ts,1} = \frac{\varepsilon_{ts,1}}{\varepsilon_{cr}},\ \beta_{ts,2} = \frac{\varepsilon_{ts,2}}{\varepsilon_{cr}} \text{ and } \beta_{ts,ult} = \frac{\varepsilon_{ts,3}}{\varepsilon_{cr}};\ \omega_{cs,1} = \frac{\varepsilon_{cs,1}}{\varepsilon_{cr}} \text{ and } \omega_{cs,ult} = \frac{\varepsilon_{cs,ult}}{\varepsilon_{cr}} \tag{6}$$

3 Moment-Curvature Response Based on Force Equilibrium

To generate the moment-curvature relationship in closed form, the section equilibrium, compatibility, and constitutive law are considered for each step according to the linear strain distribution shown in Fig. 2(b). Since the stress is expressed in a piecewise linear fashion [12], normalized compression strain, bottom skin strain, and top skin strain are given in terms of β, *k*, and *y* as defined by the distance from the neutral axis. The components of the force and moment arrays are obtained as quadratic and cubic functions. The location of the neutral axis *kh*, is obtained by the equilibrium of internal forces and solution of the quadratic equation as shown in earlier work [13] and used in the calculation of the resultant moment. The curvature is obtained from the imposed strain and neutral axis location. The curvature in terms of β is:

$$\phi = \frac{\beta\varepsilon_{cr}}{(1-k)h} \tag{7}$$

The various modes of strain distribution interact based on the material parameters which limit the boundaries of the linear segments. The closed-form solution of moment-curvature response is listed in Table 2. which the moment M_i and ϕ_i at each stage is variable within the global range of input variables $0 < \beta < \beta_{tu}$. Note that, normalized moment, m_i' and curvature, ϕ' are respected to the first cracking moment and curvature of simple rectangular section as seen in Eq. 8

$$M_{cr} = \frac{1}{6}bd^2E\varepsilon_{cr};\ M_i = m_i'M_{cr};\ \phi_{cr} = \frac{2\varepsilon_{cr}}{h_0};\ \phi = \phi_i'\phi_{cr} \tag{8}$$

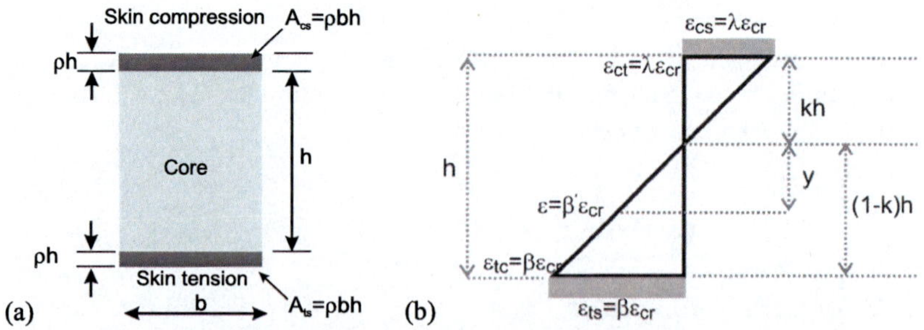

Fig. 2. (a) Sandwich section, (b) linear strain distribution along the sandwich section

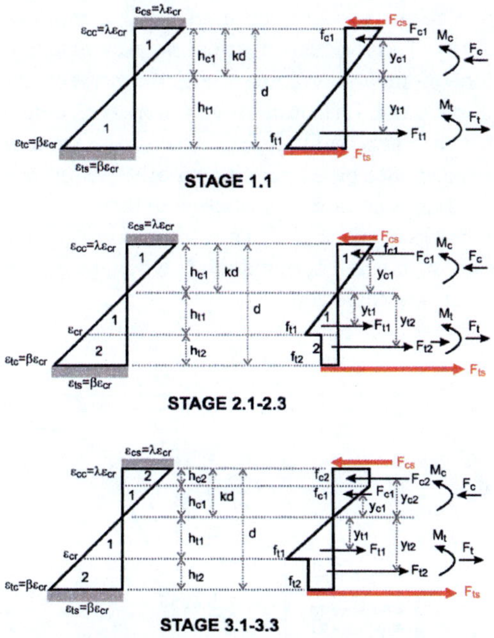

Fig. 3. The stress-strain block for various stages of variables $0 < \beta < \beta_{ult}$

The schematic stress-strain block is shown in Fig. 3. Three main stages represent the stress-strain depending on the tensile strain parameter β imposed. The first digit from 1.x to 3.x represents the compression mode of skin in elastic or plastic(see Fig. 3), and the second digit from x.1 to x.3 represented three phases of tensile stress of the skin. In zone 1.1, tensile and compressive in core and skin are elastic with a linear moment-curvature response as shown in Fig. 4(a) and (b). Initiation of tensile cracking in the core switches the response to Stage 2.1 defined as elastic compression and tension perfectly plastic core while the skin remains elastic in compression and tension. The stress in Sandwich with AAC core, (SWAAC) reduces to maintain the equilibrium since it has no residual tensile strength while Sandwich with FRAC core, (SWFRAC) remains partially loaded due to the residual strength in FRAC. The end of Stage 2.1 provides two alternatives of perfectly plastic in core compression (Stage 3.1 and then followed by Stage 3.2) or yielding of TRC skin with an elastic core in compression (Stages 2.2 and 2.3) since there are two post-crack tension stages in the skin response. The response was transferred to Stage 2.2, 2.3 by imposing material properties, and then to zone 3.3 when core compressive stress became perfectly plastic. The closed-form solution for calculating normalized flexural moment, *m'*, and neutral axis depth fraction, *k,* at different stages are listed in Table 2. When material parameters are substituted in Table 3, the only variable left to calculate *k* is β, which leads to normalized curvature, ϕ*'* and moment *m'* using Eq. 7 and 8. Hence, the envelope of the moment-curvature curve can be constructed as a benchmark in Fig. 2 when β ranges from 0 to β_{ult}.

The load-deflection response was measured using a moment-area method-based numerical approach [14]. The moment distribution was calculated by static equilibrium when the envelope moment-curvature in Fig. 4 represents the maximum moment at the mid-span. Multiple crack formation in core materials demonstrates the majority of load-carrying across the skins and induces non-uniform strain distribution [15]. The concept of the proposed model based on compatibility, equilibrium, and constitutive raw can represent an average curvature distribution in the localized zone, L_P which may assume to be equal to cracking space [16]. The closed-form simulation is based on only the deflection of flexure in skin and core materials when the core shear failure and skin wrinkling (local bucking) are neglected.

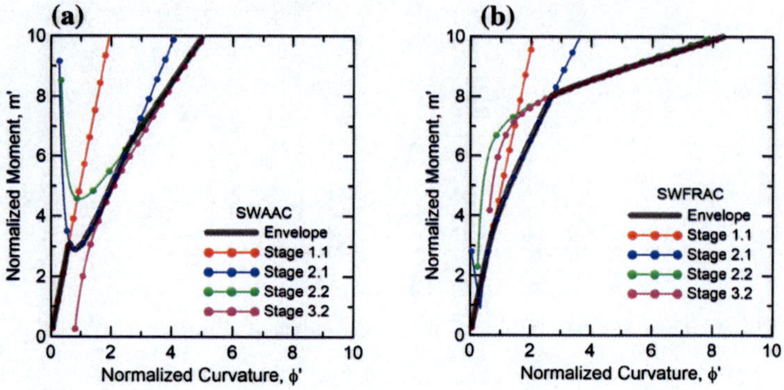

Fig. 4. Normalized moment-curvature response at each stage of β. (a) sandwich with AAC (b) sandwich with FRAC

4 Sandwich Model Validation

4.1 Material Parameters Identification

Sandwich mixtures were prepared with an AR glass skin and two different types of core using FRAC and AAC [6]. The skin layer was bonded AR glass textiles with a perpendicular set of yarns (warp and weft) as shown in Fig. 5(a) and 5(b). The textile yarn density in warp and weft directions was four yarns per centimeter with 400 filaments of fibers of diameter 13.5 μm. Tensile strength of textile was in the range of 1270 – 2450 MPa and modulus of elasticity of 78 MPa[12]. Three-point bending tests on the specimen with dimensions of (WxDxL) 50×50×250 mm for SWFRAC and 50×100×250 mm for SWAAC were performed using a servo-hydraulic testing system under a displacement control rate of 2 mm/min. Load-deflection test results and damage mechanisms of are discussed in detail in Dey et al. [17].

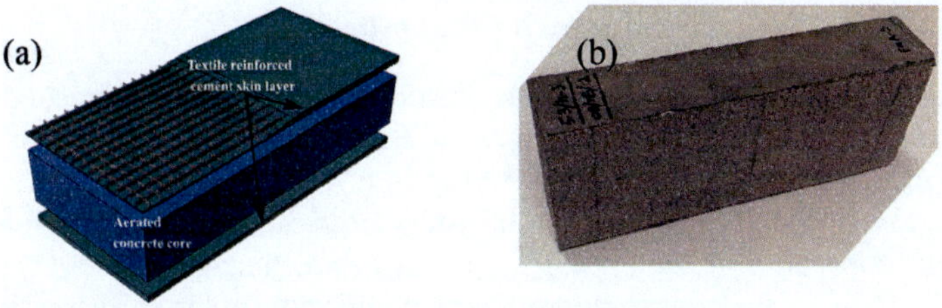

Fig. 5. Schematics of the proposed textile reinforced-aerated concrete sandwich composite system, (b) Sandwich beam specimen.

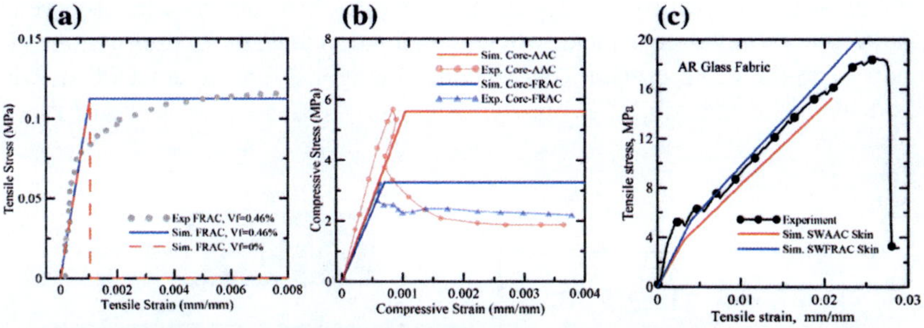

Fig. 6. Simplified stress-strain diagram of core materials, AAC and FRAC. (a) Core tensile model, (b) Core compressive model, (c) Stress-strain diagram of AR Glass fabric TRC as skin material

Tensile and compressive stress-strain as shown in Fig. 6(a) and (b), of core and skin properties were used [6]. The ratio of compressive to tensile strength is in the order of $\gamma\omega = 50$–60, hence the flexural response is dominated by tensile and shear strength. ACC is brittle with no residual tensile strength ($\mu = 0$) while FRAC experienced an elastic-perfectly plastic response. The TRC skin material made by AR Glass fabric was obtained from [17, 18] and shows typical tensile stress-strain as in Fig. 6(c). The skin material is much stiffer than the core, hence resulting in η_{ts}, and η_{cs} as 10 as it carries a majority of the uniaxial load and delays the diagonal tension failure as the core material resists transverse shear. Material parameters were reported in Table 1.

Table 1. Material parameters used to simulate tensile and compressive stress-strain diagram of core and skin domain.

Domain	Tensile					Compression		
Core	E, MPa	$\varepsilon_{cr} \times 10^{-6}$	μ	β_{ult}		γ_{cc}	$\omega_{cc}, 1$	
AAC	108	1040	0	30		50	1	
FRAC	108	1040	1	30		43	0.67	
Skin	γ_{ts}	$\beta_{ts}, 1$	$\beta_{ts}, 2$	$\eta_{ts}, 1$	$\eta_{ts}, 2$	γ_{cs}	$\omega_{cs}, 1$	$\eta_{cs}, 1$
TRC	11–12.5	1.9–3	1.9–3	0	0.2–0.55	9.5–12.5	1.92	0.20–0.55

4.2 Flexural Simulation of Sandwich Composites

The load-deflection response matches the elastic deflection of an uncracked section of Stage A and microcracks between stages A and B. The contribution of skin is after stage B in SWAAC with the load at around 500 N in SWFRAC. For SWAAC, after stage B, the core loses capacity and the load transfers from the entire sandwich to the skin. Shear crack propagation is seen in DIC analysis between stages B and C as in Fig. 7. From stages C to D, the sandwich maintains equilibrium with an increasing flexural load. The skin is the primary carrier of an axial load whereas core material shows the development of diagonal tension failure. The diagonal tension cracks in the core may not result in sudden failure. As the diagonal tension cracks coalesce during stages D to E, the SWAAC finally loses equilibrium. SWFRAC (see Fig. 8) specimens were observed to fit the model better until after point C, when delamination and shear deformation became dominant. The FRAC core materials can control diagonal tension cracks better due to the residual tensile strength.

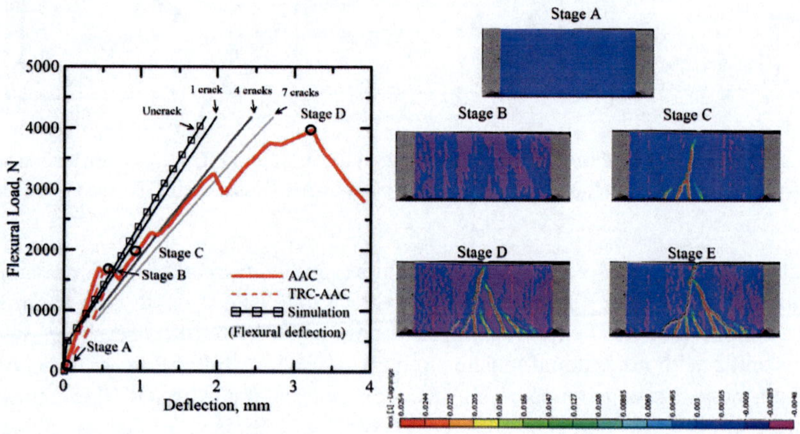

Fig. 7. Flexural load-displacement response of SWAAC with dimension 50×100×200 mm. DIC analysis on nominal strain in the x-direction are illustrated from Stage A to B

5 Conclusion

A generalized closed-form solution approach is developed and verified with the experimental flexural response of sandwich composites using an aerated concrete core and textile reinforced concrete textile used as skin membrane. The model integrates compression and tensile models of the skin and core elements using multi-component linear segments and captures the closed form flexural moment-curvature response. This response is used to model the load-deflection response and modes of behavior as affected by the material properties using a direct correlation of the mechanical response and optimization of geometrical and material properties.

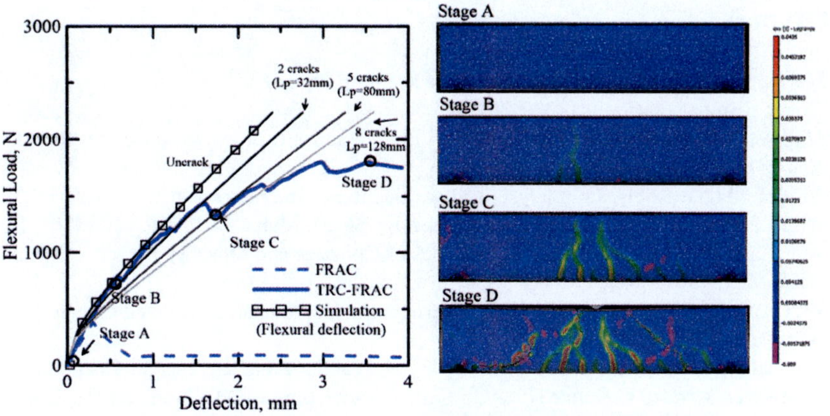

Fig. 8. Flexural load-displacement response of SWFRAC with dimension 50×50×200 mm. DIC analysis on nominal strain in the x-direction are illustrated from Stages A to D

Table 2. The closed-form solution to determine the neutral axis ratio, k and normalized flexural moment m'in terms of β and material parameters

Zone	Neutral axis ratio, k	Normalized flexural moment, m'
1	$k_1 = \frac{1}{2}$	$m_1 = \beta\left(54\rho^3 + 36\rho^2 + 6\rho + 1\right)$
2.1	$k_{21} = \frac{A_4 - \sqrt{\beta^2 A_4 (1+2\rho)}}{B_6 - \beta^2}$	$m_{21} = \frac{\left(A_2 - 2\beta^3\right)k^3 + A_1 k^2 + A_1 k + A_3 - A_2}{\beta^2 (k-1)}$
2.2	$k_{22} = \frac{B_6 + (\beta + \kappa)\rho\beta - \sqrt{\beta^2\left(B_7\rho^2 + B_8\rho + B_6\right)}}{B_6 - \beta^2}$	$m_{22} = \frac{\left(B_1 - 4\beta^3\right)k^3 + B_3 k^2 + B_4 k + B_5}{2\beta^2 (k-1)}$

$A_1 = -12\rho\beta^3 - 3A_2$; $A_2 = 3\mu\beta^2 - 3\mu + 2$	$B_5 = -B_2 - 18\rho^2\beta_{ts,1}\beta^2 - 27\rho^3\beta^3 - B_1$
$A_3 = \left(-27\rho^3 - 18\rho^2 - 6\rho\right)\beta^3$; $A_4 = 2\rho\beta^2 + B_6$	$B_6 = 2\mu\beta - 2\mu + 1$
$B_1 = 2A_2$; $B_2 = 12\rho\beta_{ts,1}\beta^2$; $B_3 = -B_2 - 12\rho\beta^3 - 3B_1$	$B_7 = \beta_{ts,1}^2 + 2\beta\beta_{ts,1} - 2\beta^2 + 3B_6$
$B_4 = 2B_2 + 18\rho^2\beta_{ts,1}\beta^2 - 36\rho^2\beta^3 + 3B_1$	$B_8 = 2B_6 + 2\beta_{ts,1}\beta$

References

1. Narayanan, N., Ramamurthy, K.: Structure and properties of aerated concrete: a review. Cem. Concr. Compos. **22**(5), 321–329 (2000)
2. Holt, E., Raivio, P.: Use of gasification residues in aerated autoclaved concrete. Cem. Concr. Res. **35**(4), 796–802 (2005)
3. Cuypers, H., Wastiels, J.: Analysis and verification of the performance of sandwich panels with textile reinforced concrete faces. J. Sandw. Struct. Mater. **13**(5), 589–603 (2011)
4. ACI 122R-02, Guide to Thermal Properties of Concrete and Masonry Systems. Farmington Hills, Michigan (2002)
5. ACI 523.4R, Guide for Design and Construction with Autoclaved Aerated Concrete Panels. Am. Concr. Inst. (2009)
6. Bonakdar, A., Babbitt, F., Mobasher, B.: Physical and mechanical characterization of Fiber-Reinforced Aerated Concrete (FRAC). Cem. Concr. Compos. **38**, 82–91 (2013)
7. Dey, V., Bonakdar, A., Mobasher, B.: Low-velocity flexural impact response of fiber-reinforced aerated concrete. Cem. Concr. Compos. **49**, 100–110 (2014)
8. Cuypers, H., Van Ackeren, J., Belkassem, B., Wastiels, J.: Impact resistance of load bearing sandwich elements with textile reinforced concrete faces. In: Proceedings of the 13th European Conference on Composite Materials (2008)
9. Djamai, Z.I., Bahrar, M., Salvatore, F., Si Larbi, A., El Mankibi, M.: Textile reinforced concrete multiscale mechanical modelling: Application to TRC sandwich panels. Finite Elem. Anal. Des., **135**, 22–35, (2017)
10. Chudoba, R., van der Woerd, J., Schmerl, M., Hegger, J.: ORICRETE: Modeling support for design and manufacturing of folded concrete structures. Adv. Eng. Softw. **72**, 119–127 (2014)
11. Ashby, M.F., Gibson, L.J.: Cellular Solids, Structure and Properties. Cambridge University Press, Cambridge, UK (1997)
12. Soranakom, C., Mobasher, B.: Closed-form solutions for flexural response of fiber-reinforced concrete beams. J. Eng. Mech. **133**(8), 933–941 (2007)
13. Pleesudjai, C., Li, A., Dey, V., Mobasher, B.: Low-velocity impact experiments and modeling of TRC skin-aerated concrete core sandwich composites. Materials (Basel) **14**(2), 390 (2021)
14. Yao, Y., Aswani, K., Wang, X., Mobasher, B.: Analytical displacement solutions for statically determinate beams based on a trilinear moment–curvature model. Struct. Concr. **19**(6), 1619–1632 (2018)
15. Dias-da-Costa, D., Valença, J., do Carmo, R.N.F.: Curvature assessment of reinforced concrete beams using photogrammetric techniques. Mater. Struct. 47(10) 1745–1760 (2014)
16. di Prisco, M., Plizzari, G., Vandewalle, L.: Fibre reinforced concrete : new design perspectives, pp. 1261–1281 (2009)
17. Dey, V., Zani, G., Colombo, M., Di Prisco, M., Mobasher, B.: Flexural impact response of textile-reinforced aerated concrete sandwich panels. Mater. Des. **86**, 187–197 (2015)
18. Soranakom, C., Mobasher, B.: Modeling of tension stiffening in reinforced cement composites: Part II. Simulations versus experimental results. Mater. Struct. Constr. **43**(9), 1231–1243 (2010)

Mix Optimisation and Bending Behaviour of Cement Composites Reinforced with 3D Textiles and Microfibres

Ciska Gielis[1](✉), Michael El Kadi[1], Tine Tysmans[1], and Didier Snoeck[2]

[1] Department Mechanics of Materials and Constructions (MeMC), Vrije Universiteit Brussel (VUB), Pleinlaan 2, 1050 Brussels, Belgium
Ciska.Gielis@vub.be

[2] Department Building, Architecture and Town Planning (BATir), Université Libre de Bruxelles (ULB), Franklin Rooseveltlaan 50, 194/02, 1050 Brussels, CP, Belgium

Abstract. In Textile Reinforced Cementitious (TRC) composites, cementitious elements are reinforced with fibre textiles instead of corrosion-prone steel reinforcement. This allows for a considerable reduction of the cross-section of the cementitious elements, and so of the required amount of cement. One of the drawbacks of brittle materials, however, is the occurrence of wide cracks, requiring proper repair. In the presented research, 3D textiles are used as reinforcement in combination with microfibres to design a smart material with self-healing features. The use of 3D textiles has demonstrated a superior flexural behaviour in comparison to 2D textiles, while the short microfibres are known to limit the width of the cracks in the cementitious matrix. These narrow cracks can be healed through the autogenous healing characteristic of the cementitious material.

In this research, the synergetic interaction between the 3D textiles, microfibres and cementitious matrix is investigated for the first time. First, the optimal material composition of 3D TRC composites with integrated polypropylene (PP) microfibres of 6 mm (PP6) and 8 mm (PP8) length was explored. This resulted in an optimal fibre content of 0.7 v% for both fibre lengths, allowing sufficient fibre penetration through the 3D textile meshes. Then, the loadbearing behaviour and crack formation of these material compositions were studied by four-point bending tests, monitored with Digital Image Correlation (DIC). The 3D TRC + PP6 samples demonstrated superior flexural properties, and showed slightly more and narrower cracks.

Keywords: Textile Reinforced Cementitious (TRC) composites · 3D textile reinforcement · Microfibre reinforcement · Four-point bending · Digital Image Correlation (DIC)

1 Introduction

In Textile Reinforced Cementitious (TRC) composites, a cementitious material is reinforced with fibre textiles to withstand tensile stresses, instead of the traditional steel reinforcement. This innovative reinforcement technique allows for a reduction of the

M. Kunieda et al. (Eds.): SHCC 2022, RILEM Bookseries 39, pp. 209–216, 2023.
https://doi.org/10.1007/978-3-031-15805-6_22

cover thickness in cementitious elements, traditionally required to protect the steel reinforcement from corroding. This results in a possible reduction of the amount of cement necessary for these structural elements, and hence a lower environmental impact [1]. As is the case for other cementitious materials, however, TRCs are still susceptible to the formation of wide cracks due to the occurring tensile stresses. These cracks act as pathways for water and harmful substances to infiltrate the structural elements, and so could cause significant damage and durability issues. Consequently, these cracks must be properly repaired, but repair works are often time-consuming and expensive, or not even possible in case of inaccessible cracks. An alternative approach is the adjustment of the cementitious material itself, with the aim to improve its durability properties a priori, and so reduce the need for repair [2].

Cementitious materials have an autogenous healing characteristic, meaning that they have the ability to heal cracks in the presence of water. By means of the further hydration of unhydrated cement and pozzolanic materials, precipitation of calcium carbonate, expansion of calcium-silicate-hydrate phases, and the closing of cracks by loose particles and impurities, narrow cracks with a width up to 30–50 μm can be entirely healed. To ensure that the cracks remain narrow enough, short microfibres, such as shown in Fig. 1a, can be added to the cementitious matrix. These fibres can limit the crack widths considerably, allowing for more narrow and healable cracks [2].

In the presented research, the addition of short microfibres to 3D-textile reinforced cementitious composites (3D TRCs) is explored, to prevent the formation of wide cracks (> 200 μm). The 3D textiles (see Fig. 1b) are composed of two continuous layers of 2D textiles, connected by knitted connections. 3D textile reinforcement has two important advantages over 2D textiles: (i) the 3D textiles form a rigid reinforcement during casting, allowing for an easy and straightforward manufacturing process, a considerable advantage in view of practical application, and (ii) 3D TRCs demonstrate superior flexural properties compared to 2D TRCs, due to the anchorage mechanism generated by the transversal connections [3].

In this paper, the mechanical behaviour and crack formation of 3D TRCs with integrated microfibres are explored for the first time. The first challenge lies within the adequate manufacturing, i.e. assuring a homogenous distribution of the microfibres when casting through the 3D textile grid. Based on a mix optimisation process, two material configurations are determined, containing respectively 6 mm and 8 mm length polypropylene fibres. Next, it is evaluated whether the addition of the microfibres positively affects the crack formation. Through four-point bending tests, monitored by Digital Image Correlation (DIC), insights are gained in the material's mechanical performance in bending, as well as the widths and patterns of the generated cracks.

2 Materials and Methods

2.1 Materials

The used mortar mixture is composed of Portland cement (CEM I 52.2 N, Holcim, Belgium), fly ash, quartz sand (M34, Sibelco, Belgium), water and superplasticizer (MasterGlenium 51, BASF, Netherlands). The short microfibres, mixed in this cementitious matrix, are commercially available polypropylene (PP) fibres with a length of

6 mm (PP6) or 8 mm (PP8) (Redco, Belgium). The selected 3D textiles (SitGrid 701, Wilhelm Kneitz AG, Germany) consist of two 2D textiles of AR-glass with a mesh size of 22.5 mm × 22.5 mm, maintained at a fixed distance of 8.5 mm from each other by means of knitted polyester connections. The textiles have a coated density of 616 g/m^2, and an uncoated density of 536 g/m^2 [3].

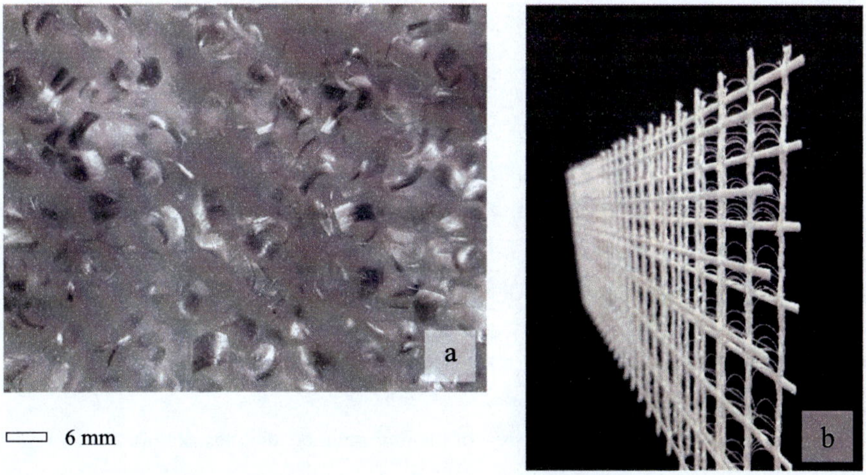

Fig. 1. (a) Short polypropylene (PP) microfibres with a length of 6 mm. (b) 3D textiles composed of two 2D textiles of AR-glass linked by means of knitted polyester connections.

2.2 Methods

For the manufacturing of the 3D TRC test specimens with integrated short microfibres, a straightforward method was applied in this research. First, the 3D textiles were horizontally placed in moulds of 500 mm × 60 mm × 15 mm. These moulds allowed for the 3D textiles to be clamped in place, to ensure an equal cover thickness on both sides. After clamping the 3D textiles in place, the fibre-mortar mixture was poured onto it, until the mould was filled, and the textiles were completely immersed in the fibre-mortar mixture (see also Sect. 3.1). The fibre volume fraction (V_f) of the 3D textiles in the test specimens was 0.7 v%. Based on a mix optimisation process, the optimal volume fraction of the PP microfibres was determined.

A first assessment of the mechanical performance of 3D TRCs with integrated microfibres was carried out by performing four-point bending tests on the produced specimens. With the PP6 fibres six test samples were produced (3D + PP6), with the longer PP8 fibres three samples were made (3D + PP8). The specimens were tested in four-point bending (see Fig. 2) with 350 mm between the supports and 100 mm between the loading pins. For the downwards displacement of the loading pins, a rate of 2 mm/min was applied, until failure of the test specimen. This procedure was monitored by means of two DIC systems, one for monitoring the side of the flexural specimens and one for monitoring the bottom. Through the DIC monitoring the deformation of the samples was examined, as well as the generated crack widths and patterns. The pixel density of the used cameras was 2448 pixels by 2048 pixels, and the applied subset, step and strain window size were respectively 21, 7 and 9.

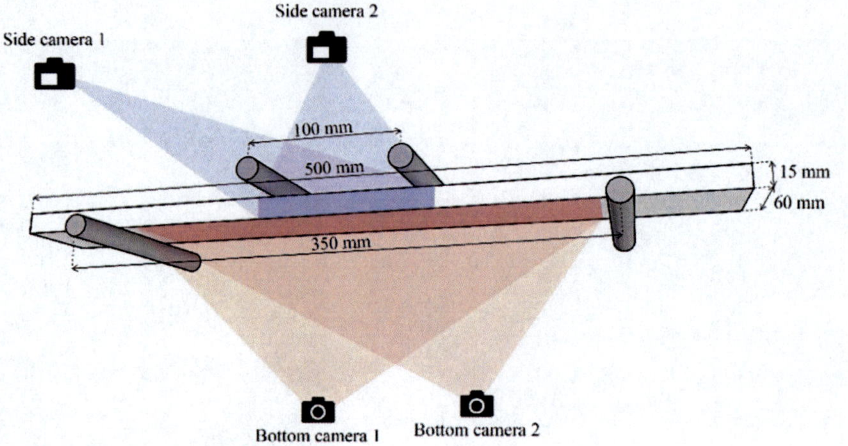

Fig. 2. Schematic overview of the four-point bending test set-up.

3 Results and Discussion

3.1 Mix Optimisation

During casting the 3D textiles ($V_{f,3D} = 0.7$ v%) form a considerable barrier for the fibre-mortar mixture, and hamper hereby the even distribution of short fibres in the TRC element. The penetration through the 3D textiles is firmly linked to the volume fraction of the added microfibres, the fibre type and length. To guarantee a sufficient fibre penetration through the 3D textiles during the casting of the test specimens, an optimal fibre volume fraction for the PP microfibres of 6 mm and 8 mm was explored by means of small-scale casting tests. These tests consisted in placing fitted pieces of 3D textiles in small moulds of 60 mm x 60 mm x 15 mm, after which little amounts of fibre-mortar mixture with varying fibre volume fractions for both fibre lengths were poured into the small moulds. By visually evaluating whether the selected fibre volume fraction allowed for a full mesh drop-through of the microfibres through the 3D textiles, an optimal volume fraction was determined. The results of the small-scale casting tests are presented in Fig. 3a, both for the PP6 and PP8 fibres. The first tested fibre volume fractions were 0.5 v% and 1.0 v%. For both fibre lengths, the volume fraction of 0.5 v% allowed sufficient fibre penetration through the 3D textile meshes, while for the 1.0 v% mixture a considerable amount of fibres remained on top of the upper mesh. By gradually increasing the fibre volume fraction starting from 0.5 v%, the optimal fibre volume fraction of 0.7 v% was determined for both the PP6 and PP8 fibres.

Next, this optimal fibre volume fraction was applied for the production of 3D TRC test specimens with integrated PP fibres. In Fig. 3b, the casting of the samples is shown. The mixture of mortar and 0.7 v% PP microfibres was fluid enough and easily penetrated the 3D textile meshes, with no large bundles of fibres remaining on top of the upper mesh. This confirmed that the optimal volume fraction of 0.7 v% also allows sufficient fibre penetration on a larger scale for the applied manufacturing method. However, it must be mentioned that 0.7 v% is low in comparison with reference works on self-healing microfibre reinforced cementitious materials, containing orders of magnitude of 2.0 v% of microfibres yet no 3D textile reinforcement [4].

Fig. 3. (a) Results of small-scale casting tests for PP microfibres of 6 mm and 8 mm length, with 0.7 v% as the optimal fibre volume fraction for sufficient fibre penetration through the 3D textile meshes. (b) Larger scale casting of 3D TRC test samples with 0.7 v% PP microfibres mixed in the cementitious matrix.

3.2 Mechanical Behaviour

After 28 days, the produced 3D TRC samples with 6 mm and 8 mm PP fibres were tested in four-point bending. Figure 4 presents the results of applied force as a function of the maximum displacement. In Table 1, a summary of the average load at first crack is given, as well as the slope of the force-displacement graphs in both the pre- and post-cracking stage. The average load at which the first crack occurred was similar between the 3D + PP6 and 3D + PP8 samples. The measured slopes in both the pre- and post-cracking stage were, however, on average higher for the 3D + PP6 samples. Therefore, the pre- and post-cracking stiffnesses of the 3D + PP6 samples were higher than the 3D + PP8 samples. The slightly superior post-cracking behaviour of the 3D + PP6 samples could be the result of a better fibre distribution in the cementitious matrix. It might also be linked to the fact that slightly more individual fibres were present for the 3D + PP6 mix, since one 6 mm fibre has an individual lower volume than one 8 mm fibre and the same total volume fraction of 0.7 v% was applied.

Table 1. Measured flexural properties of 3D TRC test samples with integrated PP microfibres of 6 mm and 8 mm long.

		3D + PP6	3D + PP8
Volume fraction PP microfibres Vf,PP	%	0.7	0.7
First cracking load	N	206 ± 16	212 ± 12
Pre-cracking slope	N/mm	458 ± 43	394 ± 18
Post-cracking slope	N/mm	13 ± 2	11 ± 1

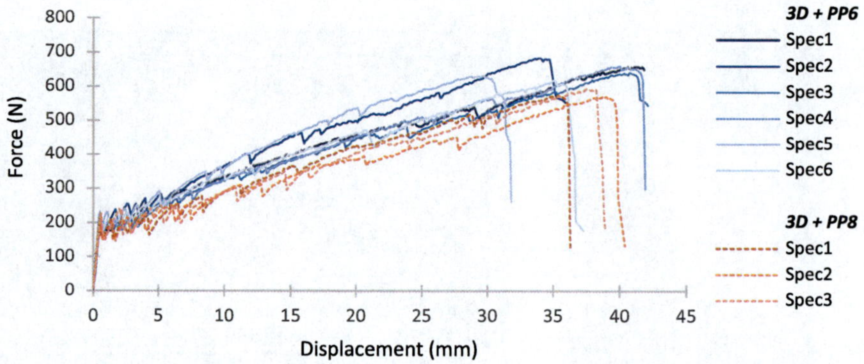

Fig. 4. Force-displacement graphs of the tested 3D + PP6 (solid, blue) and 3D + PP8 (dashed, orange) TRC samples.

3.3 Crack Pattern Analysis

Furthermore, the cracks that occurred during the four-point bending tests were analysed. Through DIC monitoring, the maximal crack widths at any displacement could be investigated. Within this study, a fixed displacement value in the post-cracking stage was selected (25 mm). It must be noted that this displacement exceeded the serviceability criteria (SLS) for these elements. As a result, the measured crack widths also exceeded the self-healing limit of 30–50 μm [2]. However, as the samples were tested up to failure (ULS), this displacement does give a valuable insight on the crack widths and patterns closer to this moment of failure.

In Table 2, the averages of the measured number of cracks, (maximum and minimum) crack widths, and crack spacings are summarised for the 3D + PP6 and the 3D + PP8 samples. The average crack spacing was similar between the two sample configurations. The cracks generally originated in the zones where the transverse fibre bundles of the lower mesh layer of the 3D textiles were located; these transverse fibre bundles are indeed discontinuities in the mortar cross-section (see Fig. 5). The measured number of cracks were, on the other hand, slightly higher in the 3D + PP6 samples, and on average the minimal crack widths were also more narrow than in the 3D + PP8 samples.

Table 2. Average values of the measured number of cracks, (maximal and minimal) crack widths, and crack spacings for the 3D + PP6 and 3D + PP8 samples.

		3D + PP6 0.7 v%	3D + PP8 0.7 v%
Number of cracks	/	12.0 ± 1.3	10.3 ± 1.2
Crack width	μm	393 ± 224	425 ± 208
Crack width between loading pins	μm	524 ± 194	503 ± 211
Max. Crack width	μm	736 ± 37	682 ± 41
Min. Crack width	μm	73 ± 31	98 ± 83
Crack spacing	mm	20.0 ± 5.5	20.3 ± 5.4

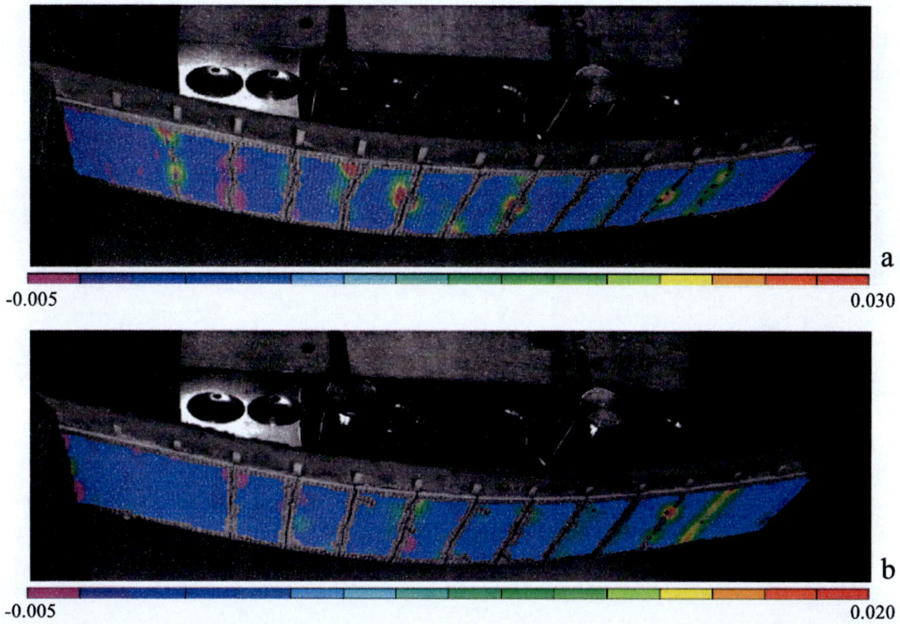

Fig. 5. DIC images of representative (a) 3D + PP6 and (b) 3D + PP8 test samples at 25 mm displacement of the midpoint, presenting the corresponding crack patterns and longitudinal strains.

4 Conclusions

This paper describes a preliminary investigation on cement composites with 3D textile reinforcement and integrated short PP microfibres, with the aim to design a novel composite material with autogenous healing features.

Based on small-scale casting tests, an optimal fibre volume fraction of 0.7 v% was determined for PP fibres with a length of 6 mm (PP6) and 8 mm (PP8). This fibre content allowed sufficient fibre penetration through the 3D textile meshes for the applied manufacturing method, however this amount of fibres is limited compared to reference works on self-healing cementitious materials with integrated short fibres (2.0 v%). An investigation of alternative manufacturing methods could allow higher fibre contents.

The 3D + PP6 and 3D + PP8 samples were tested in four-point bending. Through DIC monitoring, the mechanical behaviour and crack formation of the flexural samples were studied. The measured load at first crack was similar between the two sets of samples. The post-cracking stiffness was slightly higher for the 3D + PP6 samples than for the 3D + PP8 samples. This could be a result of a more even fibre distribution in the 3D + PP6 samples, or a higher number of individual fibres for the same fibre volume fraction of 0.7 v%. When comparing the crack patterns of the two sets of samples, a great similarity was noted, as most cracks originated at the locations of the transverse fibre bundles of the lower mesh of the 3D textiles. A slightly higher number of cracks was measured for the 3D + PP6 samples, as well as more narrow minimal crack widths.

With the aim to completely evaluate the effect of the short fibres, a comparative investigation of the flexural properties and crack formation of additional test samples with only 3D textiles on the one hand, and only short fibres on the other hand, is required in the further continuation of this research.

Acknowledgements. Financial support by the Research Foundation-Flanders (FWO-Vlaanderen) through project nr. G030721N is gratefully acknowledged.

References

1. Peled, A., Mobasher, B., Bentur, A.: Textile Reinforced Concrete, 1st edn. CRC Press, London (2017)
2. Snoeck, D., De Belie, N.: Autogenous healing in strain-hardening cementitious materials with and without superabsorbent polymers: an 8-year study. Front. Mater. **6**, 1–12 (2019)
3. El Kadi, M., et al.: Experimental study and benchmarking of 3D textile reinforced cement composites. Cement Concr. Compos. **104**, 1–10 (2019)
4. Snoeck, D., Van Tittelboom, K., Steuperaert, S., Dubruel, P., De Belie, N.: Self-healing cementitious materials by the combination of microfibres and superabsorbent polymers. J. Intell. Mater. Syst. Struct. **25**(1), 13–24 (2014)

Spacers for 3D Textiles as Reinforcement in Cement Composites: Influence on the Flexural and Cracking Behavior

M. El Kadi[1](✉), C. Gielis[1], D. Toma[1], D. Van Hemelrijck[1], H. Rahier[2], and T. Tysmans[1]

[1] Department Mechanics of Materials and Constructions (MeMC), Vrije Universiteit Brussel (VUB), Pleinlaan 2, 1050 Brussels, Belgium
Michael.el.kadi@vub.be

[2] Department Physical Chemistry and Polymer Science (FYSC), Vrije Universiteit Brussel (VUB), Pleinlaan 2, 1050 Brussels, Belgium

Abstract. Discrete spacers allow to transversally connect planar textiles and to transform them into pseudo-3D architectures. These 3D textiles can provide the optimal reinforcement in cement composites as the concrete casting process is enhanced and the textile placement within the mould is well-controlled. This research investigated the effect of discrete spacer connections on the flexural properties and cracking behaviour of TRCs and performed a comparison with equivalent 2D TRC systems (same in-plane textiles but without connections) and with knitted 3D textiles. Additional investigated parameters were the material of the reinforcement (glass, carbon) and the density of the textile (fibre volume fraction of the composite). A four-point bending experimental campaign was conducted on four different textile configurations containing (i) no connections (ref), (ii) four spacer connections distributed over the shear zone, (iii) nine spacer connections distributed over the entire length between the supports (both shear and constant moment zone) and (iv) 3D knitted connections. The spacer connections resulted in a pronounced post-cracking stiffness increase and a positive influence on the crack formation of the TRC. Their influence was however dependent on the fibre volume fraction (V_f) of the in-plane reinforcement; configurations with a higher V_f required a higher degree of connections to showing a post-cracking stiffness increase.

Keywords: 3D textile · Digital image correlation · Spacer · Textile Reinforced Cement (TRC) · Transversal connection

1 Introduction

Textile Reinforced Cement Composites (TRCs) combine a cementitious mortar (matrix) with textile reinforcement. The matrix serves as the backbone of the composite, while the textiles warrant a ductile and controlled post-cracking tensile behaviour [1–3]. Several applications using TRCs are reported in scientific literature; from ventilated façade

M. Kunieda et al. (Eds.): SHCC 2022, RILEM Bookseries 39, pp. 217–227, 2023.
https://doi.org/10.1007/978-3-031-15805-6_23

panels [4], repair and strengthening realisations [5–10] to load-bearing applications such as shell structures [11–14], sandwich panels [15, 16] and pedestrian bridges [17], which demonstrate the material's potential for the realisation of lightweight, loadbearing structures.

3D textiles as reinforcement for TRCs offer significant advantages compared to traditional, planar textile reinforcement, such as superior toughness [18] and optimal fibre exploitation in flexural applications [19]. From a manufacturing perspective, the use of a singular, three-dimensional textile reinforcement entity that can be inserted in a mould and poured over, strongly aligns with current production practices for concrete. From a mechanical point of view, different studies [19–22] have shown that the presence of transversal connections in 3D TRCs can lead to a superior flexural post-cracking behaviour of the material.

3D textiles are usually manufactured by means of a combined weaving-knitting process during which both the in-plane textiles and the transversal connections are inserted gradually [23, 24]. To modify the properties of these textiles (thickness, density, geometry), time-consuming adaptations to the weaving looms have to be undertaken, which strongly hamper the production speed. Alternatively, individual spacer connections can be transversally inserted between pre-manufactured, planar textiles to achieve pseudo-3D textile architectures. These spacer connections are however usually larger than integrated knitted connections and induce significant discontinuities in the matrix material. The effect of these individual spacer connections on the mechanical properties of the 3D TRC is currently unknown.

This research investigated the effect of discrete transversal spacer connections on the flexural behaviour of TRCs, more specifically at the level of the post-cracking stiffness and the crack formation. For this purpose, four-point bending tests were performed on TRCs with two main varying parameters: the transversal spacer configuration and the in-plane textiles. Four different transversal connection configurations were considered: (i) an equivalent 2D configuration (containing the same in-plane textiles but no connections) that served as a reference, (ii) a configuration with four spacer connections distributed over the flexural shear zone of the TRC, (iii) a configuration with nine spacer connections distributed over the entire length between the supports (both shear and constant moment zone) and (iv) a knitted 3D configuration (with homogeneously distributed transversal connections). The comparison between the four configurations will give the first insights in the, to this point, unknown influence of discretely inserted spacers on the flexural behaviour of TRC composites. In future research, more specific configurations should be investigated such as spacers exclusively in the constant moment zone, in the end anchorage zone behind the supports, etc.

At the level of the in-plane textiles used, both the fibre type and the fibre volume fraction were varied. To do this, four textiles were considered: (i) a low density 3D glass textile with similar in-plane textile layers at the top and bottom, and a squared mesh geometry, (ii) a high density 3D glass textile with different in-plane textiles at the top and bottom layers, having respectively rectangular and squared mesh geometries (iii) a glass textile with higher in-plane fibre volume fraction and squared mesh geometry and (iv) a carbon textile with squared mesh geometry. Consequently, these parameter variations resulted in 13 different reinforcement configurations, each of which tested on five identical specimens per type, resulting in 65 specimens tested in bending. Apart from the investigation at the macro-mechanical level, also the effect of the spacer connections on the crack formation and propagation was assessed using the optical Digital Image Correlation (DIC) technique [25].

2 Materials and Methods

2.1 Material Selection and Reinforcement Configurations

Textiles

Four different commercially available textiles were considered in this study. Firstly, two SBR-coated 3D glass textiles with knitted transversal connections made of polyester were selected (Fig. 1 (a) and (b)). The first 3D textile had a squared mesh configuration in both layers and will be referred to as "low density" due to its large mesh size of 22.5 mm × 22.5 mm (see Table 1 for all properties). The second 3D textile had different mesh sizes in the top and bottom in-plane layers (Table 1). It however possessed two yarns per mesh in the direction with doubled mesh size, leading to an overall balanced reinforcement distribution between both layers (similarly as for the other configurations). It will be referred to as the "high density" 3D textile in the following parts.

Secondly, two planar textiles made of respectively glass and carbon were considered (Fig. 1 (c) and (d)). These textiles were selected due to similar SBR-coating, yarn density

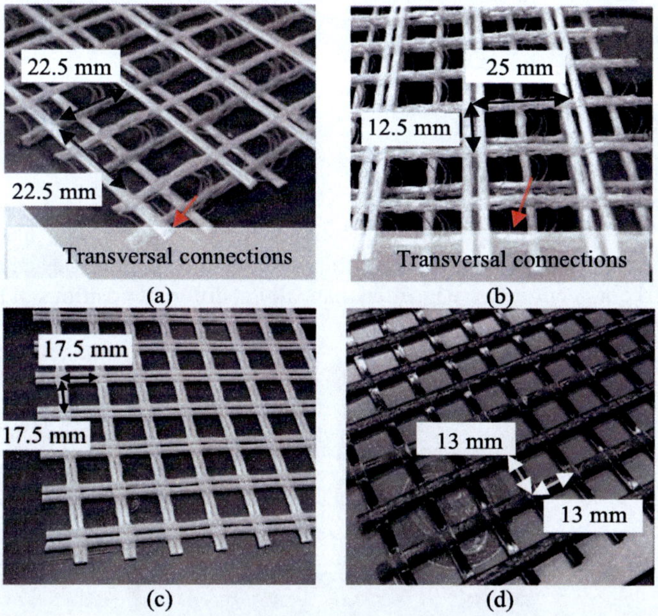

Fig. 1. (a) 3D glass textile, low density (Glass1), (b) 3D glass textile, high density (Glass2), (c) planar 2D glass textile (Glass3) and (d) planar 2D carbon textile (Carbon)

and manufacturing method as the 3D textiles. Due to their smaller mesh size, they furthermore offered a higher reinforcement ratio compared to the 3D textiles.

Table 1. Properties of the four textiles

Textile	3D glass low density	3D glass high density	2D glass	2D carbon
Name	Glass1	Glass2	Glass3	Carbon
Material	AR Glass 2400tex	AR Glass 2400tex	AR Glass 2400tex	Carbon 48K
Grid size (mm x mm)	22.5 × 22.5	12.5 × 25 (top) 12.5 × 12.5 (bottom)	17.5 × 17.5	13 × 13
Textile density coated (g/m^2)	614*	1055*	653	578
Textile density uncoated (g/m^2)	536*	917*	568	516
Yarn stiffness (GPa)	67	67	67	93
Yarn failure stress (MPa)	496	496	526	814
Coating	SBR	SBR	SBR	SBR
Transversal connection	Polyester PET	Polyester PET	/	/

* *3D textiles: the density takes into account **both** textile layers*

Connections

Apart from the knitted transversal connections (Fig. 1 (a) and (b)), discrete spacer connections were used to achieve 3D textile architectures (Fig. 2). These HF840MO polypropylene connections were inserted manually at different positions of the samples.

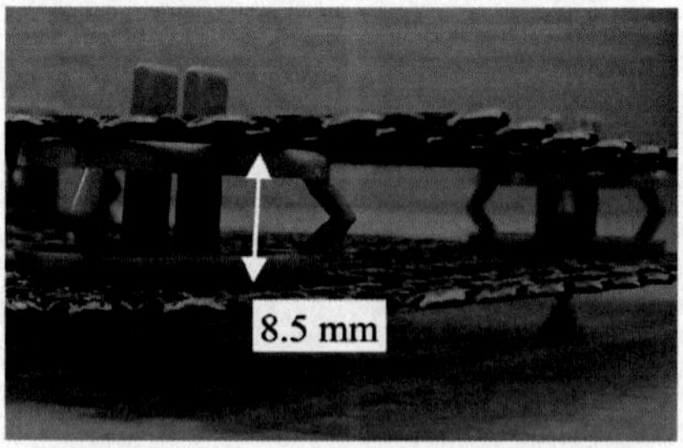

Fig. 2. Discrete spacer connections inserted between two textile layers

Geometry and Textile Configurations

The 3D textiles, Glass1 and Glass2 (Fig. 1(a) and (b)), were configured in four different ways: (i) C0(ref): an equivalent 2D configuration containing the same in-plane textiles but without connections (Fig. 3 (a)), (ii) C4: a configuration containing four discrete connections distributed over the shear zone of the specimen's length (Fig. 3 (b)), (iii) C9: a configuration containing nine discrete connections distributed in the zone between the supports (both shear zone and constant moment zone, Fig. 3 (c)) and (iv) KNIT: a 3D configuration with knitted connections at every mesh distance (Fig. 1 (a) and (b)). Table 2 gives an overview of the considered configurations for the different textiles, their respective naming and in-plane fibre volume fraction V_f in the loading direction.

All specimens had a concrete cover thickness of 2 mm at the top and bottom, two textile layers of 1.25 mm and a lever arm of 8.5 mm between the layers for a total thickness of 15 mm. The specimen length and width were respectively 450 mm and 59 mm.

Table 2. Naming of considered textile configurations

Configuration	3D glass low density, $V_f = 0.70\%$	3D glass high density, $V_f = 1.20\%$	2D glass $V_f = 1.48\%$	2D carbon $V_f = 1.94\%$
Equivalent 2D	Glass1-C0(ref)	Glass2-C0(ref)	Glass3-C0(ref)	Carbon-C0(ref)
Four connections 3D	Glass1-C4	Glass2-C4	Glass3-C4	/
Nine connections 3D	Glass1-C9	Glass2-C9	Glass3-C9	Carbon-C9
Knitted 3D	Glass1-KNIT	Glass2-KNIT	/	/

Matrix

For the matrix, a commercial grout was selected for its low viscosity before hardening to allow a manufacturing process by pouring. The mixture consists of Ordinary Portland Cement, siliceous sand and admixtures. Additionally, this grout is characterized by its compensated shrinkage and relatively small aggregate size (<1.6 mm) that allows for an optimal flow through the 3D textile architecture.

2.2 TRC Manufacturing Process

All TRC beams were manufactured in 500 mm × 500 mm × 15 mm moulds. First, an oil film was applied at the bottom of the mould. The matrix was then poured over the pre-placed textile reinforcement in the mould. After filling the mould with mortar, a plastic film was placed and the mould was sealed for 28 days to cure at room temperature. Finally, the specimens were cut from the plates in the desired dimensions of 450 mm × 59 mm × 15 mm. After cutting, the specimens' side surface was painted white and speckled for DIC monitoring of displacements and cracks during the experimental campaign.

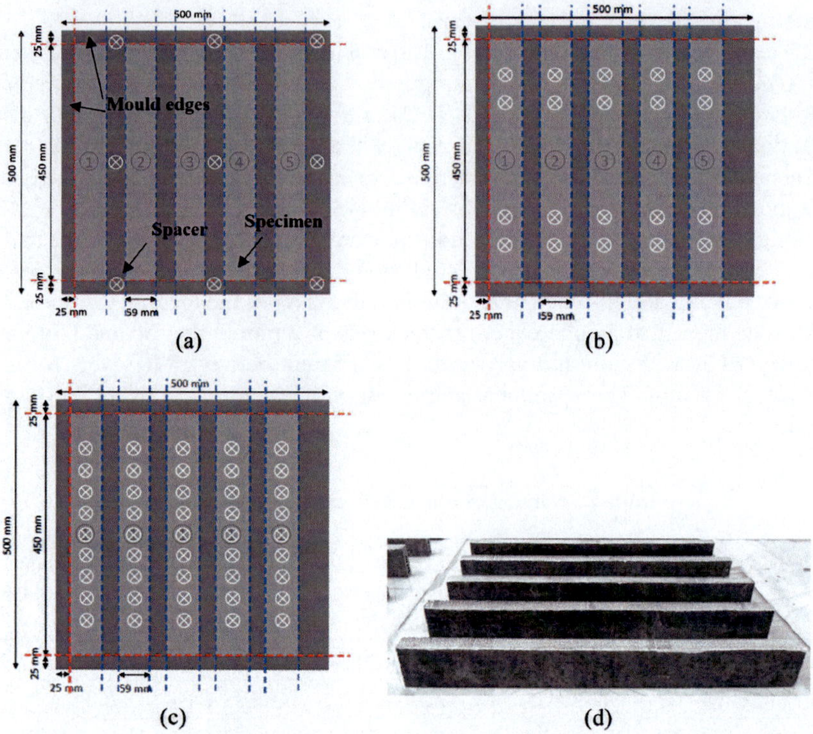

Fig. 3. Placement of discrete spacers in mould (a) C0(ref), (b) C4, (c) C9 and (d) Image of prepared specimen before testing

2.3 Experimental Test Setup

A four-point bending experimental campaign was performed. The distance between the supports and loading pins was respectively 350 mm and 100 mm. The specimens were loaded using an Instron 5885H universal electromechanical test bench with a crosshead rate of 2 mm/min. Two DIC cameras were mounted to monitor the side of the specimens. The pixel density of the cameras was 2448 pixels by 2048 pixels. The subset size, step size and strain window size were selected respectively as 21, 7 and 9. The speckle size varied between 1.75 mm and 2 mm and the resolution was 0.15 mm/pixel.

3 Experimental Results and Discussion

Figure 4 (a) to (d) show the experimental 4-point bending curves of representative specimens for each configuration. In general, all specimens of each configuration yielded repeatable results with low standard deviations. The indicated load levels on Fig. 4 (a) to (d) (horizontal dashed line) represent the levels at which the crack amounts, widths and spacings were investigated using DIC. All values and standard deviations are given in Table 3 for all the configurations.

The global macroscopic behaviour of all the specimens followed the typical three-stage response of TRCs with (i) a first, pre-cracked stage, (ii) a multiple cracking stage and (iii) a post-cracking stage dominated by fibre properties and anchorage of the textiles [3]. All specimens failed due to fibre pull-out.

Figure 4 (a) shows the experimental results of the four configurations of the Glass1 textile. These specimens had the lowest fibre volume fraction ($V_f = 0.70\%$) and therefore the lowest post-cracking stiffness of the four considered textiles (Table 3). The Glass1-C0(ref) configuration had an average crack width of 1.4 ± 1.1 mm and a post-cracking stiffness of 9.32 ± 0.51 N/mm. The configuration with knitted connections (Glass1-KNIT) showed (at the given load level of 310 N) on average smaller crack widths and a higher post-cracking stiffness (respectively 0.40 ± 0.11 mm and 20.5 ± 1.4 N/mm). These results showed that the fibre activation in the Glass1-C0(ref) configuration was limited. Compared to the Glass1-C0(ref) configuration, the configurations containing the discrete connections (Glass1-C4 and Glass1-C9) showed an increase in average post-cracking stiffness (respectively 13.59 ± 0.64 N/mm and 18.9 ± 1.2 N/mm or up to 103%). This trend suggested that the discrete spacer connections were capable of providing mechanical anchorage to the in-plane textiles and that the degree of anchorage was dependant on their placement and amount. The Glass1-C9 configuration showed a considerable decrease in amount of cracks (Table 3).

Figure 4 (b) shows the experimental curves of the Glass2 textile. Here, a smaller difference in post-cracking stiffness between the C0(ref) and the 3D configurations was observed (Table 3). This may be linked to a better inherent anchorage of the original (without spacers) in-plane textile. Glass2 had a significantly lower mesh width than the other two glass textiles. As Table 3 shows, this also resulted in a very dense and fine crack pattern: 11 cracks for Glass2-C0(ref) with an average width of 0.42 ± 0.16 mm versus 4 cracks with an average crack width of 1.4 ± 1.1 mm for Glass1-C0(ref). As a result, the addition of transversal connectors (in C0(ref), C4 or KNIT configuration) had a less pronounced effect on the anchorage and thus the post-cracking stiffness of the TRC.

Figure 4 (c) shows the experimental curves of the three considered configurations for the Glass3 textile. Differently than for the Glass1 and Glass2 textiles, the C0(ref) and C4 configurations of the Glass3 textile exhibited a similar macroscopic mechanical behaviour as well as cracking parameters (Table 3). This suggested that at higher fibre volume fractions, inserting transversal connections in the shear zone of the sample did not improve the anchorage. The Glass3-C9 configuration however clearly showed that connections placed over the entire length of the specimen did achieve a considerable post-cracking stiffness increase (from 38.5 ± 1.2 N/mm to 60.2 ± 4.5 N/mm, or an increase of 56%). These results suggested that the activation of the anchorage properties of discrete spacer connections was strongly dependant on the zone in which the connections were placed. This hypothesis was supported by the results on the Carbon textile; Fig. 4 (d) shows the experimental curves of the two considered configurations for this textile (C0(ref) and C9). Apart from an even higher post-cracking stiffness increase (from 60.1 ± 3.9 N/mm to 150.4 ± 6.2 N/mm or an increase of 150% (Table 3), both the crack amount (from six to four) and average width (from 0.20 ± 0.18 mm to 0.083 ± 0.031 mm) were strongly reduced.

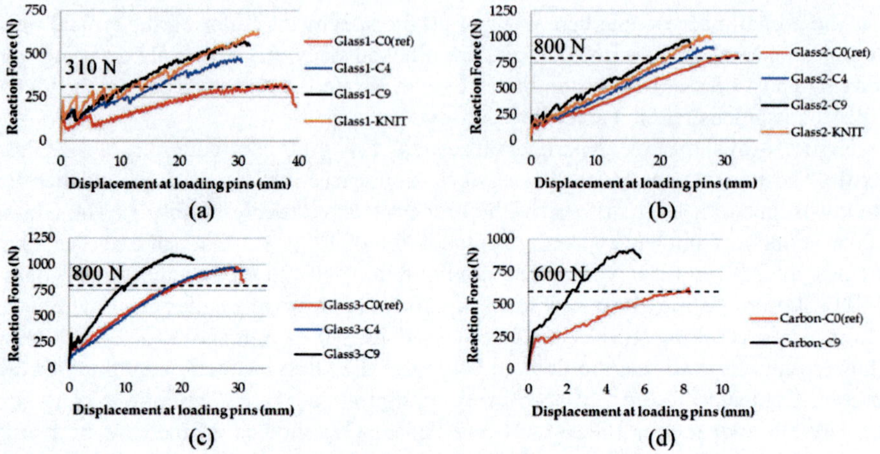

Fig. 4. Load (N) - displacement (mm) curves for the four configurations: representative curves

Table 3. Summary of experimental results for the considered textiles and configurations

Textile		**Glass1, $V_f = 0.70\%$**			
Configuration		C0(ref)	C4	C9	KNIT
Post-cracking stiffness (N/mm)		9.32 ± 0.51	13.59 ± 0.64	18.9 ± 1.2	20.5 ± 1.4
310 N	Representative crack amount*	4	9	3	5
	Crack width (mm)	1.4 ± 1.1	0.32 ± 0.14	0.58 ± 0.21	0.40 ± 0.11
	Crack spacing (mm)	33 ± 14	19.5 ± 7.4	46.1 ± 4.9	22 ± 12
Textile		**Glass2, $V_f = 1.20\%$**			
Configuration		C0(ref)	C4	C9	KNIT
Post-cracking stiffness (N/mm)		25.8 ± 1.1	30.1 ± 2.6	35.8 ± 2.3	35.3 ± 1.6
800 N	Representative crack amount*	11	14	8	17
	Crack width (mm)	0.42 ± 0.16	0.22 ± 0.12	0.46 ± 0.29	0.174 ± 0.082
	Crack spacing (mm)	16.0 ± 6.0	13.2 ± 5.4	20.4 ± 8.7	9.7 ± 3.9
Textile		**Glass3, $V_f = 1.48\%$**			
Configuration		C0(ref)		C4	C9
Post-cracking stiffness (N/mm)		38.5 ± 1.2		38.3 ± 1.8	60.2 ± 4.5
800 N	Representative crack amount*	8		7	4

(*continued*)

Table 3. (*continued*)

	Crack width (mm)	0.45 ± 0.27	0.34 ± 0.21	0.58 ± 0.30
	Crack spacing (mm)	18.4 ± 8.0	15.3 ± 5.3	26 ± 11
Textile		**Carbon, V_f = 1.94%**		
Configuration		C0(ref)	C9	
Post-cracking stiffness (N/mm)		60.1 ± 3.9	150.4 ± 6.2	
600 N	Representative crack amount*	6	4	
	Crack width (mm)	0.20 ± 0.18	0.083 ± 0.031	
	Crack spacing (mm)	22.6 ± 8.0	28 ± 14	

* *Representative crack amounts correspond to all the cracks of the representative curves*

4 Conclusions

This research evaluated the influence of discrete spacer connections on the flexural behaviour of TRC composites. A flexural experimental campaign was conducted on specimens containing four different textiles (three glass fibre textiles with increasing fibre volume fraction (Glass1–3) and one carbon textile), each containing two to four transversal connection configurations (C0(ref), C4, C9 and KNIT).

The results of the Glass1 and Glass2 textiles showed that the discrete spacer connections were able to provide similar anchorage mechanisms as the integrated, knitted connections in 3D TRCs. The anchorage was more pronounced when the connections were distributed evenly over the entire zone between the supports and present in higher amounts. A hypothesis was provided for the local anchorage mechanism provided by the spacers: these connections form local bridges between the top and bottom textile layers which hinder the (longitudinal) relative displacement of these in-plane textiles. Hence, this anchorage effect is believed to be more effective towards the supports and the edges of the samples where the relative displacement is larger. This hypothesis should still be validated in future research.

The experimental results put into evidence that discrete spacer connections can have a positive mechanical influence that is currently not considered in TRC design, even though their (original) purpose is to warrant textile spacing during manufacturing. It was also shown that this influence was dependant on the reinforcement volume fraction. At lower fibre volume fractions (Glass1 textile), both configurations with the discrete spacer connections (C4 and C9) achieved high contributions to the post-cracking stiffness due to their localized anchorage that served as discrete connections between the top and the bottom textile layers. At higher fibre volume fractions, only the configurations containing connections distributed evenly over the zone between the supports (C9) achieved a notable post-cracking stiffness increase and improvement at the level of crack formation.

Additionally, distributing the connections in both the shear and constant moment zone led to a strong reduction in amount of cracks for all considered textiles. This is generally an undesired property, since a larger amount of cracks with small crack widths are usually preferred. For the Carbon textile however (highest V_f of 1.94%), the C9 configuration exhibited a reduction of both the crack amount and width, as the post-cracking stiffness was significantly increased, which is strongly desired in TRCs.

Acknowledgments. This research was performed within the Interreg project "Beton naar hoogwaardig beton". The authors would like to thank Interreg for the provided funding of the research.

References

1. Brameshuber, W.: Textile Reinforced Concrete: State of the Art Report. RILEM Technical Committee 201-TRC, 1st edn. RILEM Publications, Aachen (2006)
2. Bentur, A., Mindess, S.: Fibre Reinforced Cementitious Composites, 601 pp. Taylor and Francis, New York (2007)
3. Aveston, J., Cooper, G.A., Kelly, A.: Single and multiple fracture. In: The Properties of Fibre Composites, pp. 15–24 (1971)
4. Hegger, J., Kulas, C., Horstmann, M.: Spatial textile reinforcement structures for ventilated and sandwich facade elements. Adv. Struct. Eng. **15**, 665–675 (2016)
5. Triantafillou, T., Papanicolaou, C.G.: Shear strengthening of reinforced concrete members with textile reinforced mortar (TRM) jackets. Mater. Struct. **39**, 93–103 (2006)
6. Brückner, A., Ortlepp, R., Curbach, M.: Anchoring of shear strengthening for T-beams made of textile reinforced concrete. Mater. Struct. **41**, 407–418 (2008). https://doi.org/10.1617/s11527-007-9254-9
7. Contamine, R., Si Larbi, A., Hamelin, P.: Identifying the contributing mechanisms of textile reinforced concrete (TRC) in the case of shear repairing damaged and reinforced concrete beams. Eng. Struct. **46**, 447–458 (2013)
8. Si Larbi, A., Agbossou, A., Hamelin, P.: Experimental and numerical investigations about textile-reinforced concrete and hybrid solutions for repairing and/or strengthening reinforced concrete beams. Compos. Struct. **99**, 152–162 (2013)
9. Verbruggen, S., Remy, O., Wastiels, J., Tysmans, T.: Stay-in-place formwork of TRC designed as shear reinforcement for concrete beams. In: Advances in Materials Science and Engineering, 9 pp. (2013)
10. Peled, A., Bentur, A., Mobasher, B.: Textile Reinforced Concrete, 1st edn., 473 pp. (2017)
11. Hegger, J., Horstmann, M., Zell, M.: Applications for TRC. In: The 15th International Congress of the International Glassfibre Reinforced Concrete Association, 10 pp. (2008)
12. Tysmans, T., Adriaenssens, S., Cuypers, H., Wastiels, J.: Structural analysis of small span textile reinforced concrete shells with double curvature. Compos. Sci. Technol. **69**, 1790–1796 (2009)
13. Scholzen, A., Chudoba, R., Hegger, J.: Thin-walled shell structures made of textile-reinforced concrete Part I: Structural design and construction, pp. 106–114 (2015)
14. Verwimp, E., Tysmans, T., Mollaert, M., Berg, S.: Experimental and numerical buckling analysis of a thin TRC dome. Thin-Walled Struct. **94**, 89–97 (2015)
15. Cuypers, H., Wastiels, J.: Analysis and verification of the performance of sandwich panels with textile reinforced concrete faces. J. Sandwich Struct. Mater. **13**, 589–603 (2011)
16. Vervloet, J., et al.: Validation of a numerical bending model for sandwich beams with textile-reinforced cement faces by means of digital image correlation. Appl. Sci. **9**, 1253–1269 (2019)

17. Helbig, T., Rempel, S., Unterer, K., Kulas, C., Hegger, J.: Fuß-und Radwegbrücke aus Carbonbeton in Albstadt-Ebingen. Die weltweit erste ausschließlich carbonfaserbewehrte Betonbrücke. Beton-und Stahlbetonbau, vol. 111. pp. 676–685 (2016)
18. Zhu, D., Mobasher, B., Peled, A.: Experimental study of dynamic behavior of cement-based composites. Constr. Build. Mater. **25**, 12 pp. (2011)
19. Amzaleg, E., Peled, A., Janetzko, S., Gries, T.: Flexural behaviour of cement based element reinforced with 3D fabric. In: VIII International Conference on Fracture Mechanics of Concrete and Concrete Structures, 10 pp. (2013)
20. Peled, A., Haik, R., Sasi, E. A.: Influence of three-dimensional (3D) fabric orientation on flexural properties of cement-based composites. Cem. Concr. Compos. **80** (2017)
21. El Kadi, M., et al.: Experimental study and benchmarking of 3D textile reinforced cement composites. Cem. Concr. Compos. **104**, 10 pp. (2019)
22. El Kadi, M., Kapsalis, P., Van Hemelrijck, D., Wastiels, J., Tysmans, T.: Influence of loading orientation and knitted versus woven transversal connections in 3D textile reinforced cement (TRC) composites. Appl. Sci. **10**, 15 pp. (2020)
23. Ciubanu, L.: Development of 3D knitted fabrics for advanced composite materials. In: Advances in Composite Materials – Ecodesign and Analysis, pp. 161–192 (2011)
24. Roye, A., Gries, T.: 3-D textiles for advanced cement based matrix reinforcement. J. Ind. Text. **37**(2), 163–173 (2007)
25. Sutton, M.A., Orteu, J.-J., Schreier, H. W.: Image Correlation for Shape, Motion and Deformation Measurements. Springer, Cham (2009). https://doi.org/10.1007/978-0-387-78747-3

Durability

Check for updates

Experimental Study on Autogenous Healing of Cracked SHCC Under Sustained Bending Loading

Yao Luan[1](✉), Keita Suzuki[1], Satoru Sakuma[2], and Katsuhiko Hirano[2]

[1] Saitama University, Saitama 338-8570, Japan
luanyao@mail.saitama-u.ac.jp
[2] East Nippon Expressway Company Limited, Tokyo 100-8979, Japan

Abstract. SHCCs have been widely reported for their distinct autogenous healing property of the multiple fine cracks because of fiber bridging effect and remarkable precipitation of secondary products in cracks. However, most of previous studies performed autogenous healing tests of SHCC without simultaneous loading, i.e., applying healing environment to specimens after removing the pre-cracking load. Experiment studies on autogenous healing of cracked SHCC under sustained loading, which is closed to serving conditions of real structures, are still limited. In this study, the authors performed autogenous healing experiment of SHCC specimens under sustained bending loading. Small SHCC beams with steel rebar reinforcement at tensile side were prepared. A setup that can offer sustained load was used to apply four-point bending to the beam. Then, the tensile side of the beam, where multiple bending cracks took place, were immersed in water for autogenous healing in more than one month during when the bending load was sustained. For the control group, the same healing operation was applied but the load was unloaded before that. The crack widths were observed using microscopy periodically to study the healing degree. The results indicated that, although all of the specimens exhibited crack healing in certain degrees, the specimens with sustained load showed slower healing process than those unloaded, leading to higher water absorption after the healing operation.

Keywords: SHCC · Autogenous healing · Multiple fine cracks · Crack width · Sustained loading · Four-point bending

1 Introduction

Strain-hardening cementitious composites (SHCC) are a class of fiber-reinforced cementitious materials characterized by superior tensile ductility and multiple cracks of tiny widths because of the bridging of fibers. SHCC has gained extensive interest in practical construction at present, being used as structural and non-structural members as well as repair and strengthening materials. Typical examples are the combination of SHCC and steel reinforcement or reinforced concrete structures (R/SHCC) [1]. In addition to the high ductility and dispersed fine cracks, it has been widely reported that the bridging

M. Kunieda et al. (Eds.): SHCC 2022, RILEM Bookseries 39, pp. 231–238, 2023.
https://doi.org/10.1007/978-3-031-15805-6_24

effect of fibers in cracked SHCC accelerates the precipitation of secondary products in the cracks, leading to an autogenous healing process [2–4], which is beneficial for long-term durability of structures.

Regarding the studies on autogenous healing of SHCC, however, many researchers performed healing operation on specimens after unloading the pre-cracking load. For the use of SHCC in practical structures, it is important to investigate the influence of sustained loading on the effect of autogenous healing. Ozbay et al. [5] performed autogenous healing tests for SHCC beams with sustained bending load, indicating less stiffness recovery of the beams than those without sustained load. On the other hand, Yildirim et al. [6] tested ultrasonic pulse velocity of SHCC specimens with and without sustained load after subjected to 150-day moist curing, reporting that recovery results were similar. In general, the studies on the autogenous healing of SHCC under sustained loading are still limited, particularly the healing degree of cracks and the influence on further durability. In the present study, the authors carried out autogenous healing experiment of R/SHCC beams under sustained bending loading by comparing with those without sustained load. Crack healing process were observed and after the healing operation water absorption test was performed.

2 Experimental Program

2.1 Materials and Mixture

The mix proportion of SHCC is shown in Table 1. PVA fibers with a specific gravity of 1.3 were added with a volume fraction of 2%, while the volume fraction of sand in mixture was 45%. The length of the fibers was 12 mm and the diameter 40 μm. The tensile strength and Young's modulus were 1,560 MPa and 41 GPa. High early strength Portland cement and the fly ash of type II in Japanese Industrial Standards (JIS) were used as cementitious powder with a ratio of 1 to 1 in weight. Water-to-binder ratio was 0.3. Superplasticizer was added with 0.8% of the powder in weight. Uniaxial tensile test was conducted to confirm the ductility and cracking performance of the SHCC using a dumbbell-shaped specimen that had a central part with the size of 13 x 30 x 80 mm. The strain-stress curve of a specimen is shown in Fig. 1.

Table 1. The mixture of the SHCC (kg/m^3)

Water	Cement	Fly ash	Sand	PVA fiber	Superplasticizer
333	555	555	500	26	8.9

2.2 Four-Point Bending Test and Autogenous Healing Operation

Small cuboid R/SHCC beams with a cross section of 40 x 70 mm and length of 550 mm were prepared for the test (Fig. 2). Bending load was applied in the center span of

150 mm and the beam was supported at both sides with the distance of 150 mm from the loading point. A ribbed steel rebar with a nominal diameter of 6.35 mm was placed longitudinally in the tensile side to disperse cracks. After casting and demolding, all the beams were cured in water at 20 °C to the age of 28 days, followed by the four-point bending test.

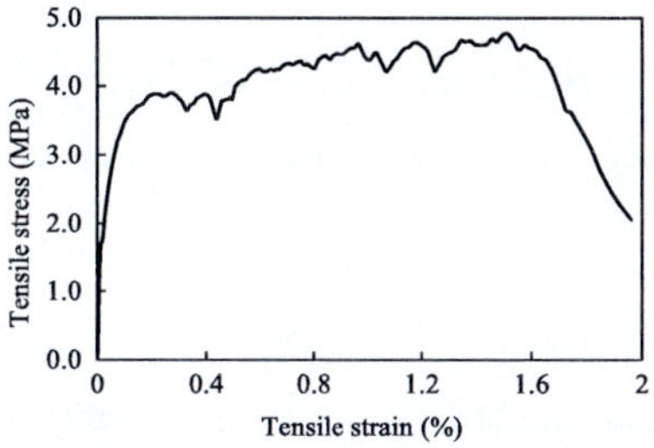

Fig. 1. The uniaxial tensile test result

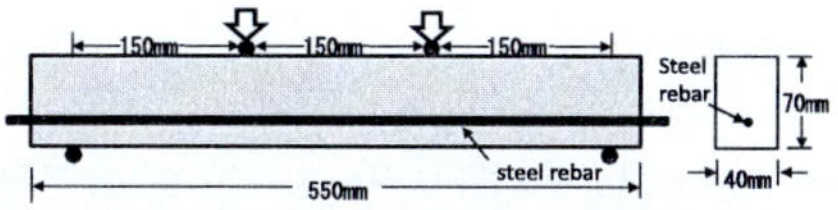

Fig. 2. The small R/SHCC beams

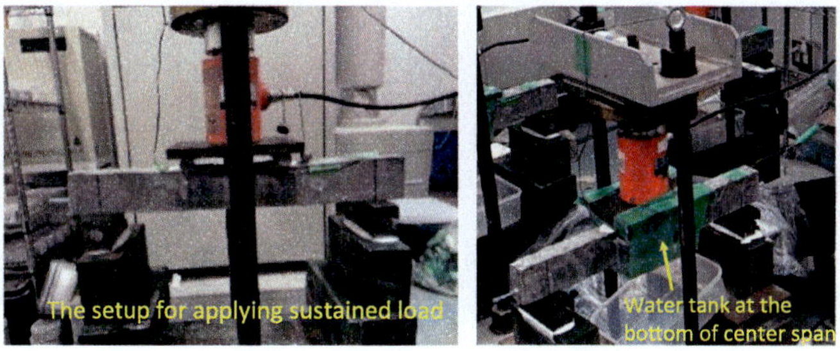

Fig. 3. The setup for applying sustained load and the operation of autogenous healing

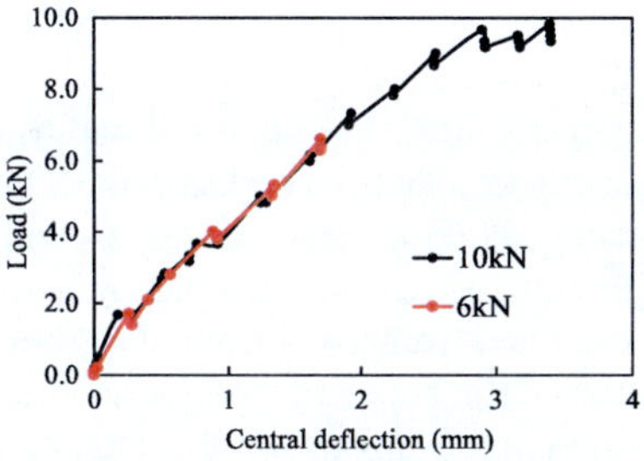

Fig. 4. Deflection-load curves

Fig. 5. Multiple cracks in a beam

A compressive creep apparatus was used to apply the load (Fig. 3). After placing the beam to the setup, the load was applied by gradually fastening the nuts on the top plate. Simultaneously, the load and the deflection at the bottom center of the beam were measured using load cell and displacement transducer. Strain at the compressive

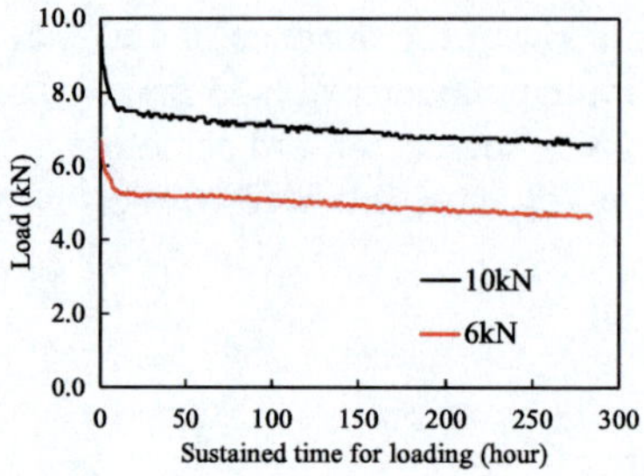

Fig. 6. Load decreasing with sustained time

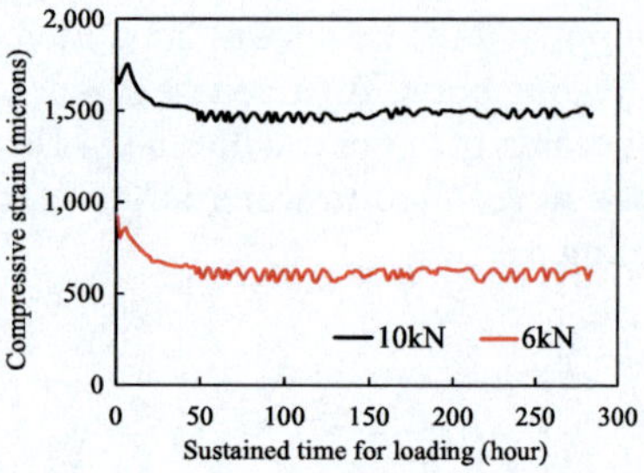

Fig. 7. Compressive strain with sustained time

edge of the beam was measured with pre-attached strain gauges. The load cell, displacement transducer, and strain gauges were connected to a data logger for continuous measurement.

Two maximum loads were applied, respectively, causing different damage degrees to the beams. One was 6 kN and the other was 10 kN, approximately. The displacement-load curves are shown in Fig. 4. At 6 kN the curve was linear, indicating that the steel rebar was in elastic state. In contrast, when the load approached approximately 10 kN, the load did not increase any longer while the deflection kept on increasing, implying the yielding of the rebar. For all the specimens applied with 10 kN, the deflection was kept as approximately 3.6 mm to achieve similar deformation, which implied similar damage degrees.

When the above loads and deflection were reached, the bottom and side surfaces of the beam were observed. At both loads, multiple cracks were confirmed in the tensile side. An example is shown in Fig. 5. In the center span between the loading points, bending cracks formed in the vertical direction. In the span between the loading point and the corresponding supporting point, some beams exhibited diagonal cracks with tiny width, but shear failure did not take place. This indicates that the tensile ductility of SHCC increased shear capacity, even though there were no stirrups. In this study, only bending cracks in the center span were studied. The cracks at bottom and side surfaces were accounted as different ones when calculating crack number. The initial widths of the cracks were measured using microscopy. For the cracks at bottom surfaces, the width was measured in the center of each crack. For the cracks at side surfaces, the largest width along each crack was approximately at the edge of bottom, which was recorded. After the measurement of initial crack widths, a water tank was fixed from the bottom of the beam across the center span using glue tape. The gap between the tank and the specimen was sealed with sealant to prevent leakage. Water was filled in the tank to two-third height of the beam to promote autogenous healing. At 7, 14, 21, 28, and 35 days, the tank was taken off from the beam and observation of crack healing was conducted using microscopy. At the same location of the initial measurement, the crack width was measured as the distance between the newly formed edges due to autogenous healing. The sustained load and compressive strain were monitored continuously to 12 days (Figs. 6 and 7) and then was stopped because of power outage. It can be seen that the sustained load kept on decreasing with increasing sustained loading time. The rate of load decreasing was particularly obvious at initial days. Consequently, the strain at the compressive edge was found to decrease with sustained loading time, but the values tended to converge after initial days in spite of slight fluctuation. The decrease of sustained load with time was probably because of both stress relaxation of SHCC and the gradual loosening of the apparatus, but the nuts were not fastened again to avoid any disturbance of the cracks and healing process.

In addition to the beams with sustained load, experiment on the control groups were performed in which the same load was applied using the setup, but the operation of autogenous healing was performed after unloading. Therefore, the study included four groups: 1) 6 kN, with sustained load; 2) 6 kN, unloaded; 3) 10 kN, with sustained load; 4) 10 kN, unloaded. In each group, two beams were prepared and tested. For the beams unloaded, initial crack widths were observed after unloading.

After the operation of autogenous healing, the part in the center span, which was a cuboid with the size of 40 × 70 × 150 mm, was cut off. The specimen was pre-dried at 40 ℃ in an over until constant weight. Subsequently, a water absorption test was carried out by immersing the bottom surface, where multiple bending cracks took place, in water and maintaining an immersion depth of 3 mm. Water absorption was measured by a gravimetric method.

3 Results and Discussion

The distributions of initial crack widths are summarized in Fig. 8. Note that the crack widths of the unloaded beams were measured after unloading. For the cases of 6 kN, all the cracks of the beams without sustained load were within 40 μm, whereas the crack widths were comparably larger for those with sustained load, ranging from 40 to 140 μm. The reason is that unloading caused the release of stress at the cross section, resulting in the decrease of crack width. Regarding the cases of 10 kN, the crack widths were remarkably larger than those of 6 kN. For the beams unloaded, the cracks principally ranged from 40 to 200 μm in width with the largest crack number around 60 μm, and localized big cracks about 320 μm were found. In comparison, the beams with sustained load exhibited most of cracks in the range of 40 to 200 μm with the largest crack number around 100 μm, and more localized big cracks were found in the range of 240 to 360 μm than the unloaded ones. Therefore, similar with the 6 kN cases, the 10 kN beams unloaded exhibited smaller crack widths than those with sustained load, which can also be attributed to the unloading and the resulted release of stress.

During the operation of autogenous healing, all the cracks were found to be healed in more or less degrees with precipitates forming in cracks. Figure 9 shows the healing process of a 20 μm wide crack in a beam with unloaded 6 kN. It can be seen that after 14 day, the crack clogged completely. The crack healing ratio was calculated by dividing the change of crack width by the initial crack width, following Zhang et al.'s definition [7] below

$$R_{heal} = \frac{W_{ini} - W_n}{W_{ini}} \times 100\% \tag{1}$$

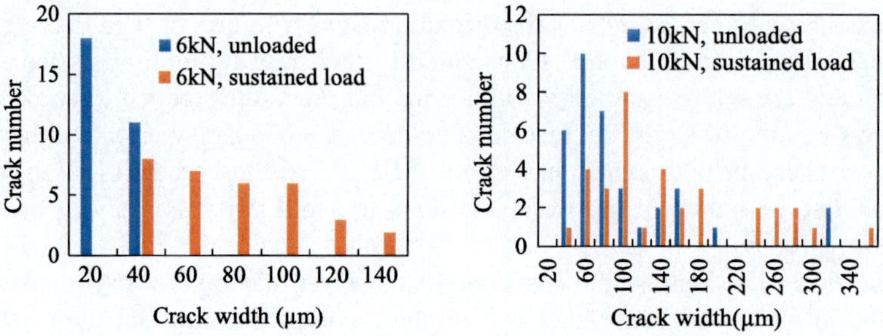

Fig. 8. Histogram of crack width distribution: (left) 6 kN; (right) 10 kN

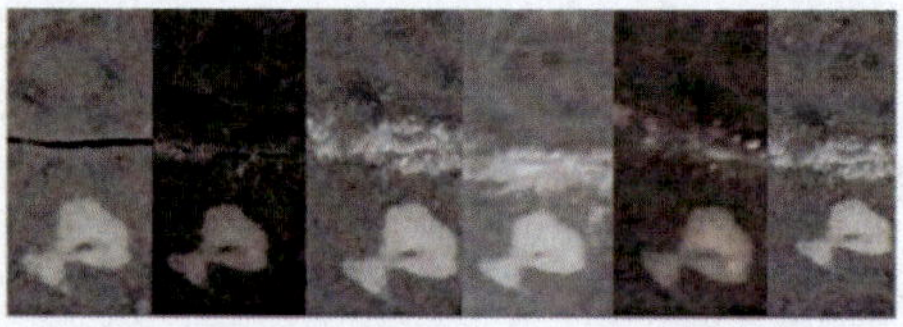

Fig. 9. Healing of a crack during immersion (6 kN, unloaded, initial crack width 20 μm)

where R_{heal} is healing ratio, W_{ini} is initial crack width, and W_n is the crack width after n days of healing.

Figure 10 shows the summary of crack healing ratios in terms of initial crack width, regardless of the maximum load and whether it was sustained load or unloading. The error bars indicate standard error of the data. The crack widths are categorized into three groups: <100 μm, 101–200 μm, >200 μm. The results clearly indicate that even though all the cracks were healed in certain degrees, smaller cracks tended to exhibit faster healing process. For the cracks small than 100 μm, the average healing ratio was above 90% after 28 days, while the average ratio of the cracks above 200 μm was less than 70% after 35 days. The dependence of crack healing ratio on initial crack width is consistent with Zhang et al.'s study [7] in which three-point bending load was applied to R/SHCC beams and autogenous healing was performed after unloading, but the healing rates in this study were faster. One possible reason is that the cracks were kept on immersed in water, but water fog or water/dry cycles were applied in their study. Another reason may be that high early strength Portland cement was used in this study, which produced more $Ca(OH)_2$ to activate the pozzolanic reaction of fly ash, while in their study ordinary Portland cement was used.

The healing ratios were also compared between the beams with sustained load and unloaded. As Fig. 8 shows, the beams with sustained load exhibited cracks in a wider range of crack widths than those unloaded. Therefore, the healing ratios were compared only for the cracks with similar widths, shown in Fig. 11. For the cases of 6 kN, the healing ratios were compared for the cracks within 40 μm, while for the cases of 10 kN it was compared within 80 μm. It is noted that for both loads, the cracks in the cases of sustained load exhibited slower healing process than those of unloading, indicating that sustained load may present a negative effect on self-healing. The mechanism has not yet been fully clarified at present. More study on this issue needs to be carried out in the future.

The result of water absorption test after autogenous healing is shown in Fig. 12. The values were calculated by averaging those of the two specimens in each group. Compared to those with 6 kN, the specimens with 10 kN exhibited significantly higher water absorption. This can be attributed to more severe damage under higher loading, which featured more cracks and larger crack widths. As indicated in Fig. 10, even after 35-day healing, large cracks, particularly those above 200 μm, exhibited only partial healing degrees. The cracks were still highly accessible to water, leading to high capillary absorption of the overall specimen. Looking at the influence of sustained load, for the cases of 10 kN, the specimens with sustained load exhibited higher water absorption than those unloaded. This can be attributed to the above-mentioned delayed healing effect, particularly for cracks with larger width. In contrast, there was only a slightly difference in absorption between the two 6 kN cases. According to Figs. 8 and 10, most of cracks in the 6 kN beams with sustained loads were below 100 μm, which had been healed completely or in high degrees after the 35-day operation. Therefore, it can be speculated that the beams achieved similar recovery with the unloaded ones.

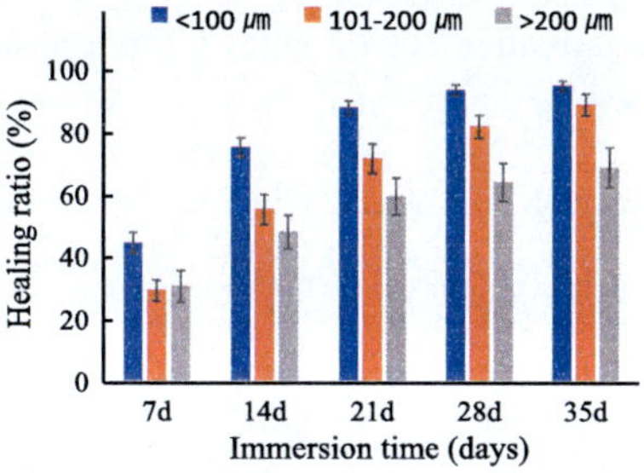

Fig. 10. Healing ratios of the cracks in terms of initial crack width

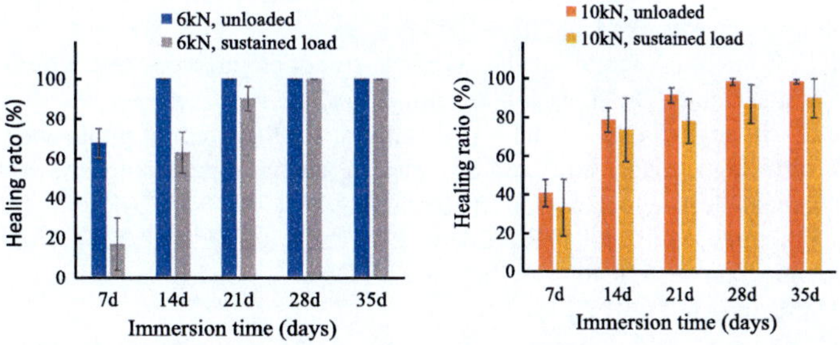

Fig. 11. Comparison of crack healing between beams unloaded and with sustained load. (left) 6 kN, initial crack widths within 40 μm; (right) 10 kN, initial crack widths within 80 μm.

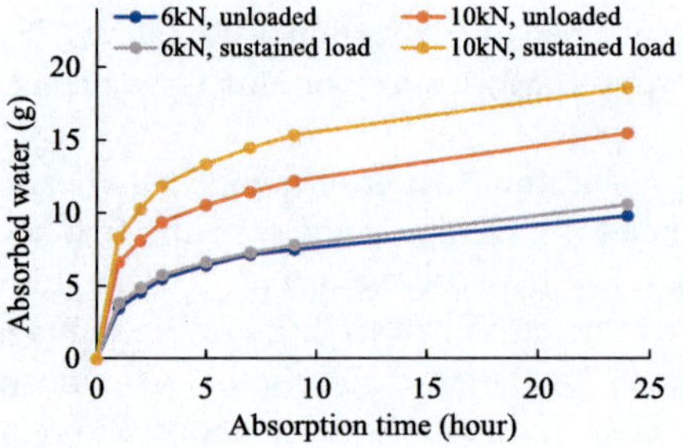

Fig. 12. The result of water absorption test

4 Conclusions

In this study, the influence of sustained bending load on the autogenous healing of SHCC was studied experimentally. Small R/SHCC beams with steel rebar reinforcement were prepared and applied with four-point bending load that was sustained during the operation of autogenous healing. The change of crack widths due to autogenous healing were observed using microscopy. The results indicated that, although all of the specimens exhibited crack healing in certain degrees, the specimens with sustained load showed delayer healing than those unloaded. The result of water absorption test indicated that the specimens with sustained load and less healing degrees exhibited higher absorption.

Acknowledgement. The authors thank Kuraray Co., Ltd., for providing the PVA fibers.

References

1. Mechtcherine, V.: Towards a durability framework for structural elements and structures made of or strengthened with high-performance fibre-reinforced composites. Constr. Build. Mater. **31**, 94–104 (2012)
2. Şahmaran, M., Li, V.C.: Durability properties of micro-cracked ECC containing high volumes fly ash. Cem. Concr. Res. **39**(11), 1033–1043 (2009)
3. Kan, L., Shi, H.: Investigation of self-healing behavior of engineered cementitious composites (ECC) materials. Constr. Build. Mater. **29**, 348–356 (2012)
4. Kunieda, M., Choonghyun, K., Ueda, N., Nakamura, H.: Recovery of protective performance of cracked ultra high performance-strain hardening cementitious composites (UHP-SHCC) due to autogenous healing. J. Adv. Concr. Technol. **10**(9), 313–322 (2012)
5. Ozbay, E., Sahmaran, M., Yucel, H.E., Erdem, T.K., Lachemi, M., Li, V.C.: Effect of Sustained Flexural Loading on Self-Healing of Engineered Cementitious Composites. J. Adv. Concr. Technol. **11**(5), 167–179 (2013)
6. Yildirim, G., Alyousif, A., Şahmaran, M., Lachemi, M.: Assessing the self-healing capability of cementitious composites under increasing sustained loading. Adv. Cem. Res. **27**(10), 581–592 (2015)
7. Zhang, P., Dai, Y., Ding, X., Zhou, C., Xue, X., Zhao, T.: Self-healing behaviour of multiple microcracks of strain hardening cementitious composites (SHCC). Constr. Build. Mater. **169**, 705–715 (2018)

How Does Self-healing Under Sustained Loadings in Aggressive Water Affect the Constitutive Response of a UHPFRC?

Salam Al-Obaidi(✉), Marco Davolio, Giovanni Recchia, Francesco Lo Monte, and Liberato Ferrara

Department of Civil and Environmental Engineering, Politecnico di Milano, Piazza Leonardo Da Vinci 32, Milan, Italy
salammaytham.alobaidi@polimi.it

Abstract. In the present study, the effects of sustained load combined with aggressive exposure on the long term performance of UHPFRC has been investigated. Three different materials have been tested: a reference one, containing a crystalline admixture as promoter of the autogenous self-healing and two further matrices, additionally enhanced with alumina nano-fibers and cellulose nano-crystals. The aim of adding these functionalizing nano-constituents is to work on the micro- and nano-structure of the matrix and enhance the durability in the cracked state. For this reason, specimens (100 × 30 × 500 mm) were pre cracked to a target crack width level. A couple of specimens was self-contained in suitable 4PBT setup to exert a constant flexural sustained load while being exposed to different exposure conditions of tap water, 3.3% chloride aqueous solution, and geothermal water. The study is proposed to elaborate different nondestructive and destructive measurements to evaluate the self-healing efficiency and its impact on the mechanical performance of the specimens, specifically the tensile constitutive response. Direct tensile test, and 4PBT have been used to assess healing efficiency by testing the conditions of the specimens before damage, after damage and after simultaneous mechanical and chemical exposure. Inverse analysis has been applied to the 4PB curves to obtain the tensile constitutive laws before and after crack localization. The obtained results proved that the autogenous self-healing of UHPFRC, as an active process, not only allows to recover the pristine condition of the cracked specimens, but also to achieve higher tensile capacity, attributed to the healing particles precipitating in the distributed cracks in the damaged area.

Keywords: Sustained loading · Self-healing · Durability performance · UHPFRC · Chemical and mechanical loadings

1 Introduction

The recent development of producing UHPFRC with strain hardening in flexural and uniaxial tensile stress that can reach 30 MPa and 10 MPa respectively [1] has oriented the intention of exploiting these materials in real case applications without using traditional reinforcements and with less thickness to account for its strength and reduce

M. Kunieda et al. (Eds.): SHCC 2022, RILEM Bookseries 39, pp. 239–248, 2023.
https://doi.org/10.1007/978-3-031-15805-6_25

the consumption of the materials [2]. Moreover, UHDC (Ultra High Durability Concrete), an emerging class of UHPFRC that incorporates nano and micro constituents to enhance the durability in cracked state and under extremely aggressive scenarios, has been introduced to the market recently [3]. This development not only solved the issue of mechanical performance but also tackled the problem of durability in the cracked state through synergistic interactions among its nano scale constituents and components (1.5% steel fibers, nano materials, and self-healing promoters). Self-healing methods can be found in the literature in two main categories: autogenous self-healing, which does not need man-made intervention, and autonomous self-healing, which can be made by injecting healing products either by dispersed capsules [4] or vascular networks [5]. Despite all the novelties of the autonomous self-healing, the autogenous self-heling is the simplest approach and draws the researchers' intention to study the effect of crack width, exposure type, exposure period, and cementitious materials on the efficiency of self-healing. The typical methodology involves induced damage to the specimens by loading them to a certain level, exposure under the desired condition in an unloaded stress-free state, and study of the self-healing mechanism and its effect on the material/specimen response after a certain time of exposure [6]. The reported results in terms of durability, transport and mechanical properties can be overestimated since the unloading state of the specimens can reduce the crack width to half of its value when under loading. The other crucial point is the creep, that can adversely interact with self-healing by continuously keeping the crack width active all along the service life. Therefore, the sustained loading influence on the self-healing efficiency needs to be investigated by simulating realistic service conditions. Despite all the interesting findings of self-healing studies, the practical demands concerning the self-healing effects in advanced materials such as UHPFRCs are more related to the evolution of mechanical performance with time, and specifically the tensile performance since it is the main attacked or beneficiary performance from cracking-healing mechanisms, and the most decisive property for design parameters. Moreover, studying UHPFRCC specimens that have been produced in the construction site would be appropriately representing the real case study.

This study investigates the performance of UHPFRC thin beams produced at a construction site and subjected to contemporary mechanical and environmental actions. The scope of the study extends to cover the synergic effects of nano and micro additives on the autogenous self-healing of cracked UHPFRC. The two main variables of the study are the exposure types and UHPFRC mix ingredients. The main objective of the study is to evaluate the performance of the cracked UHPFRC over six months exposure period based on tensile stress response, that can be extracted with two different approaches (direct test and inverse analysis from flexural test results), and associate it with types and periods of exposure.

2 Experimental Program

2.1 Materials and Basic Mechanical Properties

The H2020 ReSHEALience project has developed, as an upgrade of UHPFRC, the concept of Ultra High Durability Concrete (UHDC), defined as "a strain-hardening fibre reinforced cementitious material with macro and nano scale functionalized constituents especially added to obtain a high durability in the cracked stage under extremely aggressive exposure conditions". In the same framework, real structures made of UHDC materials have been designed and built to prove the concept of the newly adopted materials. One of these structures is the (Enel Green Power Plant) EGP pilot which is a basin to collect geothermal water in a geothermal power plant which is owned and run by the Enel Green Power company in Chiusdino, Italy.

The basin has been designed and built with three different concepts and materials, each implemented in one of the three compartments the basin is divided into; the walls of the first compartment are 100 mm thick and made of ordinary reinforced concrete; the walls of the second cell are 60 mm thick and made of UHDC and the walls of the third cell consist of 30 mm precast slabs supported by 200×200 mm^2 UHDC columns; a more detailed description of the pilot can be found in [2]. During the casting of the third compartment, extra precast slabs with dimensions of 165×150 mm^2 have been casted and shipped to Politecnico di Milano (Laboratory for Testing Materials Buildings and Structures) to perform mechanical and durability tests. Small beams with dimensions of ($100 \times 30 \times 500$ mm) have been obtained from the precast UHDC elements. Table 1 describes the mix-design of the different investigated UHDCs, distinguishing three types of batches: nano Alumina (XA-CA+ANF), nano Cellulose (XA-CA+CNC), and without nano materials, assumed as a reference (XA-CA). Mechanical properties have been evaluated for these materials at different scale and configurations, showing an appreciable ductility performance particularly in the tensile behavior [1].

Table 1. UHDC mix. components and their proportions

Constituents	XA-CA	XA-CA+ANF	XA-CA+CNC
CEM I 52,5 R	600	600	600
Slag	500	500	500
Water	200	200	200
Steel fibers Azichem Readymesh 200 ®	120	120	120
Sand 0–2 mm	982	982	982
Superplasticizer Glenium ACE 300 ®	33	33	33
Crystalline Admixture Penetron Admix ®	4.8	4.8	4.8
Alumina nanofibers NAFEN ®*	–	0.25	–
Cellulose nanocrystals Navitas ®*	–	–	0.15

* % by weight of cement

2.2 Experimental Methodology

In order to induce the controlled damages, the specimens were pre-cracked under 4 point-bending tests; tests were performed with 100 kN MTS machine and at 0.15 mm/s displacement control. All specimens were loaded to obtain crack level of 0.2 mm at the tensile face, where two 10 mm stroke LVDTs have been used to monitor the evolution

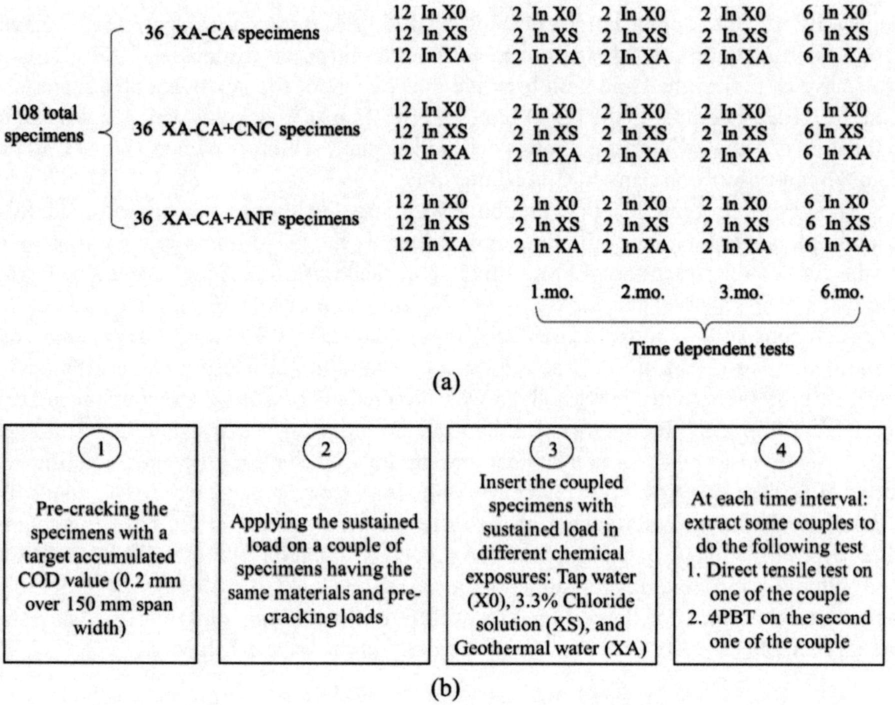

Fig. 1. The layout of the experimental plan; (a) exposure types and specimens' types and numbers, (b) the sequence of the tests

of the Crack Opening Displacement (COD) over 150 mm distance between the point loads. Upon reaching the target crack level, the specimens were unloaded with the same displacement loading rate. The recorded maximum pre-cracking load was used as the maximum sustained load level to be applied to the same specimens under exposure. Since the specimens registered different cracking loads, two "compatible" beams were chosen based on their pre-cracking load level values to perform the sustained load procedure. The target load was monitored by the load cells at each support by slightly tightening the red nuts (Fig. 2b). A pair of compatible UHDC specimens was coupled (Fig. 2c), with the cracked sides pointing outward of the cylinder rods. The steel pieces combined with rods and nuts were assembled to form a frame that can brace the specimens' pair. The load was applied on the beams by gradually tightening the top nuts, reading with the load cells up to the target specified load. Then, the load was transferred to the braced pair by tightening the nuts positioned below the load cells. As a consequence, the load cells were released, and moved to proceed with another pair. This procedure was carried out for all the specimens previously cracked (Fig. 1a) to immerse them in the aggressive exposures, represented by the chloride solution (XS exposure), geothermal water (XA exposure), and tap water (X0 exposure).

Mechanical tests have been then performed on the specimens to evaluate their mechanical recovery or deterioration. From each couple of specimens, one specimen

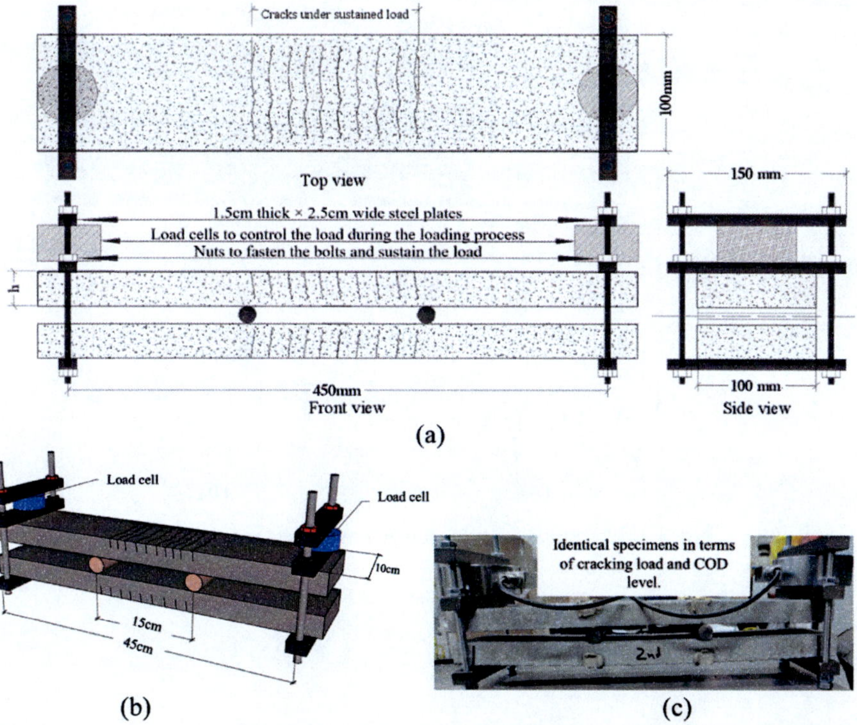

Fig. 2. Load setup used to keep the specimens under sustained loads and exposure simultaneously; (a) details sketch of the setups, (b) 3D representation of the load setup, and (c) actual beams couple stressed together by using the sustained load setup

has been tested under uniaxial direct tensile test up to failure to obtain the post-exposure direct tensile-COD curve (Fig. 3a). The other specimen of the couple has been tested under 4PBT with the same setup of the pre-cracking process to obtain flexural stress-displacement and COD curves (Fig. 3b). The aim of obtaining the tensile constitutive laws associated with flexural performance under exposures over time is to back-identify the flexural response of the UHDC materials with degradation/recovery taken into account and compare the results with the direct tensile stress-COD curves for stiffness recovery evaluation per each material type, exposure type, and exposure period. It is worth mentioning that, the pre-cracking level is 200 μm over 150 mm span. Thus, the multiple cracks are developed with very thin crack width, and sometimes they are only visible under the microscope specially after the self-healing effects, Fig. 3.

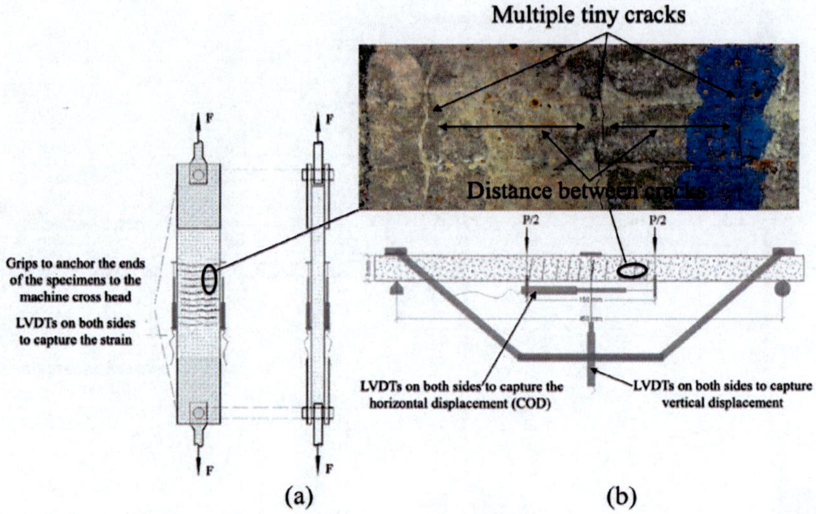

Fig. 3. Mechanical tests; (a) direct tensile test, and (b) 4PBT

3 Results and Discussions

3.1 Results of Direct Tensile Tests

This section is dedicated to the investigation of the influence of the exposure type and period on the mechanical performance of UHPC under the sustained loads and aggressive environments. Direct tensile stress versus axial displacement is presented for specimens tested after one, two, three, and six months of exposure, then compared to similar specimens that have not been under sustained load and exposure periods (Fig. 4). First of all, it is worth remarking that the tested UHPFRC specimens exhibited the multiple cracking in the most cases. Figure 4a shows the average direct tensile stress-COD evolution with time in terms of tap water exposure type, while Fig. 4b and Fig. 4c show the same curves but displayed in terms of aggressive exposure types (3.3% chloride solution, XS, and geothermal water, XA). Generally, all the materials exhibit a strain hardening response after the first cracking stress; specimens under the chloride and geothermal exposure show better performance in terms of higher tensile stress, in fact, they perfectly align with the response of the virgin specimens at the second and the sixth month of exposure, while the specimens under tap water showed the least tensile stress performance as compared to the not exposed specimens.

3.2 Results of Inverse Analysis

Inverse analysis based on five points approach adapted from [7] has been applied on the stress-displacement curves obtained from the 4PBT of the exposed and non-exposed specimens. As shown in Fig. 5, the curves have been divided into two parts, the first being before crack localization, on the left-side, containing elastic deformation and multiple cracking regime, showing strain hardening performance. The response after crack

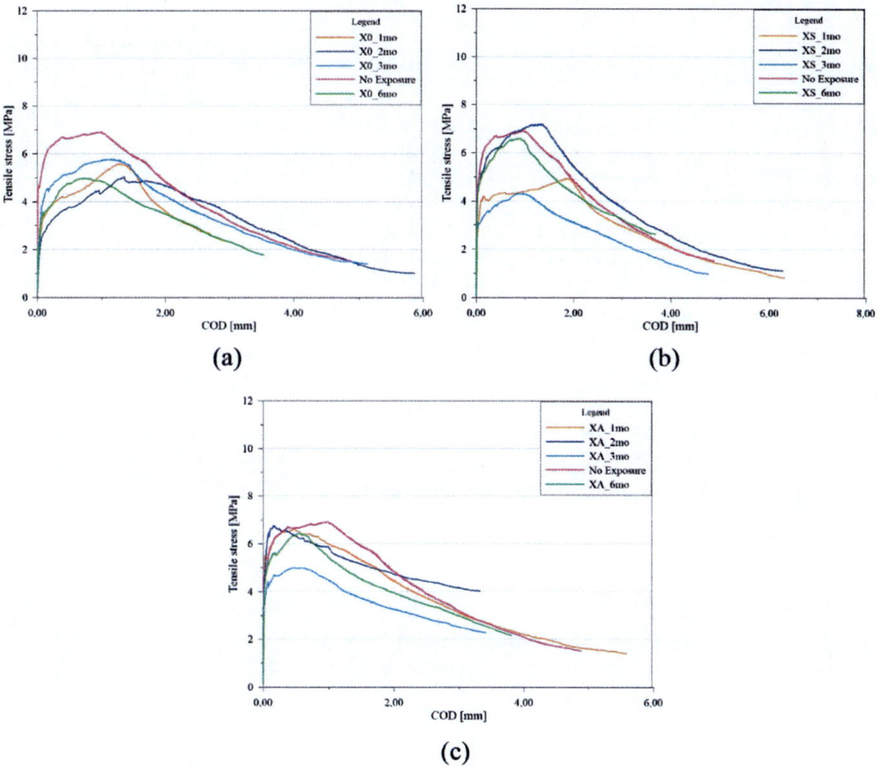

Fig. 4. Evolution of direct tensile test with time: (a) results based on exposure type (Xo), (b) results based on exposure type (XS), and (c) results based on the exposure types (XA)

localization, shown on the right-side, is mainly characterized by crack opening of the localized crack and strain softening performance. As a general observation, constitutive laws associated with specimens that went under sustained load and exposure periods show a better performance in terms of strain hardening phase than non-treated specimens (not subjected to sustained load and exposure conditions). The specimens exposed to geothermal water (XA exposure class) showed the best performance among the other exposure types; tap water (X0), and 3.3% chloride solution (XS).

This observation confirmed the aforementioned direct tensile tests' results. In fact, the specimens under the geothermal exposure showed the best performance. On the other hand, the tensile stress showed similar performance among all the specimens (different time periods of conditioned specimens and not conditioned specimens) after crack localization phase. This observation positively confirmed previously made statement that the recovered performance (either higher or equal to the pristine performance) is due to the healing particles formed in the induced smeared cracks. These particles, with the help of nano scale constituents and steel fibers for cracks at nano- and micro-scale respectively, have formed strong bonds between the damaged surfaces, which are believed to be stronger than the tensile strength of the UHPC matrix. Thus, the elastic modulus

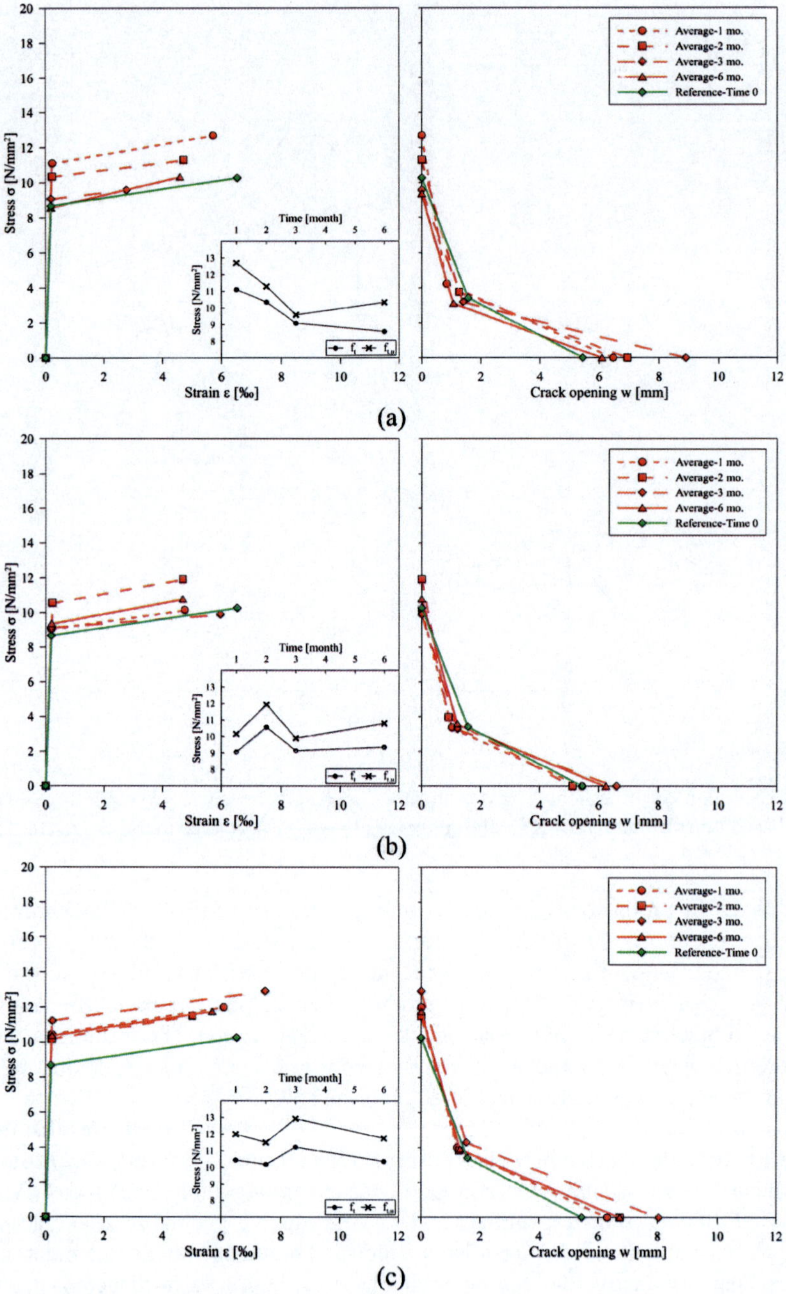

Fig. 5. Constitutive tensile laws derived from inverse analysis of 4PBT; (a) results from X0 exposure, (b) results from XS exposure and (c) results from XA exposure

and the tensile strength before crack localization of the recovered specimens showed higher performance than the virgin specimens. None the less, after crack localization the performance, characterized by the localized crack response, is similar to the virgin state specimens, also proving to some extent that the sustained load and exposure has not altered the residual bond strength of steel fiber in the localized crack and that, on the contrary, the healing has been able to restore it to a level comparable to the one which is observed in specimens without exposure and sustained loading.

4 Conclusions

Based on the experimental results presented in this study, the following concluding points can be drawn.

1. Self-healing of multiple microcracks in UHPC can occur in aggressive environments despite the high level of sustained load of as much as 80% of the ultimate flexural strength and extremely aggressive environments.
2. In addition to elastic modulus, there were associated recoveries in the strain hardening strength and deformation capacity and stiffness after exposure to aggressive environment and sustained loading.
3. Self-healing was not limited to mere sealing of the cracks, but strength recovery was also achieved, with high efficiency for the specimens with nano additions, where the damaged/healed specimens showed a better performance than the virgin specimens from the same types.
4. The healing particles formed in the damaged/healed region of the exposed specimens showed a better performance than the virgin specimens in terms of elastic modulus and tensile constitutive behaviour (obtained from inverse analysis)

Acknowledgements. The activity described in this paper has been performed in the framework of the project "Rethinking coastal defense and Green-energy Service infrastructures through enHancEd-durAbiLity high-performance cement-based materials-ReSHEALience", funded by the European Union Horizon 2020 research and innovation program under GA No 760824. The information and views set out in this publication are those of the authors and do not necessarily reflect the official opinion of the European Commission.

References

1. Lo Monte, F., Ferrara, L.: Tensile behaviour identification in Ultra-High Performance Fibre Reinforced Cementitious Composites: indirect tension tests and back analysis of flexural test results. Mater. Struct. Constr. **53**, 1–12 (2020)
2. Al-Obaidi, S., et al.: Innovative design concept of cooling water tanks/basins in geothermal power plants using ultra-high-performance fiber-reinforced concrete with enhanced durability. Sustainability **13**, 9826 (2021)

3. Ferrara, L., et al.: An overview on H2020 project "Reshealience." In: Proceedings of the IABSE Symposium: Towards a Resilient Built Environment Risk and Asset Management - Report; International Association for Bridge and Structural Engineering (IABSE), Guimaraes, pp. 184–191 (2019)
4. He, S., Zhang, S., Luković, M., Schlangen, E.: Effects of bacteria-embedded polylactic acid (PLA) capsules on fracture properties of strain hardening cementitious composite (SHCC). Eng. Fract. Mech. **268**, 108480 (2022)
5. Wan, Z., Xu, Y., Zhang, Y., He, S., Šavija, B.: Mechanical properties and healing efficiency of 3D-printed ABS vascular based self-healing cementitious composite: experiments and modelling. Eng. Fract. Mech. **267**, 108471 (2022)
6. Cuenca, E., D'Ambrosio, L., Lizunov, D., Tretjakov, A., Volobujeva, O., Ferrara, L.: Mechanical properties and self-healing capacity of Ultra High Performance Fibre Reinforced Concrete with alumina nanofibers: tailoring Ultra High Durability Concrete for aggressive exposure scenarios. Cem. Concr. Compos. **118**, 103956 (2021)
7. López, J.Á., Serna, P., Navarro-Gregori, J., Coll, H.: A simplified five-point inverse analysis method to determine the tensile properties of UHPFRC from unnotched four-point bending tests. Compos. Part B Eng. **91**, 189–204 (2016)

Acoustic Emission Technique for Monitoring Healing Induced Recovery of Mechanical Properties (HIRMP)

Avik Kumar Das[1,2](✉) and Christopher K Y Leung[2]

[1] Shenzhen International Graduate School, Tsinghua University, Shenzhen, China
akdas@connect.ust.hk

[2] Hong Kong University of Science and Technology, Hong Kong, China

Abstract. Due to bridging effect of fibers, the crack widths in strain hardening cementitious composites (SHCCs) are restricted to a small value <100 μ. Crack widths of this small size promotes autogenous healing in SHCCs. Ideally, autogenous healing should lead to noticeable reduction of crack width as well as recovery of strength and stiffness. The former leads to recovery of 'transport properties' and latter lead to recovery of 'mechanical properties' of damaged material. If we want to rely on healing-induced recovery of mechanical properties (HIRMP) are relied upon in built environment, it is important that this should be assess HIRMP in a practical scenario. Most reliable method for assessment of HIRMP is by comparing stress-strain curves of healed specimen with that of is the pristine condition. Load/stress can not be directly measured in practical scenario which render these methods impractical. Healing induced recovery stems from bonding of damages in materials, which if broken produces acoustic emission (AE) signals. Exploiting this fact, in this work potential of AE technique for assessment of HIRMP was studied. In this study, AE testing was performed during loading of 'pristine' SHCCs, which were healed and again during re-loading of these self-healed specimen. AE results were changed to power spectral entropy (PSE) to minimize adverse effect of computational parameters. PSE of healed specimen was compared with that of the 'pristine' state to assess HIRMP. The major observations of the study are 1. With increasing HIRMP the healing induced 'bonding' increases as a result cumulative PSE increases 2. With increasing HIRMP the density of PSE becomes less skewed (to the right) this is because with increasing average stiffness, lower strain is needed to reach 'hardening'. With these observations novel damage parameters are developed for practical monitoring of HIRMP.

Keywords: Autogenous healing · Acoustic emission · Healing induced mechanical recovery · Strain hardening cementitious composites

1 Introduction

Typically Strain hardening cementitious composites (SHCCs) are designed to showcase few percent of strain hardening while ensuring that cracks only open to small values [1] which lowers the transport of deleterious material [2–4] and promotes autogenous

M. Kunieda et al. (Eds.): SHCC 2022, RILEM Bookseries 39, pp. 249–259, 2023.
https://doi.org/10.1007/978-3-031-15805-6_26

self-healing [5–7]. Strategically integrating SHCCs in structures is expected to enable the development of durable civil structures with excellent crack control and earthquake resistant buildings with enormous energy absorption due to high tensile ductility [8–10]. Due to intrinsic control on the crack width previous experimental results suggest that usually under an environment with supply of water/moisture, this intrinsic crack width control in SHCCs also promotes autogenous self-healing of the structure. Interestingly, these experimental research also report that due to autogenous substantial recovery of young's modulus and strength has also been observed [11–14]. We hereby refer to this phenomenon as healing induced recovery of mechanical properties (HIRMP). As in the case of healing, Unfortunately, the amount of HIRMP depends on various factors (such as the amount of damage, environmental factors) which cannot be predicted a-priori [5, 15]. Thus, in order to make engineering decisions on the condition of a structure based on the reliance of self-healing, the accurate scoring of such a phenomenon is an essential step.

1.1 Related Work

Over the years, various techniques with different principles such as falling head [16–18], RF measurement [12, 16, 19], electrical properties [7, 20], neutron photography [21], and vision-based systems [22, 23] can be used for measurement of healing induced crack closing in SHCCs [15]. On the other hand, reloading the healed SHCC specimen under uniaxial tensile loading is a straightforward approach that is still the most reliable method to assess the mechanical recovery. In literature such as granger et al. [24], Landis et al. [25] as well as our previous result [26] have shown that results of Acoustic Emission (AE) technique is correlated to the loading response of material. Therefore, Acoustic Emission (AE) was employed in this study.

2 Proposed Method

2.1 Background

In AE testing, piezoelectric sensors are usually attached to the surface of the structural element [27, 28]. These sensors detect the acoustic emissions associated with micro-crack formation. These signals/waveforms are amplified with amplifiers and processed by computers as shown in schematic in Fig. 1. Sensors and instruments were set up following the best practices and recommendations [29, 30] as well as our previous experience with AE testing [31–33]. For this study, power spectral entropy (PSE) values of the waveforms are computed. The calculation of PSE step are as follow and also discussed in in detail in [34]. At first, the spectral power ($S(\omega_i)$) is computed and normalized following Eq. 1; and then it's power spectral entropy (PSE) values is calculated following Eq. 2 to remove adverse effect of computational parameters.

$$(\omega) = \frac{S(\omega_i)}{\sum_{i=1}^{n} S(\omega_i)} \tag{1}$$

$$SE = -\sum\nolimits_{i=1}^{n} S(\omega) * \log_2 S(\omega) \tag{2}$$

Here i is the ith frequency bin and ω is angular frequency.

2.2 Computational Steps for Assessment of HIRMP

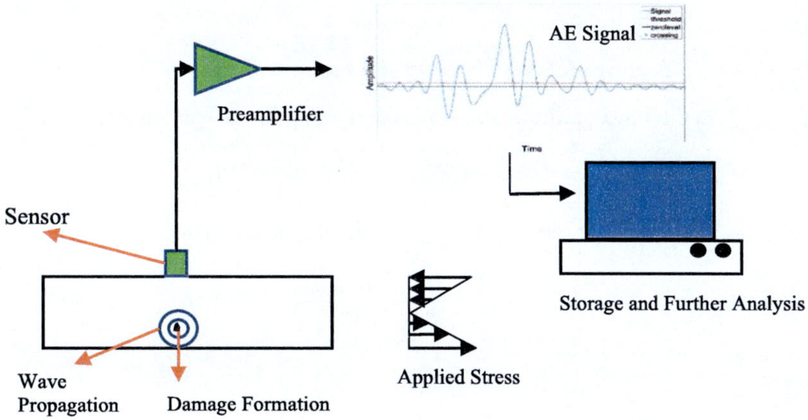

Fig. 1. Schematic for AE testing

It should be noted that fracture processes in the SHCCs are random thus, stochastic. This behavior is also expected to be observed for fracture processes within the same sample before and after healing. Therefore, to assess HIRMP we have to represent the stochastic nature of the physical process through their statistical behavior. With this consideration a data science based approach is introduced.

Data lifecycle in a data science has 4 important steps i.e. Data Collection, Feature Extraction, Feature Construction, Decision Making. With reference to these steps we will discuss the proposed technique. First step is to **data collection** of AE data in-situ by attaching sensors on the sample. AE data, which consists of AE waveforms are often interlaced with environmental noise thus, was digitally filtered. The next step in the process is **feature extraction** from the AE waveforms. To this end, for each waveform we first compute PSE values. This two steps has been discussed in the previous section. Following this, **feature construction** was performed heuristically by computing statistical behavior of *PSE vs Strain* Distribution to appropriately represent stochastic and intractable fracture process of SHCCs. To do so, 3 damage parameters are proposed- i) Cumulative of PSE (Tot _PSE), ii) Pre-Strain from PSE ($\Delta\epsilon$ _PSE), and iii) Skewness (Eq. 3) of PSE distribution (Skew_PSE). $\Delta\epsilon$ _PSE, approximates amount of pre-straining on the sample, which is the difference of strain value at the onset of hardening region (ϵ_{fs}) and strain at which test ends (ϵ_{te}). *Tot* _PSE is calculated by adding all the PSE's observed during the testing, this measures the extent of healing induced 'gluing' in damaged specimen. Skew_PSE measures the stiffness induced asymmetry within the distribution PSE.

$$Skew = \frac{Mean - Median}{Standard\ Deviation} \tag{3}$$

Finally, for **Decision Making** i.e. to assess HIRMP we calculate the ratios of these damage parameters between pristine and healed sample. η_{Tot} (Eq. 4) is the ratio of Tot _PSE of pristine and healed specimens.

$$\eta_{\text{Tot}} = \frac{\text{Tot}\_PSE_{healed}}{\text{Tot}\_PSE_{native}} \tag{4}$$

η_{Skew} (Eq. 5) is the ratio of Skew_PSE of healed and pristine specimens.

$$\eta_{\text{Skew}} = (Skew\_PSE)_{healed} / (Skew\_PSE)_{native} \tag{5}$$

And, $\eta_{\Delta\epsilon}$ (Eq. 6) is the ratio of $\Delta\epsilon\_PSE$ of healed and pristine specimens.

$$\eta_{\Delta\epsilon} = \frac{\Delta\epsilon\_PSE_{healed}}{\Delta\epsilon\_PSE_{native}} \tag{6}$$

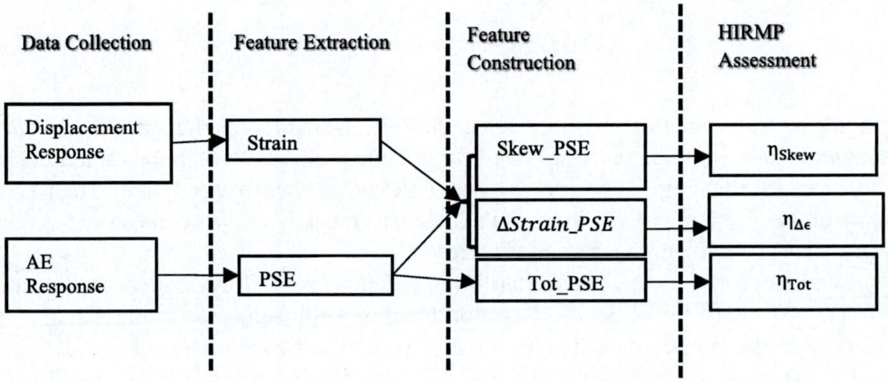

Fig. 2. A bird's eye view of the computational scheme in the proposed strategy for HIRMP

We hereby refer to this method as mechanical recovery assessment through acoustic emissions (MR_AE). As will be discussed in the later section, that these ratios-η_{Tot}, η_{Skew}, and $\eta_{\Delta\epsilon}$ were then used to HIRMP assessment. The overall data lifecycle is also schematically shown in Fig. 2.

3 Experimental Details

Table 1. Matrix design

Mix name	Fly ash	Cement	Silica fume	Sand
S	0.8	0.18	0.02	0.2

Coupon samples were cast following the matrix design shown in Table 1. These coupons were in the shape of a rectangular prism of size 350 × 25 × 8 mm as shown in Fig. 3. The binder of the mix is Ordinary Portland cement of type CEM 42.5 and class F fly ash and silica fume. The filler in this mix is especially graded silica sand with a maximum size of 0.16 mm. After mixing binder, filler, and water superplasticizer (SP) was added to ensure proper rheology required to proper dispersal of fibers. In these samples, Kuraray™ K-II REC15 Polyvinyl alcohol (PVA) was used. Nominal length, diameter, and elastic modulus of the fiber were 8 mm, 39 μ, and 42 Gpa respectively. Two different fiber concentration i.e. 1.5% and 2.5% were employed in this study. The details of the mix proportions are shown in Table 1. The direction of casting is from the top of the mold. The mix names and pre-straining value and healing conditions in reported in Table 2.

Table 2. Test series in this study

S. no	Name	No. of samples	Fiber content (%)	Healing environment	Pre-straining value (%)
1	S15A05	2	1.5	Air	0.5
2	S15W05	2	1.5	Water and Air	0.5
3	S15A10	2	1.5	Air	1
4	S15W10	2	1.5	Water and Air	1
5	S25A05	2	2.5	Air	0.5
6	S25W05	2	2.5	Water and Air	0.5
7	S25A10	2	2.5	Air	1
8	S25W10	2	2.5	Water and Air	1

approximate value

After casting, the specimens were covered with plastic wrap to prevent loss of moisture. These samples were then cured in laboratory air at a temperature of ~25 °C. for 2 days before demolding. The demolded specimens were then immediately transferred to the curing room where they were cured in water for at least 28 days age before testing. After SHCC specimen were taken out of the curing room the specimens were dried in laboratory conditions for one day. Then these samples were loaded to a pre-straining to generate damage and then transferred to healing chamber as reported in [7]. The conditioning period for all cases was fixed to 10 days for either of the cases. After conditioning period was over all the samples were kept in air for 24 h for it to dry and then they were tested again. The mechanical testing process of healed specimen is exactly the same as that of the 'pristine' specimen as described in [15, 22].

4 Results and Discussions

4.1 Qualitative Application of Damage Parameters of MR_AE

In this section, these 2 cases are described in detail to elucidate the potential of MR_AE. Based on experimental results, it could be qualitatively stated that the sample S15A10

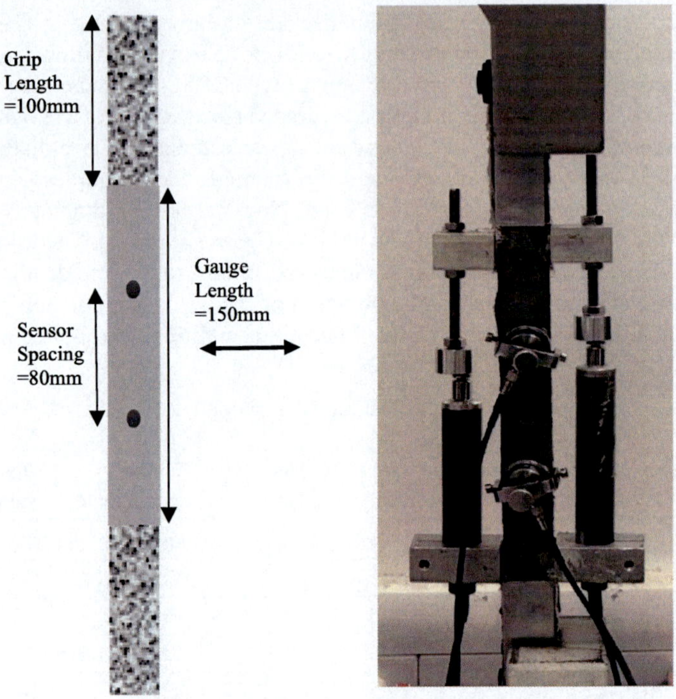

Fig. 3. Uniaxial tension test setup for SHCC

will have a much lower HIRMP as compared to the sample S25W05. We will focus the results of these samples obtained from the proposed methodologies to see if they are consistent with expectations.

The stress-strain curve for SC15A10 along with evolution of PSE for pristine state is shown in Fig. 4A and the load-strain curve after healing is shown in Fig. 4B. It can be seen that S15A10 was deliberately pre-strained to large value (shown in Fig. 4A). When samples are strained to large value it would imply that there is large number of cracks thus, increasing probability of cracks with higher widths. Moreover, this sample after testing was kept in air to discourage healing. Expectedly, this lead to negligible HIRMP. This is evident from the ratio of young's modulus of the sample which is ~0.4. This is also documented in MR_AE technique. As shown healed sample, PSE is only observed after a large 'initial region'. It is observed in Fig. 4 E,F that *strain_PSE* of the healed sample has a large gap in this region whereas the density is substantially skewed to the right as observed from the density curve Fig. 4F. Consequently, for this sample η_{Skew} value is low (0.35) as well as $\eta_{\Delta\epsilon}$ value for SC15A10 is also low approx. ~0.4 as all the PSE during reloading AE is concentrated within the small region (shown in Fig. 4D). This validates the MR_AE technique is successful in documenting HIRMP.

The stress-strain curve for S15A10 along with evolution of PSE for pristine state is shown in Fig. 5A and the load-strain curve after healing is shown in Fig. 5B. In contrast to SC15A10, for SC25W05 it is expected a significant HIRMP because the sample was

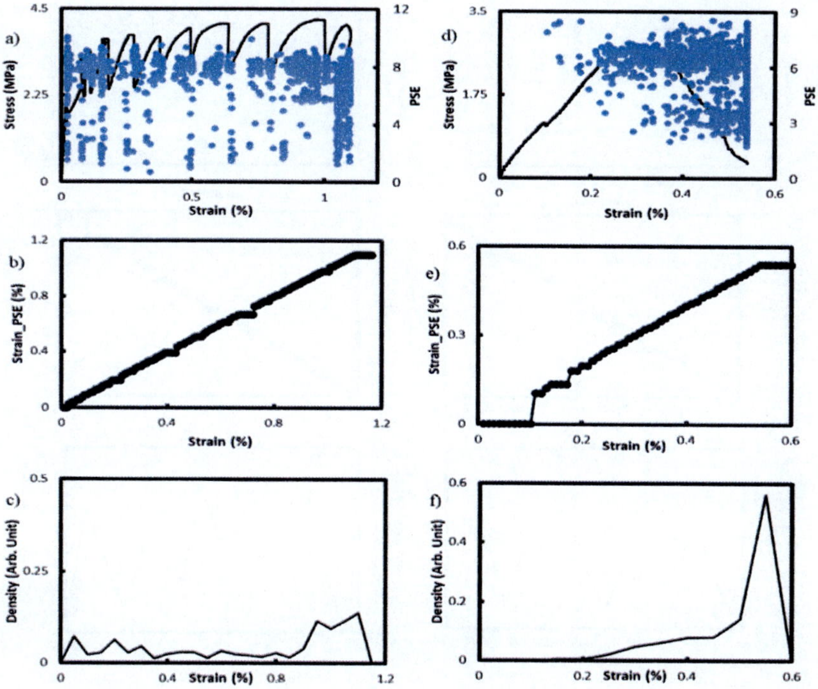

Fig. 4. Test Result of a poorly recovered sample –SC15A10 a) PSE and Load b) Strain_PSE c) Density during native state d-f are the same physical parameter during reloading

healed with water as well as pre-strained to a lower value. The ratio of young's modulus of healed and pristine sample for this case was 0.84. We expect this behavior to be documented by MR_AE technique. It is evident from the Fig. 5B that evolution of PSE of healed sample looks similar to that of the pristine sample. This behavior is in contrast to S15A10. $\Delta\epsilon$ *_PSE* as well as *density* derived from PSE evolution of healed sample is (visually) similar to that of the native sample (see Fig. 5 b.c.e.f). As a result, values of damage parameters *R_Tot_AE*, *R_Skew_AE*, *R_ΔStrain_AE* has high values (~0.8). Which suggests that autogenous healing for this sample has led to substantial HIRMP. RoA value for SC15A10 is 0.34 is much lower than SC25W05 which is 0.81. This results matches fairly well with our expectation.

4.2 MR_AE and HIRMP

The results in the previous section show that the damage parameters are capable at detecting HIRMP. In this section we will discuss the relationship of HIRMP and MR_AE for large number of cases and possibility of MR_AE to discriminate samples based on their HIRMP. Figure 6 shows the result of ratios η_{Tot}, η_{Skew}, and $\eta_{\Delta\epsilon}$ for samples in this study. The following trends were found the damage parameters η_{Tot}, η_{Skew}, and $\eta_{\Delta\epsilon}$ of MR_AE and HIRMP in the result presented in Fig. 6:

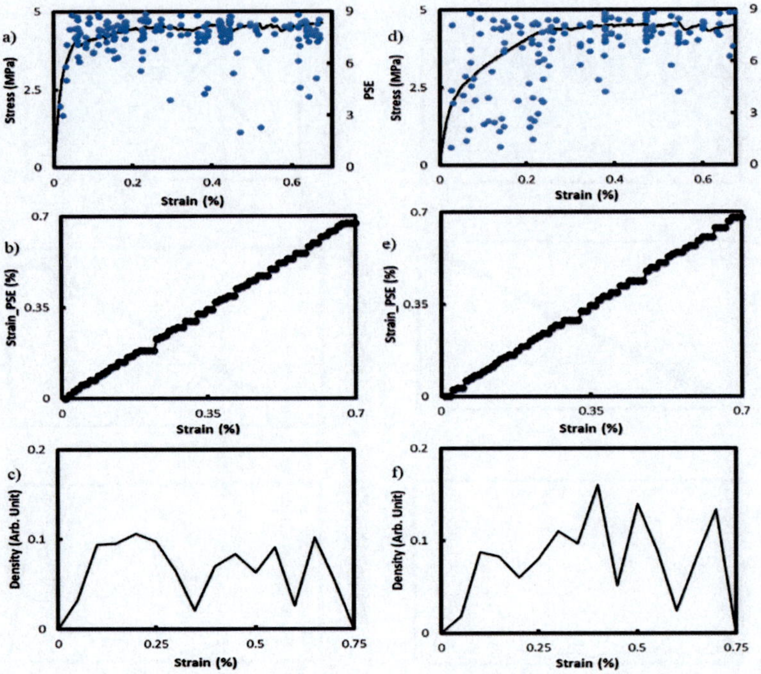

Fig. 5. Test Result of a well recovered sample –SC25W05 a) PSE and Load b) Strain_PSE c) Density during native state d-f is the same physical parameter during reloading

a) 2.5% fiber samples have higher recovery rate than 1.5% samples.
b) samples subjected to lower pre-straining have higher recovery rate.
c) samples subjected to 'water and air' conditioning have higher recovery rate.

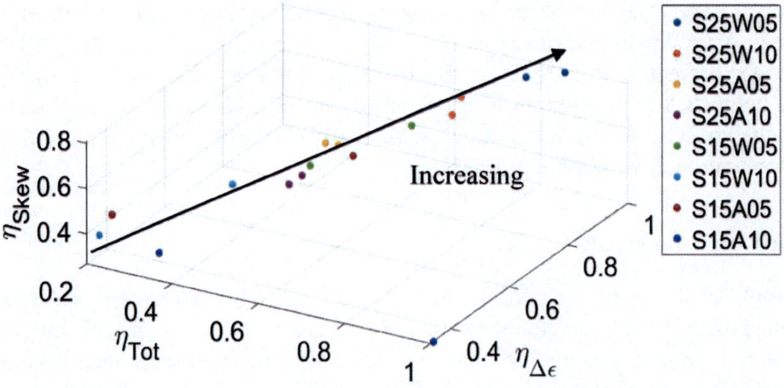

Fig. 6. Ratios - η_{Tot}, η_{Skew}, and $\eta_{\Delta\epsilon}$ for different levels of mechanical recovery

These results are consistent with our expectations and those previously reported in the literature [5, 15]. As a result of such trend observed that with increasing HIRMP the values of these damage parameters tends to follow the similar trend. The experimental results suggest that HIRMP is positively related to the damage parameters.

5 Conclusions

Ideally, autonomous healing strain hardening cementitious composites (SHCCs) should lead to recovery healing induced recovery 'mechanical' properties (HIRMP). In this work we introduce a method for assessment of HIRMP called mechanical recovery from acoustic emission technique (MR_AE) following the principle of data science. In MR_AE, 3 novel heuristic damage parameters from acoustic emission (AE) response *Tot_PSE, Skew_PSE,* Δ*Strain_PSE* were designed to stochastically represent random fracture processes of healed and pristine states. The ratio of these parameters- η_{Tot}, η_{Skew}, and $\eta_{\Delta\epsilon}$ between healed and pristine sample to corroborate HIRMP. We study this technique on HIRMP assessment for SHCC samples. Predictions of HIRMP with MR_AE match well with expected results and recovery of for a wide range of healing conditions. This result suggest that acoustic emission more specifically proposed method can facilitate monitoring of HIRMP in practical scenarios.

References

1. Li, V.C.: Engineered Cementitious Composites (ECC): Bendable Concrete for Sustainable and Resilient Infrastructure, 1st edn. (2019). https://doi.org/10.1007/978-3-662-58438-5
2. Wang, K., Jansen, D.C., Shah, S.P., Karr, A.F.: Permeability study of cracked concrete. Cem. Concr. Res. **27** (1997). https://doi.org/10.1016/S0008-8846(97)00031-8
3. Djerbi, A., Bonnet, S., Khelidj, A., Baroghel-Bouny, V.: Influence of traversing crack on chloride diffusion into concrete. Cem. Concr. Res. **38**(6) (2008). https://doi.org/10.1016/j.cemconres.2007.10.007
4. Lepech, M.D., Li, V.C.: Water permeability of engineered cementitious composites. Cem. Concr. Comp. **31** (2009). https://doi.org/10.1016/j.cemconcomp.2009.07.002
5. Herbert, E.: Development and Application of Self-healing Engineered Cementitious Composites (ECC) for Durable and Sustainable Infrastructure, ed. PhD (2016)
6. Li, V., Herbert, E.: Robust self-healing concrete for sustainable infrastructure. J. Adv. Concr. Technol. **10**(6) (2012). https://doi.org/10.3151/jact.10.207
7. Das, A.K., Leung, C.K.Y.: A strategy for in situ determination of self-healing state for strain hardening cementitious composites. Cem. Concr. Comp. (2020). https://doi.org/10.1016/j.cemconcomp.2020.103641
8. Lepech, M.D., Li, V.C.: Design and field demonstration of ECC link slabs for jointless bridge decks (2005). https://www.michigan.gov/documents/mdot/MDOT_Research_Report_RC1471_200102_7.pdf
9. Li, V.C., et al.: Durable link slabs for jointless bridge decks based on strain-hardening cementitious composites (2003)
10. Rokugo, K.: Applications of SHCC in Japan - Tools and Tips for Promoting its Use (2017)
11. Li, V.C., Yang, E.H.: Self Healing in Concrete Materials. Springer, Dordrecht (2008). https://doi.org/10.1007/978-1-4020-6250-6_8

12. Yang, Y., Yang, E., Li, V.C.: Autogenous healing of engineered cementitious composites at early age. Cem. Concr. Res. **41** (2011). https://doi.org/10.1016/j.cemconres.2010.11.002
13. Snoeck, D., De Belie, N.: Repeated autogenous healing in strain-hardening cementitious composites by using superabsorbent polymers. J. Mat. Civil Eng. **28**(1) (2016). https://doi.org/10.1061/(ASCE)MT.1943-5533.0001360
14. Ferrara, L., et al.: Experimental characterization of the self-healing capacity of cement based materials and its effects on the material performance: a state of the art report by COST Action SARCOS WG2. Constr. Build. Mater. **167** (2018). https://doi.org/10.1016/j.conbuildmat.2018.01.143
15. Das, A.K.: Development of techniques for monitoring of various aspects of strain hardening cementitious composites (2020). https://doi.org/10.14711/thesis-991012879963603412
16. Yang, Y., Lepech, M.D., Yang, E., Li, V.C.: Autogenous healing of engineered cementitious composites under wet-dry cycles. Cem. Concr. Res. **39**(5) (2009). https://doi.org/10.1016/j.cemconres.2009.01.013
17. Palin, D., Jonkers, H.M., Wiktor, V.: Autogenous healing of sea-water exposed mortar: quantification through a simple and rapid permeability test. Cem. Concr. Res. **84** (2016). https://doi.org/10.1016/j.cemconres.2016.02.011
18. Van Tittelboom, K., De Belie, N., Van Loo, D., Jacobs, P.: Self-healing efficiency of cementitious materials containing tubular capsules filled with healing agent. Cem. Concr. Comp. **33**(4) (2011). https://doi.org/10.1016/j.cemconcomp.2011.01.004
19. Das, A.K., Mishra, D.K., Yu, J., Leung, C.K.Y.: Smart Self-Healing and Self-Sensing Cementitious Composites —Recent Developments, Challenges and Prospects (2019). https://doi.org/10.1520/ACEM20190023
20. Ranade, R., Zhang, J., Lynch, J.P., Li, V.C.: Influence of micro-cracking on the composite resistivity of Engineered Cementitious Composites. Cem. Concr. Res. **58** (2014). https://doi.org/10.1016/j.cemconres.2014.01.002
21. Schröfl, C., Mechtcherine, V., Kaestner, A., Vontobel, P., Hovind, J., Lehmann, E.: Transport of water through strain-hardening cement-based composite (SHCC) applied on top of cracked reinforced concrete slabs with and without hydrophobization of cracks – Investigation by neutron radiography. Constr. Build. Mater. **76** (2015). https://doi.org/10.1016/j.conbuildmat.2014.11.062
22. Das, A.K., Leung, C.K.Y., Wan, K.T.: Application of deep convolutional neural networks for automated and rapid identification and computation of crack statistics of thin cracks in strain hardening cementitious composites (SHCCs). Cem. Concr. Comp. **122** (2021). https://doi.org/10.1016/j.cemconcomp.2021.104159
23. Fan, S., Li, M.: X-ray computed microtomography of three-dimensional microcracks and self-healing in engineered cementitious composites. Smart Mater. Struct. **24**(1) (2015). https://doi.org/10.1088/0964-1726/24/1/015021
24. Granger, S., Loukili, A., Pijaudier-Cabot, G., Chanvillard, G.: Experimental characterization of the self-healing of cracks in an ultra high performance cementitious material: mechanical tests and acoustic emission analysis. Cem. Concr. Res. **37** (2007). https://doi.org/10.1016/j.cemconres.2006.12.005
25. Landis Eric, N., Lucie, B.: Experiments to relate acoustic emission energy to fracture energy of concrete. J. Eng. Mech. **128**(6) (2002). https://doi.org/10.1061/(ASCE)0733-9399(2002)128:6(698)
26. Das, A.K., Leung, C.: Power spectral entropy of acoustic emission signal as a new damage indicator to identify the operating regime of strain hardening cementitious composites. Cem. Concr. Comp. **104** (2019). https://doi.org/10.1016/j.cemconcomp.2019.103409
27. Das, A.K.: Experimental study of cracking process in steel fiber reinforced concrete beam members, ed. M-Phil (2017). https://doi.org/10.14711/thesis-991012551768303412

28. Grosse, C., Ohtsu, M.: Acoustic emission testing: basics for research - applications in civil engineering (2008). https://doi.org/10.1007/978-3-540-69972-9
29. Ohtsu, M.: Innovative AE and NDT techniques for on-site measurement of concrete and masonry structures: state-of-the-Art Report of the RILEM Technical Committee 239-MCM (2016)
30. Ohtsu, M.: Recommendation of RILEM TC 212-ACD: acoustic emission and related NDE techniques for crack detection and damage evaluation in concrete: test method for classification of active cracks in concrete structures by acoustic emission. Mater. Struct. Materiaux et Constr. **43**(9) (2010). https://doi.org/10.1617/s11527-010-9640-6
31. Das, A.K., Leung, C.K.Y.: Fast Tomography: a greedy, heuristic, mesh size–independent methodology for local velocity reconstruction for AE waves in distance decaying environment in semi real-time. Struct. Health Monit. (2021). https://doi.org/10.1177/14759217211036881
32. Das, A.K., Leung, C.K.Y.: ICD: a methodology for real time onset detection of overlapped acoustic emission waves. Autom. Constr. 119 (2020). https://doi.org/10.1016/j.autcon.2020.103341
33. Das, A.K., Leung, C.K.Y.: A fundamental method for prediction of failure of strain hardening cementitious composites without prior information. Cem. Concr. Comp. **114** (2020). https://doi.org/10.1016/j.cemconcomp.2020.103745
34. Das, A.K., Leung, C.K.Y.: Power Spectral Entropy (PSE) as a Qualitative Damage Indicator (2018). http://hdl.handle.net/1783.1/94253

Effects of Corrosion on Bond Behavior of Reinforcing Bar in Concrete and SHCC

Ayami Naito[1], Hyun-do Yun[2], and Koichi Kobayashi[3](✉)

[1] Toyota T&S Construction, Toyota, Japan
[2] Chungnam National University, Daejeon, South Korea
[3] Gifu University, Gifu, Japan
ko2ba@gifu-u.ac.jp

Abstract. This study aimed at clarifying the bond behavior of corroded rebar in SHCC. SHCC and ordinary concrete (NC) beams with a bending deflection were exposed to chloride solution in wet-dry cycles to accelerate deterioration. Bending cracks formed by bending the beams had some influence on chloride ingress and resulted in differences in chloride penetration between NC and SHCC. SHCC showed higher corrosion proof performance than NC. Corrosion locations on the rebar corresponded to the positions of bending cracks. A double pull-out test was conducted to investigate the bond behavior of corroded rebar in NC/SHCC. NC beams suffered more longitudinal cracks than SHCC and cover spalling due to bond fracture. Crack dispersion of SHCC did not reduce with the progress of rebar corrosion. With the corrosion mass loss being not more than 1%, the bond fracture energy of SHCC beams did not reduce with the progress of corrosion.

Keywords: SHCC · Rebar corrosion · Double pull-out test · Crack distribution · Bond fracture

1 Introduction

Accurate estimation of residual performance of deteriorated structural members is essential for correct maintenance of an RC structure under a chloride attack. It is important to know not only the degree of rebar corrosion but the influence of corrosion on the bond between the rebar and the cover concrete.

Strain hardening cementitious composite (SHCC) is characteristic of numerous fine cracks formed under tensile stress. SHCC offers high corrosion proof performance even after the formation of cracks because of the fineness of the cracks [1–4], which is considered to help maintain the tensile stiffness of the deteriorated member.

This study aims at clarifying the bond behavior of corroded rebar in SHCC. After subjecting SHCC and ordinary concrete (NC) beams to chloride solution in wet-dry cycles to accelerate deterioration, the corrosion was investigated, and a double pull-out test was conducted.

M. Kunieda et al. (Eds.): SHCC 2022, RILEM Bookseries 39, pp. 260–268, 2023.
https://doi.org/10.1007/978-3-031-15805-6_27

2 Experimental Procedure

Figure 1 shows the dimensions of the beam specimen. Two D13 deformed steel bars were embedded in each beam, with D25 deformed bars welded to both ends of the D13 bars. Table 1 shows the test parameters. NC and SHCC mixtures both had a water cement ratio of 0.55. High strength polyethylene fiber (Dyneema) was mixed to SHCC in a volume percentage of 1.2%.

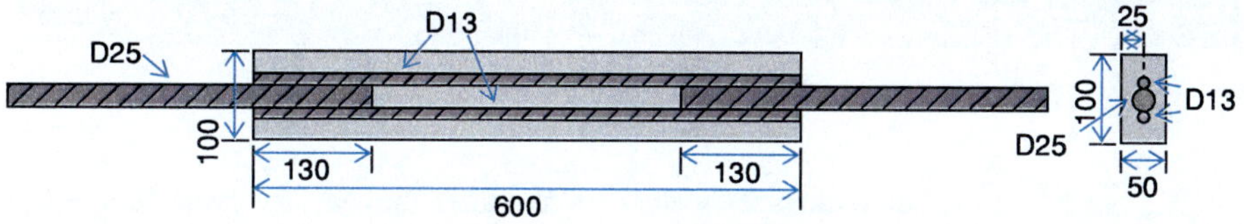

Fig. 1. Dimensions of specimen (unit: mm)

Table 1. Test parameters

Mixture	Ordinary concrete (NC), SHCC
Deflection (mm)	0.0, 3.0, 5.0
Wet-dry period (month)	0, 1, 4

After one month of wet cloth curing, the beams were fixed on an i-shaped steel beam with vises and loaded with a loading span of 500 mm. A predetermined deflection (see Table 1) was imparted to the beams by tightening the vises (Fig. 2) to cause bending cracks to form. A rectangular PVC pond (250 × 90 mm) was set on each beam and wet-dry cycles were repeated using a 3% sodium chloride solution. One cycle consisted of a two-day wet period in which the chloride solution was kept in the pond at the room temperature, and a five-day dry period in which the solution was drained and the beams were dried at high temperatures between 40 and 50 °C. This cycle was repeated for one month and for four months.

Fig. 2. Bent beam with PVC pond

After the wet-dry cycles, the beams were subjected to a double pull-out test (Fig. 3). While measuring the load with a loadcell and the entire elongation of the concrete/SHCC part with four LVDT displacement transducers, the beams were pulled until the elongation reached about 10 mm. Crack propagation was measured using four 100 mm Pi-shaped displacement meter set in a 400 mm-long mid span.

Fig. 3. Double pull-out test

The cracks on the beam surfaces were sketched after the wet-dry cycles and again after the double pull-out test. A 320 mm-long block was cut out from a center part of the beam with a concrete cutter, and split longitudinally along the rebar direction by concentrated loading. A nitric silver solution was sprayed on the split surfaces to determine chloride contaminated areas. The chloride distribution was also investigated by an electron probe micro analyzer (EPMA). After that, the rebars were taken out from the beams, and the corroded area and corrosion mass loss were examined. The influence of rebar corrosion on the bond behavior will be discussed based on the results of these tests.

3 Rebar Corrosion

Figure 4 shows the chloride mapping obtained by EPMA analysis conducted after 4 months of wet-dry cycles. As shown, chloride penetrated deeply through cracks in both NC and SHCC.

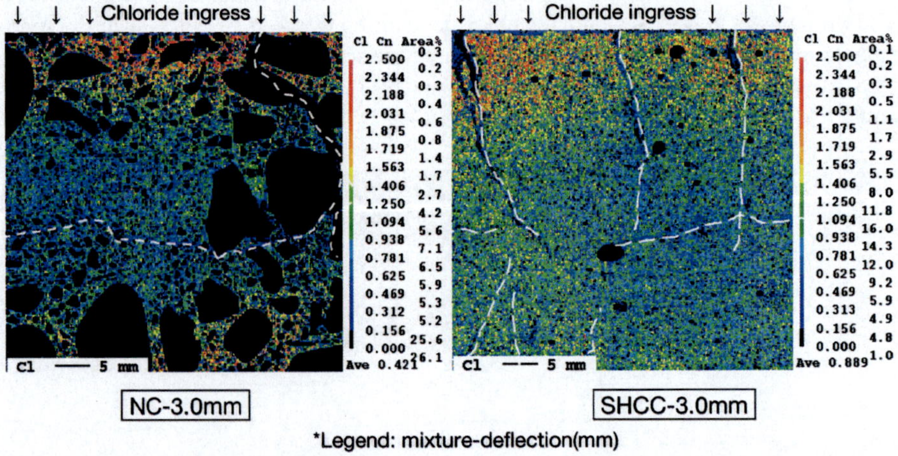

Fig. 4. Chloride mapping by EPMA (Analyzed area: 50 × 50 mm. The cross sections are parallel to the rebar. Dotted lines indicate the location of the cracks.)

Figure 5 shows the relationship between the percentage of corroded area and the corrosion mass loss. The line charts show the results obtained from the beams of the same series after one month and four months connected to each other. The corroded area of NC beams increased rapidly in the first month, after which the corrosion mass loss increased greatly. The rapid increase in surface area of rebar corrosion in NC beams at early stage is attributable to more extensive bond failure between concrete and rebar due to cracks, which allows easy access of chloride along the rebar. On the other hand, the corrosion mass loss of SHCC beams increased during the first month, after which the corroded area increased. It is assumed that bond failure occurs in smaller areas in SHCC despite numerous cracks because the rebar makes a tight bond with SHCC and that the tight bond restricted penetration of aggressive substances. Therefore, corrosion may have progressed only in limited areas around cracks at first, and spread as cracks formed on the beam surface due to rebar corrosion, resulting in the later increase in corroded area.

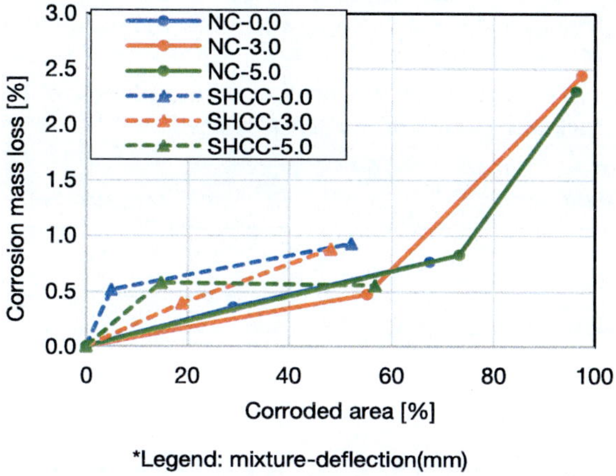

Fig. 5. Relationship between corroded area and corrosion mass loss

Figure 6 shows examples of correspondence between corroded areas of rebar and crack positions. Generally, rebar corrosion occurred near bending cracks, and the loss in cross-sectional area was large in these corroded regions. As mentioned above, the corroded area increased with time, as deterioration progressed. The degree of rebar corrosion in SHCC was distinctly lower than in NC, which is attributable to the high chloride proof performance of SHCC.

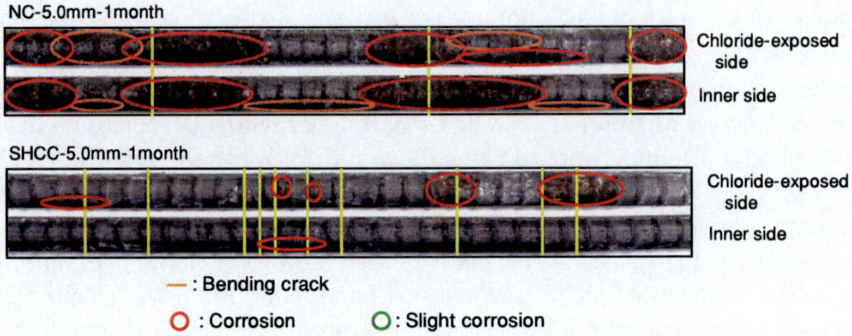

Fig. 6. Rebar corrosion and crack positions

4 Evaluation of Bond Behavior

Figure 7 shows examples of crack distributions on the beams after the double pull-out test. Longitudinal cracks and cover spalling were observed in NC beams. Severe rebar corrosion in the NC beams seems to have destroyed the bond between the rebar and the cover concrete. In contrast, there was no discernible influence of the degree of rebar corrosion on the crack distribution of the SHCC beams.

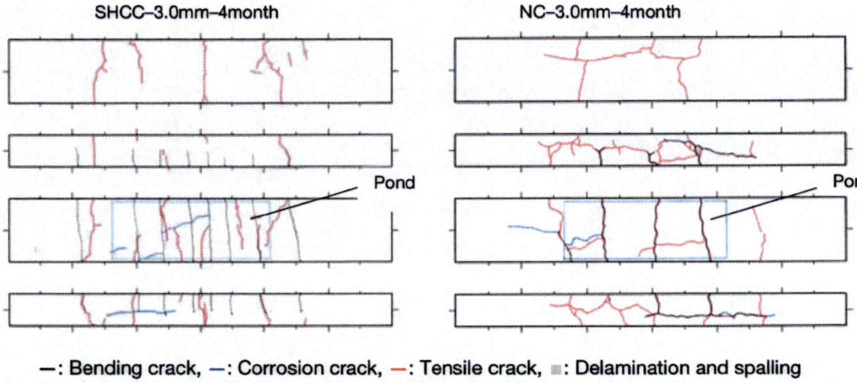

Fig. 7. Crack distributions after double pull-out test

Figure 8 shows the relationship between the corrosion mass loss and the crack dispersion ratio *R*. The crack dispersion ratio R is defined as follows:

$$R = N_{cor,n} \big/ N_{0,av} \quad (1)$$

where, $N_{0,av}$: Average number of cracks formed in sound beams after double pull-out test

$N_{cor,n}$: Number of cracks formed in each corroded beam after double pull-out test

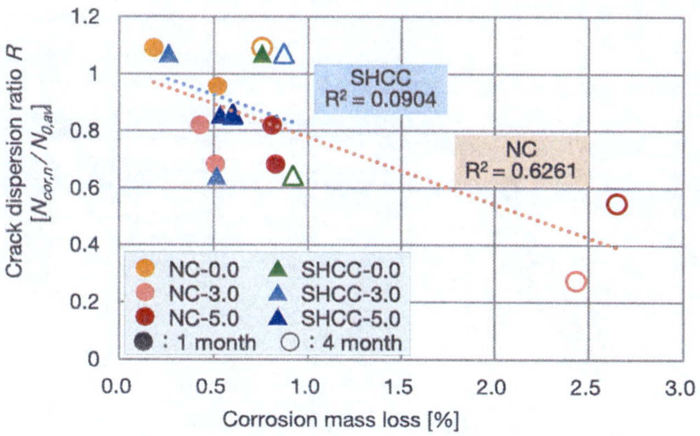

*Legend: mixture-deflection(mm)

Fig. 8. Influence of corrosion on crack dispersion ratio

The crack dispersion ratio R of NC beams with larger deflection was lower, and R reduced with the progress of deterioration. This suggests loss of bond caused by advanced corrosion, which resulted in increased crack spacing. The crack dispersion ratio R of SHCC was around 1.0, and no clear correlation between the crack spacing and the corrosion mass loss was discernible. This is probably because of the limited range of corrosion mass loss in SHCC beams. Further investigation of SHCC beams with a more extensive range of rebar corrosion is required.

Figure 9 shows load-strain curves obtained by the double-pull out test. Here, the strain is the average value of the strain in the section of 320 mm length in the central part of the beam. SHCC beams showed a higher yield load than that of rebar alone because of the characteristic high tensile strength and ductility. The larger the bending deflection, the lower the load with the same strain in both NC-series and SHCC-series beams. Both NC and SHCC beams with a deflection of 5 mm, in particular, showed lower tensile stiffness than bare steel. It is assumed that the large bending deformation destroyed the bond between the rebar and NC/SHCC near the bending cracks. There was no clear difference between the beams after one month of wet-dry cycles and after four months of wet-dry cycles.

Average bond fracture energy E_{bond} (J/m^3) was calculated from the results of the double pull out test of the beams and the tensile loading test of the rebar as follows.

$$E_{bond} = \int_0^{\varepsilon_{yield}} \frac{P - T_s}{A_c} d\varepsilon \quad (2)$$

where, P: Pull-out load on beam (N)

T_s: Tensile load on rebar alone (N, see Fig. 9)

A_c: cross sectional area of concrete/SHCC (mm^2)

ε: strain

ε_{yield}: yield strain of rebar

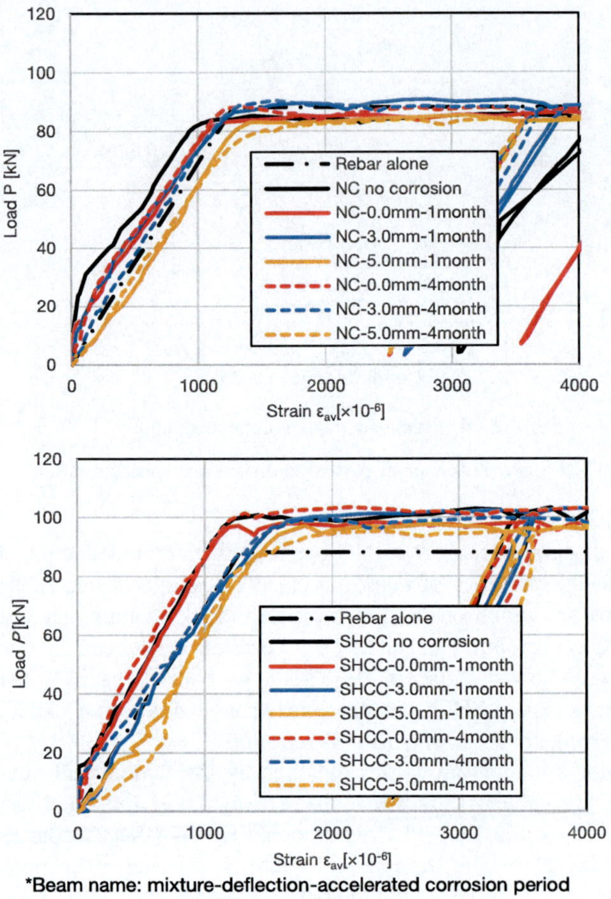

Fig. 9. Load deformation curves

Figure 10a shows the relationship between the bond fracture energy E_{bond} and the corrosion mass loss. While E_{bond} apparently decreases with the progress of corrosion in NC, no such tendency is clearly observed in SHCC. Figure 10b shows the relationship between E_{bond} and the deflection imparted to the beams during the accelerated corrosion period. This chart clearly shows that the larger the deflection, the smaller the bond fracture energy. Since the NC beams with larger deflection suffered larger corrosion mass loss as shown in Fig. 5, the decrease in bond fracture energy may have been caused by the increase in corrosion mass loss. On the other hand, as shown in Fig. 5, the amount of deflection did not directly correlate to the corrosion mass loss in SHCC. There is a possibility that the bending caused bond fracture and resulted in the lower bond fracture energy during the double-pull-out test.

Figure 10c shows the relationship between the bond fracture energy E_{bond} and the corrosion mass loss of the beams without any deflection. This chart shows a tendency of lower E_{bond} with advanced corrosion in NC, and also shows an increase in bond fracture energy of SHCC with the progress of corrosion. A volumetric expansion of rebar due to rust formation causes corrosion cracks in NC beams, and as these cracks widen, confining pressure on the rebar decreases. In contrast, the volumetric expansion of rebar must have led to the increase in confining pressure because of the crack-bridging effect of fiber in SHCC.

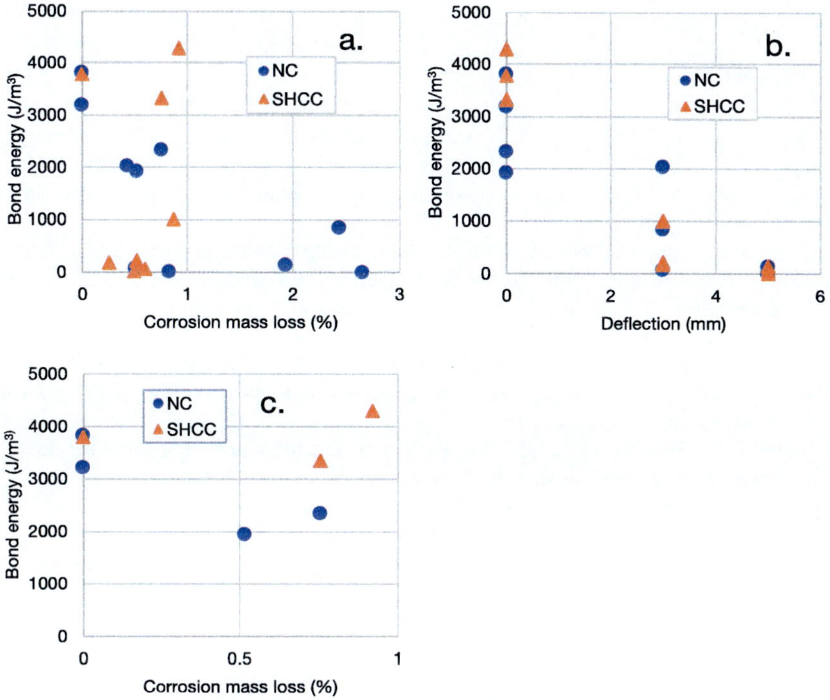

Fig. 10. Bond energy. **a.** Bond energy vs. corrosion mass loss, **b.** Bond energy vs. Beam deflection, **c.** Bond energy vs. corrosion mass loss when deflection = 0[mm].

5 Conclusions

This study aimed at clarifying the influence of corrosion on the bond behavior of rebar in SHCC. SHCC and ordinary concrete (NC) beams were exposed to chloride solution in wet-dry cycles to accelerate rebar corrosion, and the characteristics of rebar corrosion were investigated. A double pull-out test was conducted to examine the bond behavior. The results can be summarized as follows.

(1) Corrosion locations on rebar corresponded to the positions of bending cracks in the beams.
(2) Crack dispersion in SHCC did not reduce with the progress of rebar corrosion.

(3) With the corrosion mass loss being not more than 1%, the bond fracture energy of SHCC did not reduce with the progress of corrosion.

Acknowledgements. This project was partially supported by JSPS and NRF under the Japan - Korea Basic Scientific Cooperation Program JPJSBP120228807.

References

1. Wagner, C., et al.: Transfer of fluids, gases and ions in and through cracked and uncracked composites. In: van Zijl, G.P.A.G., Slowik, V. (eds.) A Framework for Durability Design with Strain-Hardening Cement-Based Composites (SHCC). RSR, vol. 22, pp. 27–58. Springer, Dordrecht (2017). https://doi.org/10.1007/978-94-024-1013-6_2
2. Kobayashi, K., Paul, S.C., van Zijl, G.P.A.G.: Reinforcing bar corrosion. In: van Zijl, G.P.A.G., Slowik, V. (eds.) A Framework for Durability Design with Strain-Hardening Cement-Based Composites (SHCC). RSR, vol. 22, pp. 147–170. Springer, Dordrecht (2017). https://doi.org/10.1007/978-94-024-1013-6_9
3. Kobayashi, K., Kojima, Y.: Effect of fine crack width and water cement ratio of SHCC on chloride ingress and rebar corrosion. Cem. Concr. Compo. **80**, 235–244 (2017). https://doi.org/10.1016/j.cemconcomp.2017.03.006
4. Kobayashi, K., Hosokawa, K., Hattori, Y., Yun, H.D.: Influence of bending cracks on the distribution of rebar corrosion in SHCC. Cem. Concr. Comp. **122**, 104146 (2021). https://doi.org/10.1016/j.cemconcomp.2021.104146

Practical Applications

Follow-Up Review of Early SHCC Applications in Japan

Keitetsu Rokugo[1,2](✉), Naoharu Morii[3], Mamoru Moriyama[3], Seung-Chan Lim[3], Masaki Seki[4], Kazuhide Shinya[5], Hideaki Hatano[1], and Koichi Kobayashi[1]

[1] Gifu University, Gifu, Japan
rk@gifu-u.ac.jp
[2] Central Nippon Highway Engineering, Nagoya, Japan
[3] Deros-Japan, Kanazawa, Japan
[4] Kanto Gakuin University, Yokohama, Japan
[5] Toyo Matelan, Kasugai, Japan

Abstract. This paper discussed some early applications of SHCC to structures in which the authors were involved, describing their outlines and evaluations to date, while presenting various technical ideas related to SHCC. SHCC has been shotcreted for surface repair of concrete hydraulic structures, since around 2005. Premixed SHCC products have since then been progressively enhanced, with the application methods being improved. It has been continuously adopted, with no defects or significant deterioration over time having been reported. In the construction of emergency bays in Hida Tunnel around 2007, multilayer shotcrete lining was adopted, and SHCC was shotcreted as the protective layer for the lining. The follow-up review of the lining conducted in 2015 confirmed that the multilayer shotcrete lining including the thin layer of SHCC provides the desired performance. SHCC has been applied to the bases of new-type noise barriers as protective mortar along Tokaido Shinkansen since 2013. No problems have arisen regarding the renewed new-type noise barriers and the protective mortar (SHCC) for the bases.

Keywords: SHCC · HPFRCC · ECC · Tough mortar · Application · Follow-up

1 Introduction

A considerable period is required until a new material turns out to be good. Follow-up reports on the results of new material applications are not fully available, though they are strongly desired.

Strain-hardening cementitious composites (SHCC), which demonstrate strain-hardening and multiple fine-cracking behavior under tension, emerged in the 1990s as an attractive new cement-based material developed by substantially enhancing the performance of normal concrete and conventional steel fiber-reinforced concrete. Research on SHCC was early on conducted by Li, V.C. [1] and Kanda, T., et al. [2]. Note that SHCC are referred to in Japan as HPFRCC, ECC, tough mortar, etc. The term SHCC is used in this report as a generic term.

M. Kunieda et al. (Eds.): SHCC 2022, RILEM Bookseries 39, pp. 271–280, 2023.
https://doi.org/10.1007/978-3-031-15805-6_28

In the 2000s, production and performance assessment technologies related to this material progressed with the organization of committees in Japan Concrete Institute (JCI) and Japan Society of Civil Engineers (JSCE), with trial application to structures being initiated. In 2003, the authors, together with Kanda, T., et al., carried out trial shotcreting of SHCC (ECC and tough mortar) for restoring the appearance of a gravity-type retaining wall with ASR cracks on the premises of a facility managed by Gifu Prefecture.

In 2007, JSCE published the Recommendations for Design and Construction of High Performance Fiber Reinforced Cement Composites with Multiple Fine Cracks (HPFRCC) (Draft) [3], facilitating their application to structures. The authors were involved in the application of SHCC to repair of agricultural hydraulic structures, construction of emergency bays in Hida Tunnel, large-scale renovation to Tokaido Shinkansen, repair of bridge girder ends, and so forth [4].

This paper reports on the outline of SHCC applications and their follow-up review, while describing technical ideas related to SHCC.

2 Trial Application of SHCC for Restoring the Appearance of ASR-Cracked Concrete Retaining Wall (2003 and 2008) [5]

2.1 Outline of Retaining Wall and Initial SHCC Shotcreting (April 2003) [6]

This gravity-type retaining wall (approximately 18 m in width and 4 to 5 m in height) constructed around 1970 suffered cracking due to ASR. In April 2003, surface repair was conducted by shotcreting SHCC and repair mortar to a thickness of 50 to 70 mm for the purpose of restoring. Figure 1 shows the block layout of the wall.

Block Nos. 1 to 4 were shotcreted with tough mortar, a type of SHCC, with compressive strength (f_c) of 37.6 MPa. The tough mortar contained polyvinyl alcohol (PVA) fibers 12 mm in length and high strength polyethylene (PE) fibers 12 mm in length (1.5 vol % in total). Block Nos. 5 to 8 were shotcreted with ECC containing 12-mm long high-performance PVA fibers (2.1 vol %) with f_c of 54.2 MPa. Block No. 9 was shotcreted with normal repair mortar with f_c of 59.3 MPa. The bottom area to a height of 2 m was additionally coated with acrylic. Prior to shotcreting, the wall surface was polished by waterjetting. Welded wire mesh (D 6, SD 295 at 100 mm pitch) was set on Block Nos. 1, 5, and 9. Expanded metal (75 by 203 mm, XS-82) was set on Block Nos. 2 and 6, both at around 10 mm from the wall surface.

2.2 Thin Layer Shotcreting of SHCC (October 2008)

In April 2008, the range from the bottom of the wall to a height of 2.5 m including the acrylic coating was polished by waterjetting, as stain on the acrylic became noticeable. In October 2008, shotcreting of SHCC (tough mortar) on the polished area was demonstrated to participants of an international convention held at this time nearby. The shotcreting thickness was approximately 20 mm from the bottom of the wall to a height of 1 m and 10 mm in the range of 1 to 2 m in height (Fig. 2).

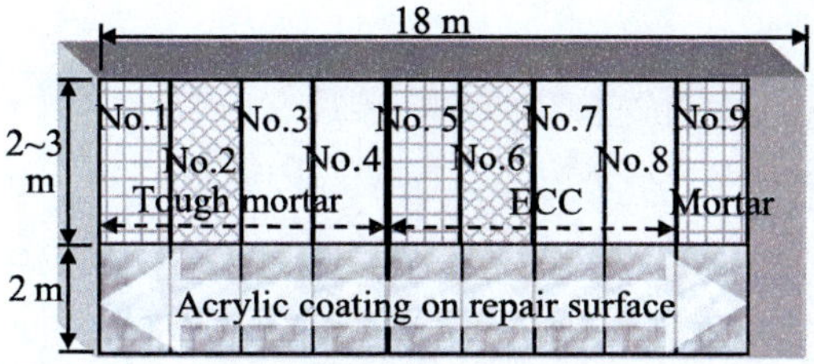

Fig. 1. Block layout of the wall.

Fig. 2. Thin layer shotcreting of SHCC.

2.3 Follow-Up Survey (September 2020)

Figures 3 and 4 show normal and thermal images, respectively, taken with a drone-mounted camera in September 2020, which is 17 years after the initial application. These were made by synthesizing 23 images each into orthomosaic images of the wall frontal.

Fig. 3. Image of the retaining wall.

Stain due to efflorescence, etc., was less evident on Block Nos. 1, 2, 5, and 6 where reinforcement was laid out in the SHCC layer in the initial application (Fig. 3). In Block No. 9 (thin-layer SHCC on normal repair mortar), a wide range of delamination was recognized by hammering test and thermal imaging analysis (Fig. 4) under the thin-layer SHCC (10 to 20 mm in thickness), which was shotcreted five years after the initial application. Delamination was also found mainly near the boundaries between Block Nos. 5 to 9. On the other hand, no delamination was found in the initially shotcreted SHCC (the area above the level 2 m from the bottom, 50 to 70 mm in thickness) by hammering test and thermal image analysis.

2.4 Technical Discussion

SHCC was found superior to repair mortar in suppressing crack widening and inhibiting stain due to efflorescence. It was proven appropriate to select the thickness of 50 to

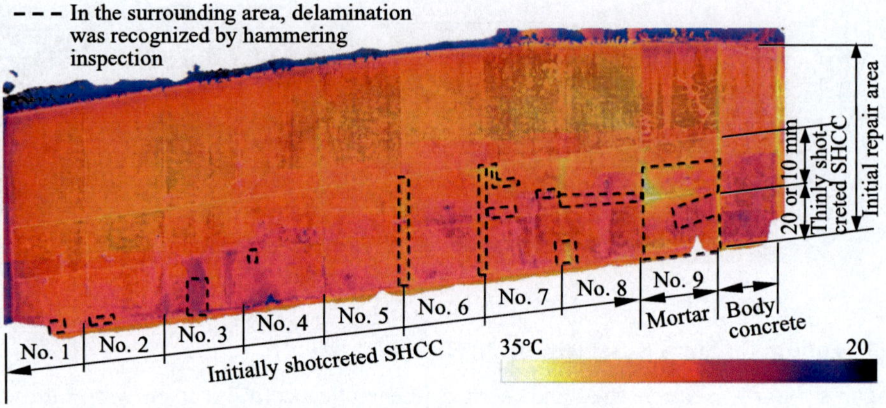

Fig. 4. Thermal image of the retaining wall.

70 mm, instead of 10 to 20 mm, in the initial shotcreting of SHCC onto this size of retaining wall, causing little delamination. The placement of reinforcement (welded wire mesh or expanded metal) was proven more effective, than without, in suppressing the crack width and inhibiting stain due to efflorescence. The restoring effect of acrylic coating over the lower half area after the initial application lasted only around two years.

3 Application of SHCC to Concrete Hydraulic Structures as a Surface Repair Material (2005–)

3.1 Repair of Concrete Hydraulic Structures

Construction of agricultural hydraulic facilities including waterways began in the 1950s in Japan. Some of them have been suffering from functional degradation with age, leading to the launch of a new subsidy system for repair in 2003. Hydraulic structures made of concrete are generally repaired by surface coating and patching, utilizing the existing concrete body, unless it is structurally deteriorated. The performances required for such repair materials include resistance to abrasion, resistance to frost attack, steel corrosion protection, and adhesion to the structural body. SHCC has therefore been used as a promising repair material.

3.2 Premixed SHCC Products

Premixed SHCC ready for mixing on site is generally used in wet shotcreting for surface repair of hydraulic structures. It incorporates 1.7 vol % PVA fibers to provide ultimate tensile strain (s_t) of around 1.0% and f_c of around 40 MPa. During the early stage of application for 10 years since 2005, only the mortar fraction was premixed while the short fibers were added during mixing in a mixer. However, premixed products with short fibers have become available since 2015. While a high-speed mixer rotating at more than 60 revolutions/min was initially used to ensure fiber dispersion, since 2015 the premixed products with short fibers can be mixed using a general-purpose low-speed mixer. In the production of the premixed products with short fibers at factories, in addition to adjustment of flowability and viscosity using water reducers and viscosity enhancing agents, there have also been changes in the aggregate size (from 0.5 mm or less to 1.0 mm or less), and changes in the fiber length (from 12 mm to 8 mm). Even after these changes, the maximum tensile strain remains at around 1.0%.

3.3 Repair Methods

Application of SHCC begins with removal of the deteriorated surface to a depth of approximately 1 mm by water jetting, and then follows patching or surface coating with SHCC by troweling (small areas) or shotcreting (Fig. 5). A squeeze pump is used for shotcreting, with the pumping distance being around 50 m at the maximum. After shotcreting with SHCC, the surface is sprayed with an acrylic emulsion at a spread of approximately 150 g/m^2 before troweling. The emulsion improves the trowel slidability, enhancing the efficiency of finishing work, while increasing the initial curing effect due to the formation of the emulsion coating on the finished surface. A method is also adopted for waterway tunnels by which fiber grid reinforcement is placed, onto which SHCC is shotcreted to form a 15 mm coating, to enhance the load-bearing capacity of the existing lining and prevent spalling (Fig. 6).

(a) Waterjetting (b) Coating with SHCC (c) Spraying with emulsion (d) Troweling

Fig. 5. Application of SHCC to agricultural waterway.

(a) Fiber grid reinforcement (b) Coating with SHCC (c) Completed

Fig. 6. Application of SHCC to waterway tunnel.

3.4 SHCC Applications to Concrete Hydraulic Structures and Follow-Up Review

Application of SHCC for repair of concrete hydraulic structures totaled more than 400 structures including open waterways, intakes, dams, and waterway tunnels made of concrete, with the area of application being more than 280,000 m^2. No report has been received regarding defects or significant age deterioration of SHCC for 16 years since its initial application. Rather, its crack-suppressing performance and resistance to frost damage are highly rated, with its application being continuously in demand.

4 Application of SHCC to Emergency Bays in Hida Tunnel as a Protective Layer for Multilayer Lining (2007)

4.1 Hida Tunnel

Hida Tunnel, which opened in July 2008, is a long tunnel spanning 10.7 km on the Tokai-Hokuriku Expressway, being located to the south of Shirakawago, a World Heritage site. It consists of the main tunnel and escape tunnel (pilot tunnel) with a cross-sectional area of 130 m^2 and 16 m^2, respectively. Excavation of the escape tunnel began in October 1996 prior to the main tunnel.

4.2 Multilayer Shotcrete Lining Adopted for the Emergency Bays [7]

After being completed in January 2007, the main tunnel was widened at 33 points to construct emergency bays. Multilayer shotcrete lining (Fig. 7) was adopted for the walls of 16 of the 33 emergency bays instead of conventional cast-in-place concrete lining, for the first time among domestic expressway tunnels.

Fiber-reinforced shotcrete with f_c of 36 MPa incorporating 0.75 vol % or 1.0 vol % polypropylene (PP) fibers 32 mm in length was used as the primary support layer to a thickness of around 150 mm after excavation to ensure stability of the excavation site and the surrounding ground.

Steel fiber-reinforced shotcrete with f_c of 36 MPa incorporating 0.75 vol % or 1.0 vol % steel fibers 30 mm in length was then applied as the secondary layer in two phases to a thickness of around 250 mm in total, to increase the structural performance.

As the tertiary layer, SHCC with f_c of 40 MPa incorporating 2.0 vol % PVA fibers 12 mm in length was applied to prevent spalling of the secondary layer, prevent corrosion of steel fibers, enhance water tightness, and improve the appearance (smooth the surface irregularities of the secondary layer). The SHCC was shotcreted with a thickness of 10 to 20 mm and then smoothed by surface rolling (Fig. 8).

4.3 Follow-Up Review of Multilayer Shotcrete Lining

Follow-up review of the multilayer shotcrete lining was conducted in 2015 based on the results of visual inspection and strength tests [8]. This section introduces the content of this review. Note that SHCC are characterized by their capability to form many fine cracks under tensile forces.

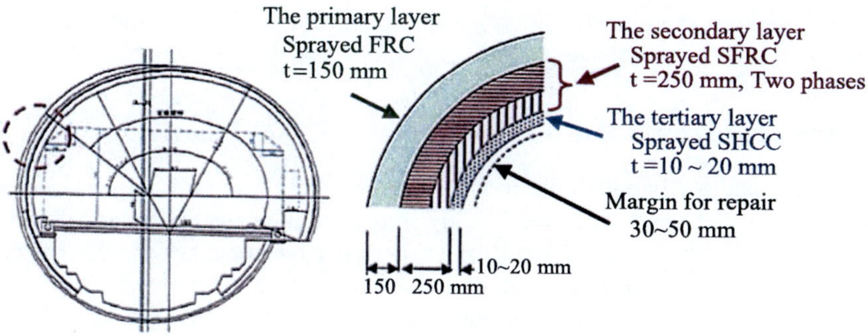

Fig. 7. Multilayer shotcrete lining.

Fig. 8. Completed shotcrete lining [7].

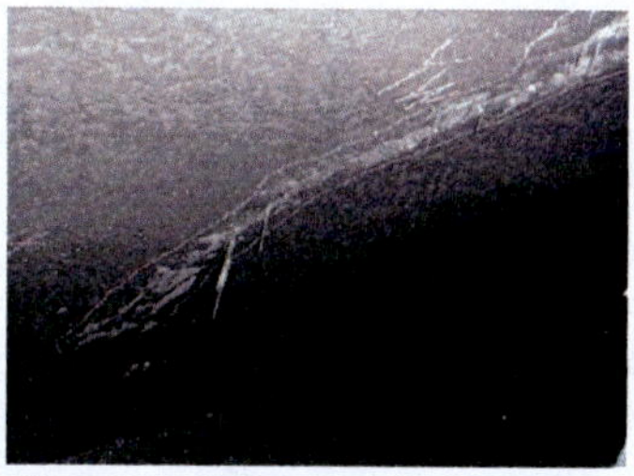

Fig. 9. Typical efflorescence [8].

Figure 9 shows a photograph of typical efflorescence at emergency bay No. 16 where efflorescence and cracks are extensively found. In this bay, the lengths of efflorescence and cracks totaled 20,450 mm and 17,045 mm, respectively. Despite their long extension, no water leakage was found, and the crack width was small, ranging from 0.2 to 0.5 mm, with no sign of progressive growth. Three core samples in total were drilled from emergency bays Nos. 12, 16, and 32 to be subject to compression tests, with the results of all cores exceeding 50 MPa with no delamination at the layer boundaries. Accordingly, the multi-layer shotcrete lining revealed no structural problem about 10 years after application, though with efflorescence and cracks [8].

Since it was confirmed by the follow-up review that the multi-layer shotcrete lining is providing the desired performance, SHCC forming the tertiary layer is also considered to provide the performance initially anticipated.

5 Application of SHCC to Viaducts of Tokaido Shinkansen as Protective Mortar for the Bases of Noise Barriers (2013–)

5.1 Large-Scale Renovation of Tokaido Shinkansen

Tokaido Shinkansen, a bullet train connecting Tokyo, Nagoya, and Shin-Osaka, has been in service since 1964. Along its overall length of 515 km, a large-scale renovation project has been under way since 2013 at steel bridges (total extension of 22 km), concrete bridges (148 km), and tunnels (69 km). Among the concrete bridges under renovation, those totaling 98.2 km are viaducts [9].

5.2 Application of SHCC to the Bases of New-Type Noise Barriers

Tokaido Shinkansen initially had no noise barriers, but they were progressively constructed thereafter, where required, mostly in residential areas. The loads due to the addition of such barriers were therefore not expected in the initial design of the viaducts. In the large-scale renovation project, noise barriers have therefore been replaced with a new type to minimize the effect of wind loads generated by the noise barriers on the cantilever parts of viaducts. In the new-type barriers, anchor bolts that pierce through the cantilever concrete together with the top and bottom steel plates (jackets) strengthen the cantilever members by making them composite structures [10]. As part of the large-scale renovation project of Tokaido Shinkansen, SHCC with f_c of 40 MPa, containing 2.0 vol % PVA fibers 12 mm in length, has been basically used since 2013 to the bases of the new-type noise barriers for the purpose of suppressing crack width, preventing steel corrosion, and securing inspection passages (Fig. 10). To date, the usage of SHCC on this line has exceeded 10,000 m^3.

(a) Before and after renovation of viaducts with new-type noise barriers

(b) Casting of SHCC to the bases

Fig. 10. Casting of SHCC to the bases of new-type noise barriers on viaducts.

5.3 Technical Discussion and Follow-Up Review

Prior to selecting the protective material for the bases of the new-type noise barriers, trial application was conducted using SHCC, high-early-strength concrete, PP fiber-reinforced concrete, ultraquick-setting non-shrink concrete, etc. These tests led to the selection of SHCC, which is less prone to cracking. The steel bases for the new-type noise barriers were coated with an epoxy primer to ensure unification with SHCC. It was therefore important to place SHCC within the setting time of the primer.

Replacement with the new-type noise barriers and SHCC placement in the bases are mostly carried out in a short time from midnight to early morning when the trains are out of operation. Experiments have confirmed that vibration equivalent to train running does not adversely affect the bond strength development between steel and SHCC [11]. This substantially relaxed the time limits of SHCC placement.

Ten years having passed since the start of replacement with the new-type noise barriers and placement of SHCC in their bases, no problems have arisen regarding the new-type barriers, the jackets of steel plates of the cantilever segments, or protective mortar (SHCC) for the bases.

6 Conclusions

This paper discussed some early applications of SHCC to structures in which the authors were involved, describing their outlines and evaluations to date, while presenting various technical ideas related to SHCC.

SHCC was shotcreted onto an ASR-cracked concrete retaining wall for restoring on a trial basis to a thickness of 50 to 70 mm in 2003. Five years later, a thin layer (10 to 20 mm) of SHCC was shotcreted on this wall. In an inspection conducted in 2020, delamination was discovered over a wide range of the thin layer of SHCC shotcreted on reference mortar. It was therefore judged appropriate that SHCC was initially shotcreted to a thickness of 50 to 70 mm, with reinforcement being laid out.

SHCC has been shotcreted for surface repair of concrete hydraulic structures including waterways, etc., since around 2005. Premixed SHCC products have since then been progressively enhanced, with the application methods being improved. It has been continuously adopted, with no defects or significant deterioration over time having been reported.

In the construction of emergency bays in Hida Tunnel around 2007, multilayer shotcrete lining was adopted, and SHCC was shotcreted as the protective layer for the lining. The follow-up review of the lining conducted in 2015 confirmed that the multilayer shotcrete lining including the thin layer of SHCC provides the desired performance.

SHCC has been applied to the bases of the new-type noise barriers as protective mortar along Tokaido Shinkansen since 2013. No problems have arisen regarding the renewed new-type noise barriers, the jackets of steel plates of the cantilever segments, and the protective mortar (SHCC) for the bases.

When cracks occur in concrete, the smaller their width, the more desirable in terms of both durability and aesthetic appearance. In the application cases stated above, SHCC is used as a concrete less prone to cracking and with smaller crack width. Meanwhile, despite abundant studies and various suggestions to make the most of its high tensile resistance and fracture toughness (energy-absorbing capacity), there are still few application examples to take advantage of these features, raising hopes for the future.

For a concrete-based material including fiber-reinforced concrete, deterioration is known to be accelerated under the combined effects of cracking, water, and member deformation/pressure. It is therefore important to take sufficient measures for waterproofing and drainage, suppress deformation by combining with structures with high stiffness, and ensure the bond between concreting lifts and at boundaries with other materials. With these points in mind, it is desired that SHCC will be extensively used, making bold and careful steps forward.

References

1. Li, V.C.: From micromechanics to structural engineering - The design of cementitious composites for civil engineering applications. J. Struct. Mech. Earthq. Eng. Jpn. Soc. Civil Eng. **10**(2), 37–48 (1993)
2. Kanda, T., Li, V.C.: Multiple cracking and saturation in fiber reinforced cementitious composites. Concr. Res. Technol. **9**(2), 19–33 (1998) (in Japanese)
3. Japan Society of Civil Engineers. Recommendations for design and construction of high performance fiber reinforced cement composites with multiple fine cracks (HPFRCC). Concr. Eng. Ser. **82** (2008)
4. Rokugo, K.: Applications of SHCC in Japan – Tools and tips for promoting its use. In: Proceedings of the 4th International RILEM Conference on SHCC (SHCC4), pp. 671–680. Dresden (2017)
5. Rokugo, K., Lim, S-C., Hatano, H., Matsudaira, T.: Thermal imaging research on retaining wall landscaped by shotcreting of HPFRCC 17 and 12 years ago. In: Proceedings of the 9th International Conference of Asian Concrete Federation (ACF2020/2021), pp. 46–53. Thailand (2021)
6. Rokugo, K., Morii, N., Lim, S-C., Kanda, T., Sakata, N.: Landscaping of ASR-cracked retaining wall using HPFRCC shotcretes and observation over 5 years. In: Proceedings of the 7th International Conference on Fracture Mechanics of Concrete and Concrete Structure (FraMCoS 7), pp. 1676–1682. Jeju (2010)
7. Moriyama, M., Kobayashi, S., Itagaki, K., et al.: Design of multilayer shotcrete lining for emergency parking lots of Hida tunnel. In: Japan Society of Civil Engineers 2008 Annual Meeting, pp. 6–290, 579–580 (2008) (in Japanese)
8. Adachi, Y., Tsuno, Y., Onoyama, T., et al.: Evaluation of multilayer shotcrete lining for emergency parking lots of Hida tunnel after 10 years. In: Japan Society of Civil Engineers 2017 Annual Meeting, VI–611, pp. 1221–1222 (2017) (in Japanese)
9. Seki, M.: Large-scale renovations of the civil engineering structures along the Tokaido Shinkansen. Jpn. Railw. Transp. Rev. **62**, 16–31 (2013)
10. Mitsumune, H., Seki, M., Rokugo, K., et al.: Performance evaluation and application of ductile fiber reinforced polymer mortar for large scale renovation to extend service life of the Tokaido Shinkansen. In: Proceedings of the 9th RILEM International Symposium on Fiber Reinforced Concrete (BEFIB 2016), Vancouver (2016)
11. Lim, S-C., Shinya, K., Ohkubo, H., Morii, N., Rokugo, K.: Test method for bond strength between steel plates and an ultraquick-hardening SHCC subjected early vibration. In: Proceedings of the 9th RILEM International Symposium on Fiber Reinforced Concrete (BEFIB 2016), Vancouver (2016)

Strain-Hardening Cement-based Composites (SHCC) for Impact Strengthening of Buildings: Recent Advances in the DFG Research Training Group 2250

Cesare Signorini(✉) and Viktor Mechtcherine

Institute of Construction Materials, Technische Universität Dresden, 01062 Dresden, Germany
cesare.signorini@tu-dresden.de

Abstract. Concrete is by far the most widespread construction material worldwide for buildings and infrastructures. While offering wide range of advantages, concrete structures are vulnerable to impact loading such as collisions, rock fall or explosions. This can be traced back to the intrinsically brittle nature of the material. Against this background, the Research Training Group (RTG) 2250 funded by the German Research Foundation (DFG) focuses on the development of strengthening overlays made of strain-hardening cement-based composites (SHCC) and other quasi-ductile mineral based materials capable of drastically enhancing the impact resistance of existing concrete structures. Multidisciplinary collaborative work is carried out by three renowned research institutions in Dresden with nine departments involved. In this contribution, an overview of the recent achievements in the RTG 2250 work are presented, spanning from the design of new sustainable SHCC as high-ductility matrices for textile-reinforced strengthening layers to the structural performance of such layers under impact loading. The latter is assessed by means of customized real-scale test protocols. Furthermore, some insights into the advanced techniques of data acquisition and management, numerical modeling as well as sustainability and resilience assessment are provided.

Keywords: SHCC · ECC · Textile reinforcement · Strengthening · Impact resistance

1 Introduction

Concrete is nowadays the most widespread construction material worldwide, owing to its high compressive strength and economic affordability. The use of steel reinforcement has amended the weakness of concrete against tensile loading as well as reduced brittleness of concrete structures. However, concretes are still highly vulnerable to high strain-rate loading, like e.g., impact, blast, earthquakes and/or explosions. Such deleterious events can be regarded as exceptional during the service life of a building, but when suddenly occurring, they may be prejudicial for human safety, often bearing dramatic societal impact. Given the urgent need to protect structures of systemic importance against exceptional events, the Research Training Group 2250 - "*Mineral-bonded*

M. Kunieda et al. (Eds.): SHCC 2022, RILEM Bookseries 39, pp. 281–290, 2023.
https://doi.org/10.1007/978-3-031-15805-6_29

composites for enhanced structural impact safety" was established at the TU Dresden in 2017. Currently, the second cohort of doctoral projects is ongoing, and the third one will start in May 2023. The main goal of the RTG 2250 consists in developing high-performance composite materials, fulfilling requirements to withstand impact loading, fostering resilience of buildings, increasing the safety degree of critical infrastructure, and hence preserving human life and the built environment. Strain-hardening cementitious composites (SHCC) represent an attractive opportunity in this regard. SHCC are well-established fiber-reinforced materials able to dissipate high amounts of impact energy, thus, turning out to be an excellent candidate to build protective strengthening layers against dynamic loading [1]. Furthermore, exploiting synergistic effects of short fibers and woven textiles leads to hybrid composites with further improved response against impact loads. Within the RTG 2250, a multidisciplinary vision of the topic is warranted by joining the expertise of several institutes, spanning from the development of new materials with exceptional dissipation capability, to the assessment of the structural behavior of strengthened members and to a comprehensive analysis of material behavior and structural performance by means of advanced computational methods. Finally, sustainability assessment and analysis of resilience play an essential role in developing efficient holistic solutions. Figure 1 gives an overview of the individual doctoral and post-doctoral projects. The article at hand presents the structure and goals of RTG 2250 as well as a concise summary of the main results achieved so far in the framework of the second cohort.

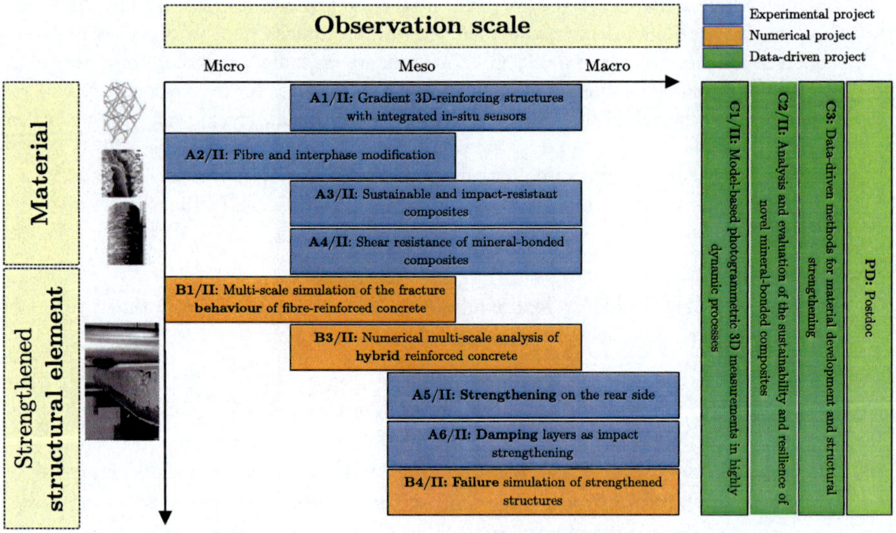

Fig. 1. The research framework the Research Training Group 2250, cohort II.

2 Structure and Goals of RTG2250/II

The second cohort of the RTG 2250 accounts for a dense network of knowledge from several fields, taking advantage of the outcomes achieved by the first cohort. Currently, 13 individual projects bring together various experimental, numerical and data-driven approaches, as briefly presented in the following sections.

2.1 The Experimental Approach: SHCC Development and Assessment

Development of the Composite Constituents. The experimental projects are mainly devoted to developing new sustainable mineral-based composites with exceptional impact resistance as well as exploring their performance as strengthening layers applied on existing concrete structures. The main focus is on composites with hybrid reinforcement, namely textile-reinforced SHCC, hereafter regarded as TR-SHCC. Project A1 focuses on textile reinforcement and pursues the twofold aim of (i) developing optimized textile meshes made of high tenacity continuous fibers in two- and three-dimensional (cellular) configurations with optimized topology [2], and (ii) including smart diagnostic capabilities by using embedded sensors. The non-destructive techniques (NDT) for monitoring damage are being implemented by exploiting the intrinsic conductive nature of fibers made of carbon or steel. Embedded sensors are capable of monitoring the integrity of the reinforcement, thus providing a comprehensive picture of the health condition of the composite while subjected to quasi-static or impact loading [3]. Continuous reinforcement can be applied both on the impact and on the opposite side of structural elements. When applied on the rear side with respect to the impact occurrence, textile reinforcement delivers a membrane action and enables to retain scabbing portion of the concrete substrate after impact. The performance is improved by using SHCC as the matrix. When TR-SHCC is applied on the impacted side of the concrete element, both the textile and short fiber mitigate the punching effect associated with high local shear stresses. In parallel to the development of continuous reinforcement, project A2 tackles the design of short fibers possessing engineered adhesion to cementitious matrices. The main focus is on modifying polyethylene (PE) and polypropylene (PP) fibers by chemical treatment or by introducing micro- and nano-particles during the spinning of the filaments or through coating. The fine particles interact either chemically or mechanically with the surrounding mineral-based matrix and enhance the bond. It has been proved that higher surface roughness of PP fibers generally fosters better adhesion properties, even more remarkably under high displacement rates, resulting in notable increase in the energy absorption capability [4]. The development of environmental-friendly SHCC compositions with highly reduced Portland cement (PC) content is the central point in project A3. Among the alternatives to PC, the use of limestone calcined clay cement (LC^3) is under investigation as a viable option for matrices in strain-hardening composites. LC^3 implies a 50% reduction in OPC content by using widely available mineral materials, namely clay and limestone [5]. Wang et al. [6] showed how different LC^3 binders can be used to produce mortars with compressive strength up to 150 MPa. Strain-hardening calcined clay cement composites ($SHLC^4$) exhibit a pronounced deflection-hardening behavior in bending, with first cracking formation observed at approximately 10 MPa to 15 MPa, and failure at

20 MPa to 25 MPa, depending on the binder composition. Thorough rheological characterization is currently ongoing to optimize SHLC[4] for various application scenarios such as automated shotcreting. Figure 2 shows the first steps in that direction.

(a) Manual wet lay-up of TR-SHCC (b) Shotcreting SHCC

Fig. 2. Application techniques for SHCC composites (LC^3 mortar with UHWPE fibers and carbon biaxial textile)

Mechanical Characterization of SHCC and TRC Under Impact. The assessment of the mechanical efficiency of quasi-ductile SHCC composites under impact is extremely challenging due to highly transient conditions during dynamic loading. Finely tailored testing setups at different scales need to be developed to accurately characterize materials properties and to gain comprehensive insights into the structural behavior of strengthened structures subjected to impact. Materials parameter quantification under dynamic regimes is also crucial for the calibration of advanced analytical and numerical models. At the material scale, split-Hopkinson pressure and tension bars (SHPB/SHTB) are sound facilities to test material behavior under impact compression and impact tension, respectively. Generally, SHB consists of an incident bar that hits specimen with a high-rate loading wave, passing through the sample and being then channeled into a transmitter bar. The loading wave can be either a compression or a tension one and it is partly transmitted and partly reflected. The characteristics of the incident and transmitted waves are measured by means of strain gauges attached to the bars. In RTG 2250, a novel gravitational SHTB device has been developed. It is easy to operate and offers a notable versatility in the testing protocols for SHCC and TRC [7]. Recently, in Project A4, the device has been extended to enable testing shear response of strain-hardening composites subject to impact loading [8]. For this purpose, a customized gripping system has been developed based on comprehensive numerical analysis of SHTB experiments in the shear configuration; see Fig. 3. On the structural level, projects A5 and A6 focus on the behavior concrete elements strengthened with overlays made of TRC, SHCC or TR-SHCC. The major facility for that is a 11 m-high drop-tower complimented by a hydraulic equipment to control the impactor velocity and energy before the collision with the structural element. This setup enables to tailor testing methodology (i) to determine crucial performance features of the strengthening layers, as presented by Hering et al. [9], and, (ii) to assess the structural behavior of real-scale structural members under impact [10]; see Fig. 4. For the first purpose, usually small-scale plates made of strengthening composite and measuring 60 cm × 60 cm × 3 cm are tested. The changes in the impactor velocity due to impact serves is one of the major parameters to assess the performance of such protective layers. Eventually, concrete slabs strengthened with mineral-bonded composites are studied. The key damage parameters here include the extent of concrete spalling and scabbing, the punching cone angle, and the variation of the free vibrations of the plates before and after the test. In project A5, it was shown that steel stirrups serving as shear reinforcement in the substrate slab strongly affect the force distribution and therefore also the punching cone slope within the strengthened element. As a consequence, the loading and the response of the strengthening composite layer positioned at the rear side of the slab are significantly influenced as well.

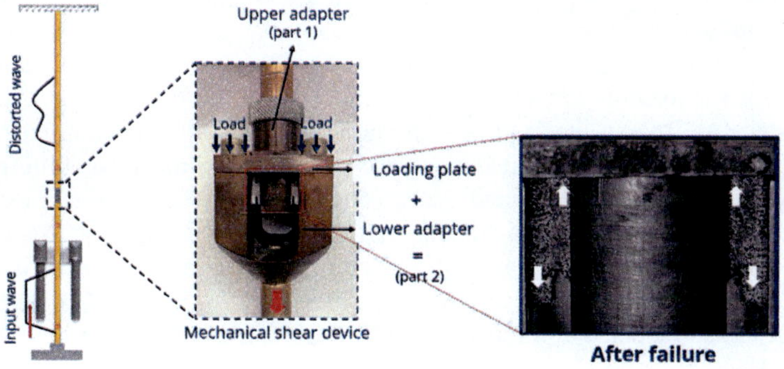

Fig. 3. Gravitational SHTB [7] with novel adapters to characterize SHCC behavior under shear impact loading

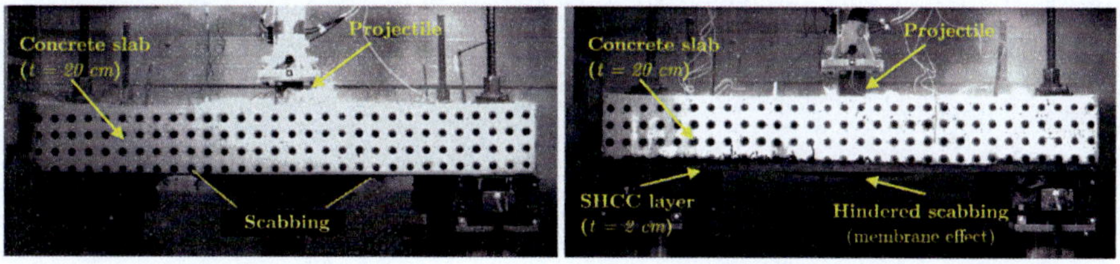

(a) Unreinforced slab (b) Slab reinforced with TR-SHCC

Fig. 4. Real-scale impact tests on concrete slabs (photos: Marcus Hering)

In project A6, Leicht et al. [11] investigated the impact response of reinforced concrete slabs and correlated the punching cone generated by the impactor with the overall damage level. Next, the effect of damping layers on the impact side has been investigated. Such layers may induce a remarkable energy dissipation, reducing the maximum load acting on the structure. For this purpose, new blends of cementitious materials and rubber aggregates are currently under development to maximize the damping effect, but also taking into account the ecological aspect by using recycled materials.

2.2 Numerical Modeling Techniques for Optimizing Impact Resistance

The development of new technologies requires handling multiple material and process parameters, whose combination would entail a huge amount of experimental work to obtained necessary characteristics. Thus, a purely empirical approach based on the trial-and-error principle is clearly not affordable. Rather, the development of robust simulation methods is necessary as a tool for acquiring information from selected experiments and eventually providing a robust predictive framework in which various parameters can be finely tuned. Individual projects encompass different scales of observation, from material level to the structural one. Recently, Tamsen et al. [12] developed a multiscale framework to take into account inertia in dynamic tests on SHCC. Inertia related phenomena can lead to a significant overestimation of the dynamic magnification factor stranding for the material response at high strain rates. SHTB experiments on SHCC were modeled by tuning the loading condition via piece-wise polynomial functions. The current work in project B1 also pursues a multiscale approach combining a multiscale projection method (MPM) and the so-called extended finite element method (X-FEM). The goal is to simulate nucleation and propagation of cracks at the meso- and macroscales. Indeed, complex wave patterns propagate and reflect as a consequence of impact, heavily affecting the fracture behavior of SHCC. In the previous work in RTG 2250, SHCC was modeled at the mesoscale through the strong discontinuity approach (SDA) by Häußler-Combe et al. [13]. Single fibers were explicitly represented as truss elements having nodes independently placed within the matrix continuum. The crucial role played by interphase bond was considered by connecting fibers and embedding medium through special bond elements. At a larger scale, crack propagation has been studied by applying novel, robust algorithms based on the variational approach to modeling fracture. The purpose of this approach is to formulate a material model representative for the physics behind the phenomena observed. In this context, the eigenerosion approach has been proved a numerically efficient and stable realization of the variational eigenfracture scheme [14, 15]. The material description developed in project B4 is suitable to predict a wide range of concrete material characteristics at high loading rates [16], but should be further adapted to describe SHCC in an effective manner, as the crack-bridging effects exerted by the fibers need to be duly taken into account. A major step forward for the variational eigenfracture scheme consists in the introduction of a mesh-free spatial discretization in terms of the Material Point Erosion scheme, whose implementation in the FEM should enable to simulate concrete plates strengthened with SHCC. Finally, neural networks are regarded as an attractive approach to modeling complex materials like SHCC, since they can be trained by means of selected datasets, drastically reducing the computational expense, with the chance of attaining a tremendous increase of the modeling potential [17]. The recent paper by Stöcker et al. [18] prepared in the framework of project B3 reports on a novel training procedure that make use of adversarial examples to improve the robustness of recurrent neural networks adopted to simulate the constitutive behavior of inelastic composite materials. Pilot simulations on fiber-reinforced polymer endorse the validity of the proposed model, which is currently under development to capture the behavior of mineral-based composites containing both short and continuous fibers. The investigation is generally performed on a modular component of the composite system, regarded as representative volume element (RVE), as elaborated in a previous publication by Fuchs et al. [19]. In this case, a numerical material testing is conducted on TRC, where textile yarns are regarded as elastic isotropic unidirectional layers, while the fine-grained concrete matrix is modeled with a gradient enhanced microplane model which properly accounts for plasticity and damage.

2.3 Data-driven Research: Sustainability, Photogrammetry and Design Optimization

The vision of RTG 2250 includes a high degree of cross-disciplinary, fostering holistic solutions. The key approaches are: (i) to understand underlying phenomena governing the mechanical response of complex materials and strengthened structures through photogrammetry, (ii) to outline a robust feasibility assessment of the technologies under development at a large scale, accounting for sources availability as well as for environmental and societal impact, and (iii) to develop reliable design tools necessary to boost the use of innovative materials on a regular basis in the practice of construction. The assessment of the mechanical behavior of structural elements under impact loading is extremely challenging, as dynamic processes occur within time-frames of few microseconds. Advanced photogrammetric algorithms developed within project C1 enable to capture and interpret deformation, cracking and failure behavior of materials and structural members subject to impact loading. For instance, optical techniques offer the possibility of monitor large surfaces in impact experiments with micrometric accuracy and temporal resolution beyond 100 kHz, when stereo high-speed camera systems are employed. In this manner it is possible to track crack development, which is a key indicator of damage evolution in SHCC and TRC. Furthermore, formation of cracks, their growth and width can be reliably computed using a new algorithm presented by Liebold and Maas [20]. Crack propagation speed for SHCC, which is approximately 800 ms^{-1}, can only be monitored with high-frequency imaging combined with automated photogrammetric algorithms for image processing; see Liebold et al. [21] and Fig. 5. On the basis of time-resolved sets of 3D point coordinates, deformations, cracks and even wave patterns can be determined. Another paramount goal in the RTG 2250 consists in designing materials and technologies able to comply with high sustainability and resilience demands. In this framework, sustainability and resilience assessment plays a fundamental role in the compliance of impact safety of strengthened concrete structures with economic, environmental and societal requirements [22]. In a recent review paper, the refurbishment potential of concrete structures through various techniques was systematically tackled by means of relevant case studies and a thorough assessment of the methodological tools for pursuing best practices in the construction industry [23]. Ongoing investigations within project C2 concentrate on the economic product perspective, with particular consideration of mineral-based composite materials as such and their performance in representative structural elements. Finally, a new transversal research project C3 initiated for the second cohort of RTG 2250 is dedicated to a sustainable and synergistic handling of the massive amount of experimental and simulations data produced over the entire 9-year duration of the RTG. This concept strictly follows the so-called FAIR principles which stand for Findability, Accessibility, Interoperability, and Reuse. FAIR principles promote the widest accessibility of data and digital asset in line with the DFG policy. In RTG 2250, research methodologies and data types are extremely heterogeneous, so that very different formats need to be processed and by means of a multiplicity of software and procedures. To be able to handle such datasets in a comprehensive way, also beyond the framework of the Research Training Group, a holistic and rigorous methodology for scientific data and metadata management and exploitation is under development. The investigation involves the formulation of automated computational procedures and the synergistic cross-checking of multiple data sources, both from experiments and simulations, aiming to achieve optimal design and a predictive view of the performance of concrete elements strengthened by SHCC or TRC layers.

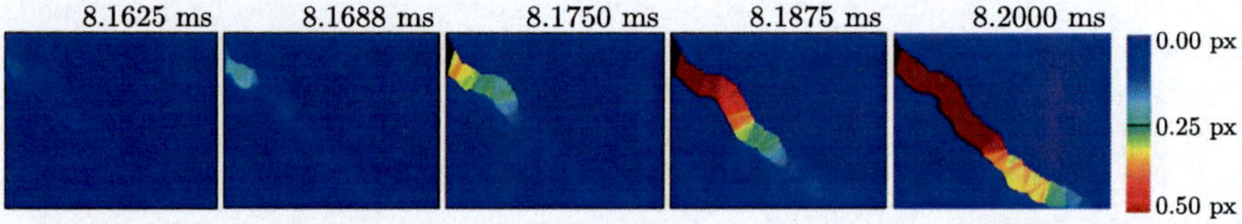

Fig. 5. Color maps showing the crack evolution by plotting the relative translation vector lengths [21]

3 Conclusions and Outlook

This contribution offered a concise overview of the most recent achievements in the Research Training Group 2250 funded by the German Research Foundation (DFG). After having investigated the fundamental mechanisms behind the performance of SHCC and TRC under impact loading in the first RTG cohort, the research group is now focusing on the development of novel, more sustainable composites while still targeting a very high-performance of the strengthening layers. Furthermore, experimental characterization methods, advanced computational tools and photogrammetric techniques are under continuous development to provide a comprehensive understanding of the mechanical response of complex composite materials and strengthened structured, so that eventually their designs can be purposefully optimized. Cutting-edge results and important achievements have been obtained so far in these topics, thanks, among other points, to close collaborations established with a number of renowned international experts and research institutions. A further crucial step currently under investigation consists in the implementation of newly developed materials using automated application technologies such as automated shortcreting or 3D printing.

Acknowledgments. The financial support of the German Research Foundation (DFG) within the Research Training Group 2250/2 "*Mineral- bonded composites for enhanced structural impact safety*", project number 287321140, is gratefully acknowledged.

References

1. Li, V.C.: On engineered cementitious composites (ECC) a review of the material and its applications. J. Adv. Concr. Technol. **1**(3), 215–230 (2003)
2. Vo, D.M., Sennewald, C., Hoffmann, G., Cherif, C.: Fiber-based 3D cellular reinforcing structures for mineral-bonded composites with enhanced structural impact tolerance. Int. J. Civil Environ. Eng. **12**(5), 582–586 (2018)
3. Gaben, M., Goldfeld, Y.: Self-sensory carbon-based textile reinforced concrete beams-characterization of the structural-electrical response by AC measurements. Sens. Actuators A: Phys. **334**, 113322 (2022)
4. Wölfel, E., Brünig, H., Curosu, I., Mechtcherine, V., Scheffler, C.: Dynamic single-fiber pull-out of polypropylene fibers produced with different mechanical and surface properties for concrete reinforcement. Materials **14**(4), 722 (2021)
5. Sharma, M., Bishnoi, S., Martirena, F., Scrivener, K.: Limestone calcined clay cement and concrete: a state-of-the-art review. Cement Concr. Res. **149**, 106564 (2021)
6. Wang, L., et al.: On the use of limestone calcined clay cement (LC3) in high-strength strain-hardening cement-based composites (HS-SHCC). Cement Concr. Res. **144**, 106421 (2021)
7. Heravi, A.A., Curosu, I., Mechtcherine, V.: A gravity-driven split hopkinson tension bar for investigating quasi-ductile and strain-hardening cement-based composites under tensile impact loading. Cement Concr. Comp. **105**, 103430 (2020)
8. Tawfik, A., Curosu, I., Alsous, G., Mechtcherine, V.: A testing device to investigate the properties of strain-hardening, cement-based composites (SHCC) under impact shear loading. Int. J. Impact Eng. **167**, 104280 (2022)
9. Hering, M., Kühn, T., Curbach, M.: Small-scale plate tests with fine concrete in experiment and first simplified simulation. Struct. Concr. **22**(2), 637–649 (2021)
10. Hering, M., Curbach, M.: Strengthening of RC plates with mineral-bonded composite layers for enhanced impact safety. In: MATEC Web of Conferences. vol. 323, pp. 01015 EDP Sciences (2020)
11. Leicht, L., Beckmann, B., Curbach, M.: Influences on the structural response of beams in drop tower experiments. Civil Eng. Design **3**(5–6), 192–209 (2021)
12. Tamsen, E., Curosu, I., Mechtcherine, V., Balzani, D.: Computational micro-macro analysis of impact on strain-hardening cementitious composites (SHCC) including microscopic inertia. Materials **13**(21), 4934 (2020)
13. Häussler-Combe, U., Shehni, A., Chihadeh, A.: Finite element modeling of fiber reinforced cement composites using strong discontinuity approach with explicit representation of fibers. Int. J. Solids Struct. **200**, 213–230 (2020)
14. Pandolfi, A., Ortiz, M.: An eigenerosion approach to brittle fracture. Int. J. Num. Methods Eng. **92**(8), 694–714 (2012)
15. Qinami, A., Bryant, E.C., Sun, W., Kaliske, M.: Circumventing mesh bias by r-and h-adaptive techniques for variational eigenfracture. Int. J. Fracture **220**(2), 129–142 (2019)
16. Qinami, A., Pandolfi, A., Kaliske, M.: Variational eigenerosion for rate-dependent plasticity in concrete modeling at small strain. Int. J. Num. Methods Eng. **121**(7), 1388–1409 (2020)
17. Böttcher, M., Fuchs, A., Leichsenring, F., Graf, W., Kaliske, M.: ELSA: An efficient, adaptive ensemble learning-based sampling approach. Adv. Eng. Softw. **154**, 102974 (2021)

18. Stöcker, J., Fuchs, A., Leichsenring, F., Kaliske, M.: A novel self-adversarial training scheme for enhanced robustness of inelastic constitutive descriptions by neural networks. Comput. Struct. **265**, 106774 (2022)
19. Fuchs, A., Curosu, I., Kaliske, M.: Numerical mesoscale analysis of textile reinforced concrete. Materials **13**(18), 3944 (2020)
20. Liebold, F., Maas, H.G.: Strategy for crack width measurement of multiple crack patterns in civil engineering material testing using a monocular image sequence analysis. PFG-J. Photogram. Remote Sens. Geoinform. Sci. **88**(3), 219–238 (2020)
21. Liebold, F., A Heravi, A., Mosig, O., Curbach, M., Mechtcherine, V., Maas, H.G.: Crack propagation velocity determination by high-speed camera image sequence processing. Materials **13**(19), 4415 (2020) (2020)
22. Ilg, P., Hoehne, C., Guenther, E.: High-performance materials in infrastructure: a review of applied life cycle costing and its drivers-the case of fiber-reinforced composites. J. Cleaner Prod. **112**, 926–945 (2016)
23. Scope, C., Vogel, M., Guenther, E.: Greener, cheaper, or more sustainable: reviewing sustainability assessments of maintenance strategies of concrete structures. Sustain. Prod. Consump. **26**, 838–858 (2021)

Repair of a Hydraulic Structure with Different Strain-Hardening Cement-Based Composites

Steffen Müller(✉), Michaela Reichardt, and Viktor Mechtcherine

Institute of Construction Materials, TU Dresden, Dresden, Germany
steffen.mueller@tu-dresden.de

Abstract. This article reports on the repair of an approximately 100-year-old hydraulic structure with different strain-hardening retrofitting materials under special consideration of application technology, labour costs, material properties and durability. The lock structure was characterised in places by very poor concrete quality (compressive strength below 5 MPa) and wide separating cracks (crack width over 2 cm), while the load-bearing capacity did not cause any concern. The aim of the repair campaign was to restore the impermeability of the lock with the minimum possible changes in the cross-section of the existing structure. Depending on the section, the repair was carried out with strain-hardening cement-based composite (SHCC) or with textile reinforced concrete (TRC), whereby TRC was applied in four variants which included different types of grids and numbers of layers. Epoxy resin, SBR and mineral-impregnated carbon fibre yarns were used. None of the areas repaired was smaller than 35 m^2.

The entire structure was monitored with regard to its deformation behaviour for half a year before the repair and regularly monitored after the retrofitting for up to two years. The observations showed that all applied retrofitting variants can effectively bridge the cracks despite the low layer thicknesses of max. 3 cm and that the permeability of the entire structure decreased considerably.

Keywords: SHCC · ECC · Textile reinforced concrete · Retrofitting · Water tightness · Monitoring

1 Introduction

The transport infrastructure in many countries has been increasingly ageing. Both space limitations and narrow time frames for construction significantly restrict the possibilities for replacement or new construction. In addition, the need to use resources as efficiently as possible makes maintenance and repair to sustainable measures. Particularly in the case of superficial repairs, solutions are required with low demands to the existing substrate, the thinnest possible overlays, high crack-width controlling properties and fast application times. These requirements can be met by thin, strain-hardening material layers as shown in different studies [1–3], but often fail due to existing regulations and standards, which is why practical demonstrators are urgently needed. In the following, a hydraulic engineering demonstrator, a water lock, with the objective of restoring the

M. Kunieda et al. (Eds.): SHCC 2022, RILEM Bookseries 39, pp. 291–301, 2023.
https://doi.org/10.1007/978-3-031-15805-6_30

impermeability of the structure, is presented and its rehabilitation by means of strain-hardening cement-based composite (SHCC) and textile reinforced concrete (TRC) is examined in detail.

2 The Structure

2.1 General

The reference structure, the *Schustergraben* lock, is located in north-eastern Germany in the Lower Oder Valley National Park in Polder 10 near Schwedt/Oder. The lock, built in 1926, connects the river Oder (river km: 702.58) with the *Gatower Graben.* It used to serve to make the water areas on Polder 10 navigable for fishing despite different water levels between the polder and the Oder. Today it serves to equalise the water levels between the Oder and the nature conservation area. The lock is 26.93 m long, 5.25 m deep and 6.5 to 7 m wide. It is oriented in an east-west direction. The north and south walls are mirror images of each other in the division of the wall sections and the use of built-in parts. Expansion joints are located at a distance of about 7.30 m from the centre of the walls. Design documents for the lock are no longer available, but it can be assumed that it is a heavyweight structure.

2.2 Structural Condition

Due to its age and visible damage to the concrete structure, it was classified by the owner as being in need of renovation. During the summer 2019, as well as after the dewatering in autumn of the same year, preliminary investigations were carried out to characterise the old concrete and the general condition of the structure. The following works were performed:

- Crack width monitoring, accompanied by recording of the ambient temperature
- Indirect determination of the compressive strength by means of a rebound hammer
- Crack mapping and damage assessment
- Determining the compressive strength on drill cores and in-situ adhesive tensile strength.

Vertical separation cracks several millimetres wide were found in the centre of the main walls of the structure. In the gate chamber areas, the cracking was even more pronounced and consisted of horizontal and vertical crack paths, each several millimetres wide. The crack sensors showed particular activity at the vertical joints of the north wall (changes in crack width of up to 0.6 mm) and the crack behind the stairs on the south side. It was also deduced that the horizontal cracks were hardly subject to movement (changes in crack width < 0.1 mm) and that the north wall was exposed to significantly higher structural movements, probably due to a greater solar radiation. The extent of crack width changes served as a starting point for the design of the repair system.

The concrete samples examined showed very large differences between the samples taken from the air side and those from the water change zone. According to the applicable regulations [4, 5], the characteristic values were so low that, normatively, a large-scale removal of old concrete and application of thick repair layers made of steel reinforced concrete supplement would be necessary. The measured characteristic values for compressive strength and adhesive tensile strength are given in Table 1.

Table 1. Compressive and tensile strengths in MPa of the existing structure (standard deviations are given in brackets)

Extraction area	Compressive strength [MPa]	Tensile strength [MPa]
Top of structure	25.7 [0.5]	1.21 [0.59]
Water change area	15.0 [7.3]	0.67 (Minimum: 0.09) [0.41]

As can be clearly seen, the values scatter very strongly, in the water change area at the first place. The concrete was damaged by water absorption and freeze-thaw cycles. Further reasons for the high scatter are, on the one hand, the clearly visible concreting sections with corresponding variable material properties. On the other hand, binder was permanently leached out of the surface of the concrete due to the constant water contact. Spalling and washout were particularly prominent at the water change zone and in the wall-base transition area. Figure 1 shows cracks and local washouts as well as damaged areas.

Fig. 1. Appearance of the lock before retrofitting: a) overview, b) deep erosion cavern, c) extensive washout, d) cracks with large crack openings

Due to the partially very low load-bearing capacity and unknown structural stability, large-scale and deep concrete removal was avoided, but instead the thinnest possible rehabilitation layer was used to restore the water-tightness of the structure.

The old concrete walls were sand-blasted to remove loose adhesions and dirt. Then, washed-out areas/cavities were reprofiled before the actual retrofitting in order to avoid hollow layers.

3 Retrofitting

3.1 Surface Sectioning and Materials

Due to the size of the existing structure and the operator's willingness to experiment, different material combinations could be used for the repair. Based on the undefined main crack direction, PVA short-fibre reinforced SHCC was applied in the gate chambers. The remaining areas were divided into four sections of approximately equal size, whereby the south wall was reprofitted with a soft, thermoplast-polymer-impregnated reinforcement (Section A: 3 layers; Section B: 1 layer). The north wall was retrofitted with a 2-layer stiff, thermoset-polymer-impregnated textile reinforcement in Section C, while in Section D a mineral-impregnated variant (MCF = Mineral Coated Fibre) was used. Referring to former studies the type of impregnation influences the composite behaviour to a great extent [6, 7]. The distribution of the materials and the location of the main damages are shown in the sketch in Fig. 2. The product data of the reinforcement is given in Table 2.

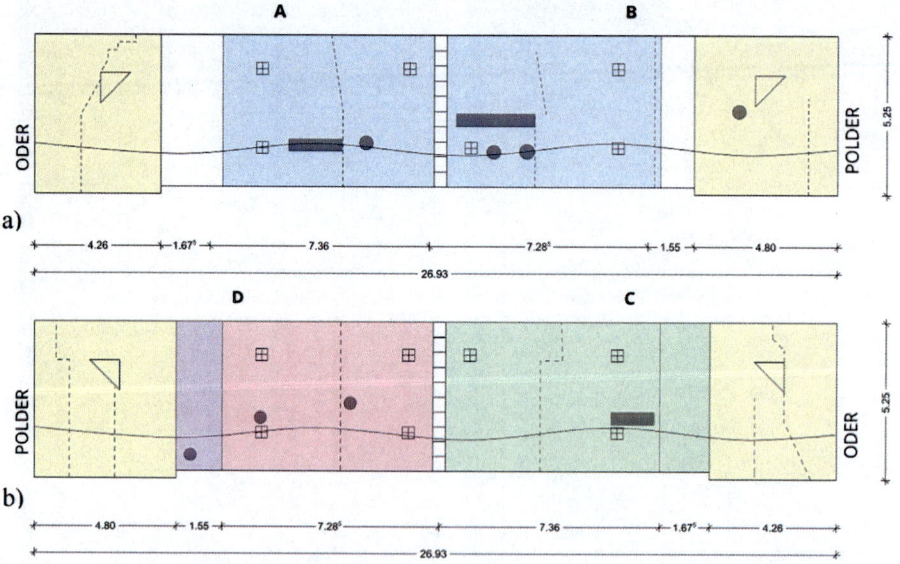

Fig. 2. Overview of the vertical concrete structure of the lock with zones for the retrofitting (blue: TRC with soft polymer impregnation, green: TRC with stiff polymer impregnation, red: TRC with MCF (Mineral Coated Fibre), yellow (Gate chambers): SHCC), major cracks (dashed line) and areas of caverns (grey zones): a) south wall, b) north wall, (unit = m) (Color figure online)

The matrix material for the TRC sections was a Portland cement based fine grained shotcrete containing hollow microspheres to reduce the modulus of elasticity and to enhance the freeze-thaw-resistance of the repair layer. The detailed mix design was part of the industrial partner development and therefore cannot be published here. The following mechanical properties could be determined after the spraying process: compressive strength of 50.1 MPa (standard deviation 5.4 MPa) and flexural tensile strength (measured on the 160 mm × 40 mm × 40 mm prisms) of 1.73 MPa (standard deviation 0.2 MPa). The density of spayed material was 2.13 kg/dm^3.

The SHCC under investigation has already been used in several projects [8–12] and studies, among others as a spray material in the field of hydraulic engineering [13]. Its composition is given in Table 3.

Table 2. Allocation of reinforcement to the test surface.

Section	Reinforcement type	Manufacturer	Cross section per layer longitudinal I rectangular	Number of layers
A	Soft impregnated carbon mesh	TUDATEX	141 I 28	3
B	Soft impregnated carbon mesh	TUDATEX	141 I 28	1
C	Stiff impregnated carbon mesh	SOLIDIAN	85 I 85	2
D	Mineral impregnated carbon rebars	made at TUD	187 I 0	1
Gate	PVA fibres, l = 12 mm, Ø = 40 μm	–	–	–

Table 3. Composition of SHCC in kg/m^3

Cement CEM I 42.5R	Fly ash	Quartz sand	Water	Superplasticizer	Viscosity agent	PVA fibres
505	621	536	338	10	4.8	26

All TRC applications were performed using dry spraying method, which has advantages in terms of the amount of material required and the cleaning effort, as residual material can remain in the hose package overnight. However, the spray quality and material properties depend heavily on the spraying operator, as he or she adjusts the dosage of water to be added by experience. In the process sequence, a first mortar layer was sprayed onto a moistened substrate, on which the corresponding textile was lightly pressed and then sprayed over. This eliminated the need for any work to fix the reinforcement to the old concrete structure. This procedure was repeated for any subsequent layers on the following day until the final number of layers was applied.

For the SHCC, a wet spraying process with SHCC mixed on site in 60l batches with respective quality control (lump formation, spreading diameter, etc.) was planned. A machine failure forced to change the procedure to application by hand.

To provide a smooth finish, a layer of TRC mortar with a thickness below 2 mm was finally sprayed onto the SHCC surfaces. No further smoothing measures were used. Curing was carried out by regularly moistening the surfaces.

3.2 Functionality of the Rehabilitation Layers

The mode of action of all material combinations used to restore the tightness of the lock structure is based on the principle of reducing the widths of the existing cracks at the structure surface. For this purpose, the wide cracks in the substrate are to be "distributed" over as many narrow cracks in the repair layer as possible. To facilitate such behaviour, it is necessary to introduce a decoupling area along each major crack in the substrate. This decoupling area creates a free length in the repair layer, which is a basic requirement for the formation of a multiple crack pattern, see Fig. 3. For this purpose, the main cracks were covered with a 14 cm wide adhesive tape to create a sufficiently smooth surface. The functionality of this practical solution was confirmed in previous laboratory tests on system test specimens, in accordance with [14].

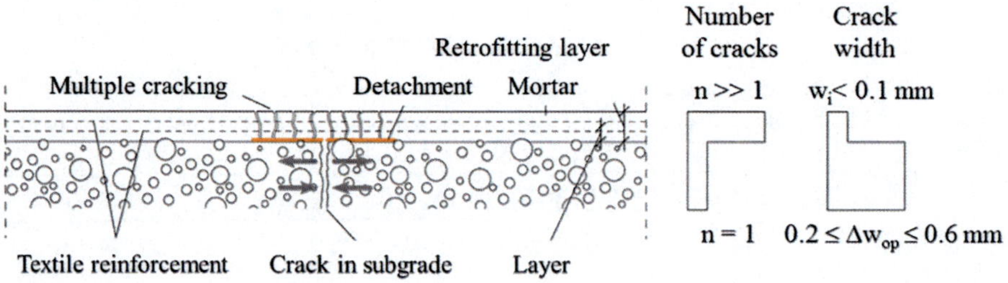

Fig. 3. Functionality and structure of a multiple cracking repair layer including decoupling area

4 Implementation

4.1 Accompanying Laboratory Tests

To assess the quality and performance of the repair measure, a large number of accompanying tests were carried out. In this section, the on-site sprayed specimens for uniaxial tension test on the textile concrete layer in Section C will be considered in particular. The focus of the investigation was to prove that the applied material exhibits multiple cracking with the narrow crack openings.

The preliminary investigations on the unrestored lock showed a maximum crack changes in the crack width of 0.68 mm. The decoupling area over this crack was designed with a width of 14 cm, which corresponds to a maximum strain of 0.48%.

The specimens produced under site conditions (cf. Fig. 4a) were tested in a direct displacement-controlled tension test at 0.3 mm/min until failure. The cracks forming on the surface were photogrammetrically recorded and could thus be evaluated with respect to crack width, number of cracks and crack spacing. Fig. 4b shows the cracks that form on an exemplary test specimen, each represented by one particular colour, and their crack widths in relation to the total elongation of the test specimen. Furthermore, the maximum measured strain on the object described at the beginning is entered. Table 4 shows the characteristic values for crack formation for the condition of the maximum strain of 0.48% expected on the basis of the preliminary investigations on the unrestored lock. It can be seen that although the repair material yielded an average crack width well below 100 μm, the maximum crack width was approximately double as high as this value. Nevertheless, the basic effectiveness of such a system could be proven by

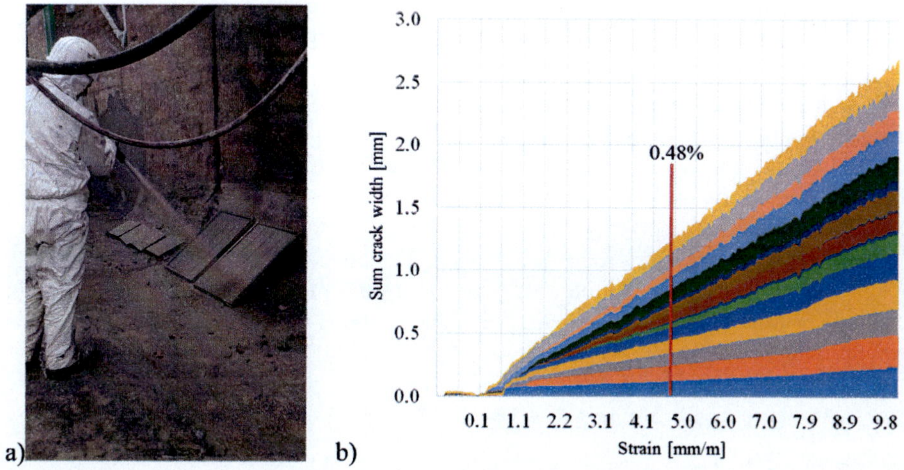

Fig. 4. a) Production of uniaxial tension test specimens under site conditions and b) sum crack-width diagram over total strain of an exemplary test specimen

Table 4. Characteristic values for the crack development in lab specimen comparable to Section C

Characteristic	Value
Average crack width at 4.8 mm/m [standard deviation]	79.1 μm [47.5 μm]
Max. crack width	191.1 μm
Cracks/meter	55.5

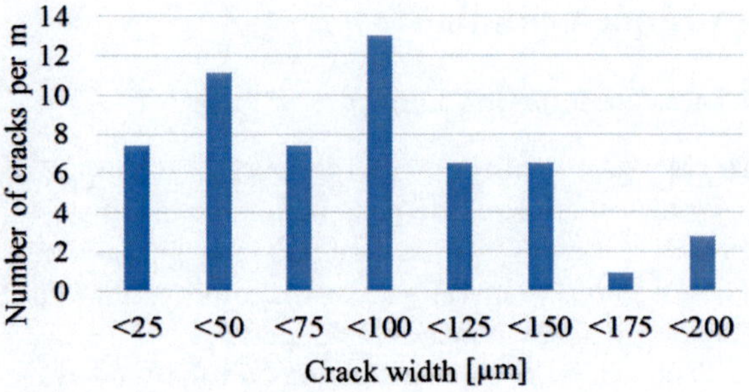

Fig. 5. Number of cracks per m according to their size from specimen comparable to Section C

the tests. Further changes or crack width reduction can result from a different textile geometry, which, however, requires an additional systematic investigation. An overview over the crack width and the distribution can be found in Table 4 and Fig. 5.

4.2 Monitoring

A monitoring campaign was initiated to assess the effectiveness of the repair. This included both the measurement of strains in the reinforcement plane by means of strain gauges applied directly above the crack on the reinforcement and the recording of displacements on the surface in the area of the previously identified main cracks. In addition, influencing variables such as the water level in the lock and temperature were recorded. In total, data of 7 strain gauges (3x Section C, 2x Section D, 2x SHCC), 3 crack opening gauges (Section C, Section D and ladder area), 4 temperature sensors (2x air, 1x in concrete above water level, 1x concrete under water level) and the measurement of the water level in the lock was recorded over a period of 1.5 years at the time of the reporting. The measurements are still ongoing.

The results showed that the strains of the reinforcement are independent of the external conditions such as temperature and water level and are only very small (<0.2 mm/m).

The displacement measurements at the surface indicate a clear decrease in the crack width changes over time. While changes of up to 0.5 mm were detected at the main crack of Section C before the repair, this value dropped to less than 0.1 mm after the rehabilitation; see Fig. 6.

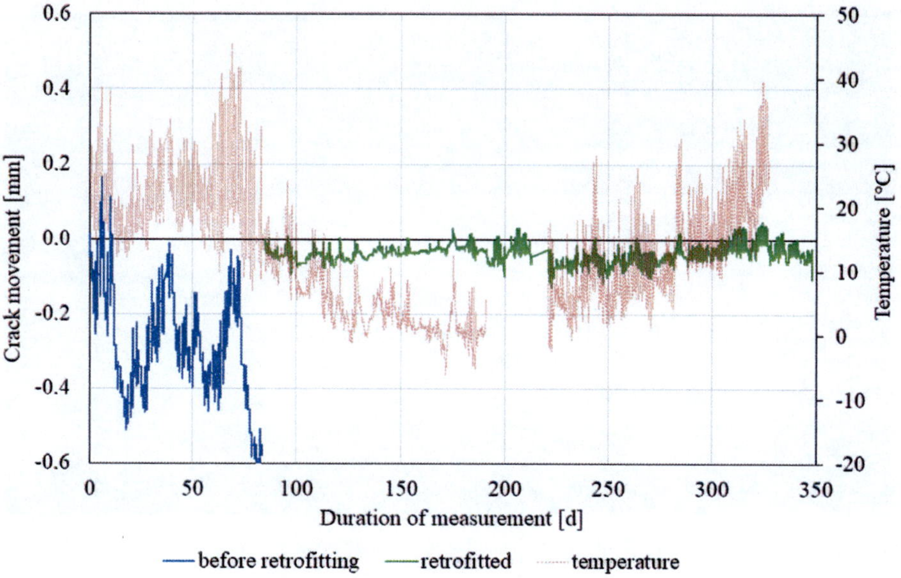

Fig. 6. Changes in crack opening in Section C with changing temperature; blue: before retrofitting; green: after retrofitting (Color figure online)

5 Conclusion and Outlook

The presented large-scale application of textile reinforced concrete and SHCC as a repair layer on hydraulic engineering structures has demonstrated the feasibility of this lean and material-efficient approach. Accompanying investigations on material samples have proven that the values determined in preliminary tests can also be achieved under construction site conditions. The observed reduction in the characteristic values has not been significant.

The appearance of the repair surfaces is still very good. After 2.5 years with 3 frost periods no degradation could be detected. The substrate cracks have not penetrated to the surface in their previous extension, but rather very narrow, finely distributed, and actually hardly detectable cracks occurred in the repair layer; see Fig. 7.

The repaired lock offers the possibility of a long-term investigation/monitoring. The remarkable performance of the repair measure serves as proof of its practical suitability for further approval and licensing procedures. For this reason, the monitoring will be continued in the coming years and the findings will be published accordingly. The construction parameters not mentioned in this article, but determined during the execution, allow the executing companies to create initial planning bases for an economic calculation of such applications.

Fig. 7. Repaired Section C with a surface detail

References

1. Mechtcherine, V.: Towards a durability framework for structural elements and structures made of or strengthened with high-performance fibre-reinforced composites. Constr. Build. Mater. **31**, 94–104 (2012)
2. Mechtcherine, V., Lieboldt, M.: Permeation of water and gases through cracked textile reinforced concrete. Cem. Concr. Compos. **33**, 725–734 (2010)
3. Büttner, T., Morales Cruz, C., Raupach, M.: Durable protective layers of textile concrete for structural surfaces in hydraulic engineering. Bautechnik **90**(8), 485–490 (2013). https://doi.org/10.1002/bate.201300037(inGerman)
4. DAfStb-Richtlinie: Instandsetzung von Betonbauteilen (2001). (in German)
5. Bundesministerium für Verkehr und digitale Infrastruktur (BMVI): ZTV-W LB 219: Zusätzliche Technische Vertragsbedingungen – Wasserbau (ZTV-W) für Schutz und Instandsetzung der Betonbauteile von Wasserbauwerken (Leistungsbereich 219) (2017)
6. Kruppke, I., Butler, M., Schneider, K., Hund, R.-D., Mechtcherine, V., Cherif, C.: Carbon fibre reinforced concrete: dependency of bond strength on Tg of yarn impregnating polymer. Mater. Sci. Appl. **10**, 328–348 (2019)
7. Mechtcherine, V., Michel, A., Liebscher, M., Schneider, K., Großmann, C.: Neue Carbonfaserbewehrung für digitalen, automatisierten Betonbau. Beton-und Stahlbetonbau **114**(12), 947–955 (2019)
8. Müller, S., Mechtcherine, V.: Fatigue behaviour of strain-hardening cement-based composites (SHCC). Cem. Concr. Res. **92**, 75–83 (2019). https://doi.org/10.1016/j.cemconres.2016.11.003
9. Jun, P., Mechtcherine, V.: Behaviour of strain-hardening cement-based composites (SHCC) under monotonic and cyclic tensile loading-Part 1–experimental investigations. Cem. Concr. Compos. **32**, 801–809 (2010). https://doi.org/10.1016/j.cemconcomp.2010.07.019
10. Ranjbarian, M., Mechtcherine, V.: A novel test setup for the characterization of bridging behaviour of single microfibres embedded in a mineral-based matrix. Cem. Concr. Compos. **92**, 92–101 (2018). https://doi.org/10.1016/j.cemconcomp.2018.05.017

11. Ranjbarian, M., Mechtcherine, V.: Influence of loading parameters in cyclic tension-compression regime on crack-bridging behaviour of PVA microfibres embedded in cement-based matrix. Construct. Build. Mater. **228**, 116760 (2019). https://doi.org/10.1016/j.conbuildmat.2019.116760
12. Junger, D., Storm, J., Müller, S., Kaliske, M., Mechtcherine, V.: Increasing the fatigue resistance of strain-hardening cement-based composites (SHCC) by experimental-virtual multi-scale material design. Materials **14**(19), 5634 (2021). https://doi.org/10.3390/ma14195634
13. Müller, S., Mechtcherine, V.: Use of strain-hardening cement-based composites (SHCC) for retrofitting. ICCRRR 2018, MATEC Web Conf. **199**, 09006 (2018). https://doi.org/10.1051/matecconf/201819909006
14. Bundesanstalt für Wasserbau: BAW Merkblatt Flächige Instandsetzung von Wasserbauwerken mit textilbewehrten Mörtel-und Betonschichten (MITEX) (2019). (in German)

Overlay of RC Bridge Deck Deteriorated by ASR Using an Ultra High Performance-Strain Hardening Cementitious Composite (UHP-SHCC)

Minoru Kunieda(✉) and Koki Banno

Gifu University, 1-1Yanagido, Gifu 501-1193, Japan
kunieda@gifu-u.ac.jp

Abstract. A case of overlaying on ASR-deteriorated reinforced concrete bridge deck is introduced for structural use in repair of Ultra High Performance-Strain Hardening Cementitious Composites (UHP-SHCC). The material is a type of ultra high strength fiber reinforced mortar. Results of investigation are presented as well, focusing on the restraint stress of substrate concrete as one of the subjects to consider when selecting a repair material. Specifically, the author highlights the "cracking potential" calculated as a ratio of restraint stress, which is measured by pseudo perfect restraint testing, to the cracking strength at the same age. By this method, the cracking potential of the material under study turns out to be 70% at 5 days.

Keywords: ASR · Overlay · Reinforced concrete bridge deck · UHP-SHCC

1 Introduction

A number of Ultra High Performance Fiber Reinforced Cement Composites (UHPFRC) have been developed, with their applications progressively increasing [1].

These now include research anticipating shear strengthening [2] and enhancement of fatigue resistance [3], in addition to tensile strengthening using short fibers. Based on their results, applications of UHPFRC are increasing, including to road bridge decks. A number of studies have also been conducted on combinations of different materials for reinforced concrete bridge decks to enhance their performance [1]. Regarding Ultra High Performance Strain Hardening Cementitious Composite (UHP-SHCC) [4] has a dense matrix, and also exhibit stain hardening behavior and multiple fine cracks in tension. There are repair applications using the UHP-SHCC to enhance protective performance against penetration of deterioration factors, such as moisture, oxygen, carbon dioxide, and chloride ions.

This paper introduces a case of an overlay on an ASR-deteriorated reinforced concrete bridge deck as a potential use of UHP-SHCC for structural repair/retrofitting. Test results focused on the restraint stress by substrate concrete are also reported as a point to consider for their application.

M. Kunieda et al. (Eds.): SHCC 2022, RILEM Bookseries 39, pp. 302–309, 2023.
https://doi.org/10.1007/978-3-031-15805-6_31

2 Overlay on ASR-Deteriorated Reinforced Concrete Bridge Deck

2.1 Concept of Repair

The bridge in this study, completed over 50 years ago and located in Gifu Prefecture, is a simple non-composite steel box-girder bridge spanning approximately 60 m, as shown in Fig. 1. Since ASR-induced deterioration was found on its reinforced concrete deck during inspections conducted up until 2008, mainly the sidewalk and wheel guards were repaired. Only several years later, however, map cracking occurred again in the wheel guards, and defects in the pavement including potholes arose. Drastic measures were therefore considered. A countermeasure, such as re-placement of the slab concrete, was included in the initial proposals, but was dismissed in view of the necessity to retain the bridge's function on the relatively important route. With all things considered, including the soundness of the piers, it was judged unlikely that the bridge would be in use for another 50 or 100 years. Consequently, it was decided to place an overlay on the deck between A1 and P1 using UHP-SHCC, placing upmost priority on securing safety for the coming 20 to 30 years. The pavement was also to be replaced with the UHP-SHCC in expectation of enhancing the stiffness of the entire deck and durability of the pavement. Note that the deck between P1 and A2 was overlaid using another material, the details of which are omitted herein.

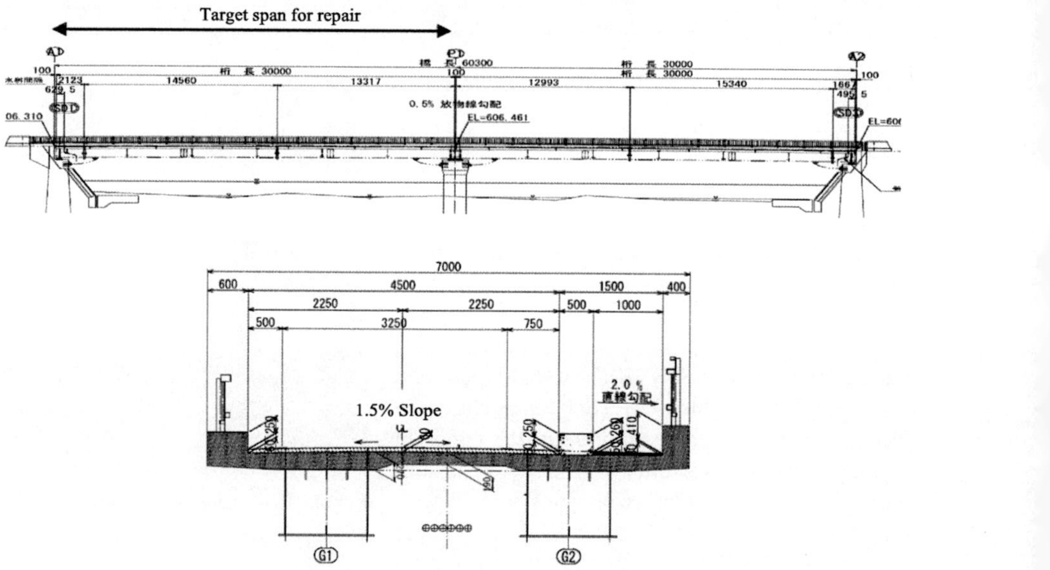

Fig. 1. Bridge configuration (upper: side view, lower: cross sectional view).

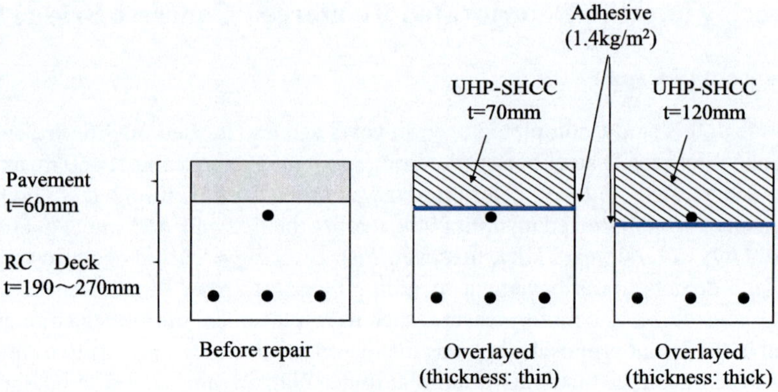

Fig. 2. Concept of repair.

Table 1. Mix proportions of UHP-SHCC.

Unit content (kg/m^3)						SP(kg)	PE fiber(kg)
Water	Low-heat cement	Silica fume	Limestone powder	Expansive agent	Silica sand		
347	1017	233	264	40	156	23.3	14.6

The repair was made to an area of approximately 134 m^2 representing one span of the bridge. The existing deck was paved with 60 mm-thick asphalt over a minimum deck thickness of 190 mm. It was planned to remove the asphalt pavement, roughen the concrete surface by water-jetting to approximately 10 mm in depth, and then place 70 mm-thick UHP-SHCC, as shown in Fig. 2. However, as the condition of the concrete surface on the road shoulder (on the sidewalk side) was worse than anticipated as shown in Fig. 3, the surface was roughly broken with a concrete breaker and then further chipped by waterjetting until the reinforcing bars became exposed.

Fig. 3. Condition after treatment for damaged concrete.

2.2 Materials

Fig. 4. Placement of adhesive.

Table 1 shows the mix proportions of the UHP-SHCC used for the repair. A UHP-SHCC is a composite of a cementitious material with a low water-binder ratio reinforced with high-strength polyethylene short fibers (1.5 vol % and 6 mm in length). It provides high mechanical performance such as a compressive strength of over 80 N/mm^2 and tensile strength of 6 to 8 N/mm^2 at 28 days, as well as high deformation performance of around 1–2% tensile strain, while exhibiting multiple fine cracks under tensile stress. With its dense matrix highly resistant to the penetration of deterioration factors, such as moisture, oxygen, carbon dioxide, and chloride ions, the UHP-SHCC is a material suitable for concrete repair where surface-protecting performance is particularly required.

Table 2. Mechanical properties of UHP-SHCC.

Mechanical properties	Averaged values
Compressive strength	85.6 N/mm^2
Flexural strength	19.8 N/mm^2
Tensile strength	8.5 N/mm^2
Strain at tensile strength	1.0%

The material was mixed using two pan-type mixers with a capacity of 100 L each, with the batch size being approximately 45 L. The as-mixed mixture was pumped to the placing point by a squeeze-type pump, spread evenly to the specified thickness, and surface-finished. Prior to placing the UHP-SHCC, a dedicated epoxy adhesive was applied at a spread of 1.4 kg/m^2 (Fig. 4). The mortar flow value is ranging from 160–180 mm.

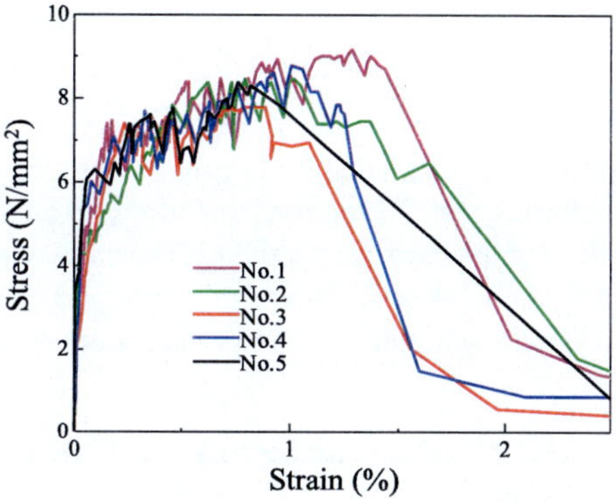

Fig. 5. Tensile stress-strain curves.

The repair work was executed over two days by dividing the 4.5-meter-wide traffic lane area into two parts.

2.3 Strength Tests and Condition After Execution

Table 2 shows the strength test results at the age of 28 days. Note that the compressive strength was determined by tests using sections of mortar bars (JIS A 1171, six specimens); the flexural strength, by tests specified in JSCE G 552 (five specimens 150 by 150 by 550 mm); and the tensile strength, by uniaxial tension tests using dumbbell shaped specimens. The compressive, flexural, and tensile strengths at the age of 28 days were approximately 86 N/mm^2, 19.8 N/mm^2, and 8.5 N/mm^2, respectively. Figure 5 shows the tensile stress-strain curves obtained by the tension tests, revealing that all specimens have sufficient tensile strength and demonstrate strain-hardening behavior.

(a) 1day later (b) 1 month later (c) 6 months later

Fig. 6. Condition after execution.

Figure 6(a) shows the state of the surface immediately after repair. Note that the surface was finished by raking at right angles to the bridge axis. Figures 6(b) and 6(c) show the state of the road surface with a lane marker at the beginning of its service and six months later, respectively. The surface condition six months later was good without any visible cracks. However, some parts of the lane marker were found to have peeled off presumably due to the dense nature of the UHP-SHCC surface.

3 Evaluation of Cracking Potential

3.1 Introduction

For the above-mentioned overlay repair, the ultra high strength material was selected to prevent the penetration of deterioration factors through the bridge face and to enhance the fatigue endurance. In the meantime, the use of a material with such a high strength can raise the risk of cracking in the repair material when restrained by the substrate concrete, due to restraint stress induced by drying shrinkage and autogenous shrinkage. Repair materials with excellent dimensional stability are generally selected. However, the selection should be based not only on free shrinkage of material but also on the evaluation of the risk of shrinkage cracking, taking account of such material properties as tensile strength, elastic modulus, and creep, which change over time and depend on the degree of restraint. Although it is desirable, when evaluating the internal stress generated by shrinkage of the repair material, to incorporate the degree of restraint depending on the thickness, shape, and size of the overlay, it is not realistic to take all these factors into consideration with every test. In this study, therefore, safe side evaluation by conducting evaluation in terms of uniaxial tensile stress under pseudo perfect restraint conditions was carried out. In such evaluation, if the margin against cracking can be quantified in addition to crack detection, then the reliability of the material can be enhanced even in the case of no cracking.

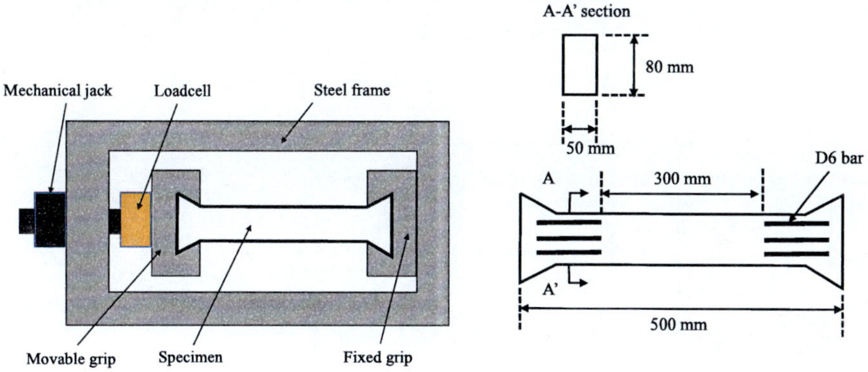

(a) Overview (b) Specimen geometry

Fig. 7. Test setup of pseudo perfect restraint test.

This paper introduces an accessing approach where the shrinkage cracking potential (i.e. an index to the degree of proneness to cracking) of the above-mentioned UHP-SHCC was evaluated by pseudo perfect restraint test.

3.2 Experiment Overview

The pseudo perfect restraint testing machine used in this study is based on the temperature stress testing machine developed to measure the internal stress generated from changes in the temperature of mass concrete [6] (Fig. 7). The specimen is fabricated by directly placing a mixture in a mold 500 mm in length and 80 mm in width with enlarged ends ("dog-bone" shaped) installed in the machine. To simulate the behavior of the repair material completely restrained by the substrate, the thickness of specimens is set to 50 mm. The specimen is restrained by grips at both ends. The system consists of movable and fixed grips, and the high-rigidity framing. The test is controlled by designating a measurement zone of 300 mm in the center of the specimen, measuring the gauge mark distance, which changes with the shrinkage of the repair material, using displacement transducers with a precision of 1/1000 mm, and then moving the movable grip to resume the datum length of 300 mm. A loadcell is connected to the movable grip side to measure the load.

As described above, the UHP-SHCC as shown in Table 1 is used. Specimens are fabricated in a constant temperature room at 20 °C and 40 to 50% R.H. by lining the mold in the machine with a PVC sheet 0.05 mm in thickness and placing the UHP-SHCC mixture in the mold. Each end of the specimen is reinforced with three 100-mm-long D6 bars to prevent cracking and breakage outside the measurement zone during testing. After placing the mixture, the specimen is wrapped with a PVC sheet to prevent rapid drying. The mold and PVC sheet are removed 12 h after placing, and measurement is started, with all surfaces except the bottom of the specimen being kept in a dry condition. During the test, the restraint load and the measurement zone displacement are measured, with the displacement being offset by moving the movable grip each time the strain value changes by 20 μ. Note that the measurement is continued up to an age of 5 days.

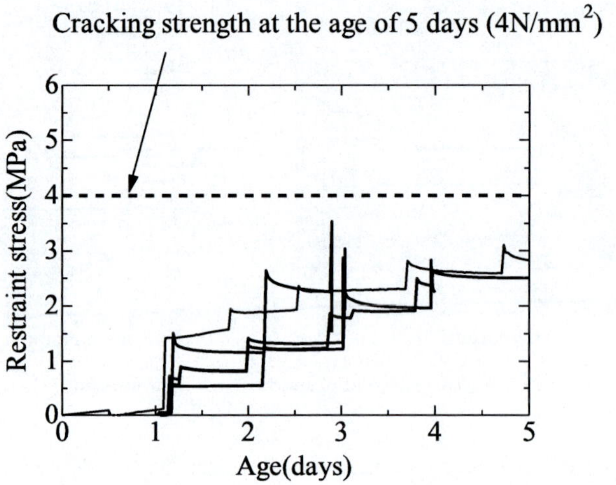

Fig. 8. Induced stress at early age.

3.3 Results and Discussion

Figure 8 shows the restraint stress which is determined by dividing the load measured with the loadcell by the apparent cross-sectional area. Relatively large restraint stress develops by around the age of 3 days, increasing to 2.8 N/mm^2 by 5 days. Meanwhile, the free shrinkage strain was approximately 200 μ at 1.5 days, increasing to 380 μ by 5 days. This can cause cracking in an actual structure in terms of free shrinkage strain alone. Table 3 shows the cracking strength and the restraint stress at 5 days, as well as the cracking potential, which is determined by dividing the restraint stress by the cracking strength [7]. The restraint stress is found to be 70% of the UHP-SHCC's cracking strength of 4.0 MPa, revealing that a certain margin remains against cracking.

When selecting and applying a repair material, it is highly beneficial to confirm beforehand that no crack occurs at early age and to quantitatively ascertain the extent of the restraint stress accumulating by 5 days with respect to the cracking strength as described above.

Table 3. Calculated cracking potential.

Cracking strength (N/mm^2)	Restraint stress (N/mm^2)	Cracking potential (restraint stress/cracking strength)
4.0	2.8	0.7

4 Concluding Remarks

In this paper, a repair case where the UHP-SHCC was used for overlaying an ASR-deteriorated reinforced concrete bridge deck was introduced. Also, to reduce the risk of shrinkage cracking inherent to ultra high strength overlay materials, preliminary investigation into the materials by a method in which the cracking potential is determined by measuring the internal stress induced by material shrinkage was conducted.

Acknowledgement. The author expresses his sincere gratitude to Takayama Civil Engineering Office, Gifu Prefecture, for its outstanding cooperation.

References

1. Brühwiler, E., Denarié, E.: Rehabilitation and strengthening of concrete structures using ultra-high performance fibre reinforced concrete. Struct. Eng. Int. **23**(4), 450–457 (2013)
2. Lim, W.-Y., Hong, S.-G.: Shear tests for ultra-high performance fiber reinforced concrete (UHPFRC) beams with shear reinforcement. Int. J. Concr. Struct. Mater. **10**(2), 177–188 (2016). https://doi.org/10.1007/s40069-016-0145-8
3. Makita, T., Bruehwiler, E.: Tensile fatigue behaviour of ultra-high performance fibre reinforced concrete (UHPFRC). Mater. Struct. **47**, 475–491 (2014)
4. Kunieda, M., Denarie, E., Bruehwiler, E., Nakamura, H.: Challenges for strain hardening cementitious composites-deformability versus matrix density. In: Proceedings of the 5th International RILEM Workshop on High Performance Fiber Reinforced Cement Composites, pp. 31–38 (2007)
5. Hamada, H., Kunieda, M., Habuchi, T., Amino, T., Onoe, K.: Development of mortar lining method using UHP-SHCC for steel pipe piles in the tidal zone. In: Proceedings of the 4th International Conference on Construction Materials, pp. 1378–1383 (2009)
6. Kovler, K.: Testing system for determining the mechanical behaviour of early age concrete under restrained and free uniaxial shrinkage. Mater. Struct. **27**, 324–330 (1994)
7. Kunieda, M., Asai, K., Sasaki, K.: Cracking resistance of UHPFRC for repair application, In: The Proceedings of the International Conference on Bridge Maintenance, Safety, Management, Life-Cycle Sustainability and Innovations, pp. 2805–2809, Sapporo (2021)

Full-Scale Experiment of AFt-UHPFRC for Overlay of Bridge Deck

Yoh Arakawa(✉) and Yuji Watanabe

Kajima Technical Research Institute, Tokyo, Japan
arakawyo@kajima.com

Abstract. In the expressway in Japan, large-scale renewal projects of the bridge in which various changes were actualized by the progress of the deterioration were advanced. In concrete slab bridges, deterioration due to salt damage or freezing and thawing, or fatigue deterioration due to traffic loads are the majority. In steel slab, the fatigue degradation by heavy traffic has become obvious. From such background, research and development have begun on upgrading bridge decks by utilizing UHPFRC in ways of increasing the thickness of the members, replacing surfaces, and increasing the rigidity. The UHPFRC used in this study which is called "AFt-UHPFRC" is characterized by a matrix densified by ettringite (AFt) formation. In this paper, the authors report the results obtained from the Full-scale construction experiment. In this experiment, fields simulating concrete slab and steel slab were prepared. In addition, the AFt-UHPFRC was casted using construction equipment and its workability and quality were evaluated on the assumption that it would be applied under the two-lane regulation. Specifically, the formability in the condition with the gradient, filling situation around the reinforcing bar, adhesion with the existing slab, orientation of the fiber, etc., were confirmed. In addition, the strength development, shrinkage behavior, pore size distribution, and chloride penetration resistance of casted AFt-UHPFRC were confirmed by this experiment, and the feasibility of the developed method was confirmed.

Keywords: UHPFRC · Strengthening · Durability enhancement · Bridge

1 Introduction

1.1 Background

Over the last decade, damaged and deteriorated bridges have been increasingly observed in expressways of Japan. Typical examples of these damaged and degraded bridges are reinforced concrete (RC) void slab top concrete degraded by antifreeze agents and welded joints of steel deck slabs damaged due to heavy traffic. Bridge deck overlays are most often used when rehabilitation is required on an existing bridge. Overlays using Steel fiber Reinforced Concrete (SFRC) can also be used as a preventative deterioration of sound existing decks. However, there is a possibility that the top surface of bridge decks may perhaps deteriorate again due to the repeated loading of heavy traffic, severe

M. Kunieda et al. (Eds.): SHCC 2022, RILEM Bookseries 39, pp. 310–320, 2023.
https://doi.org/10.1007/978-3-031-15805-6_32

environmental conditions and improper construction methods. In addition, the SFRC overlay thickness could reach to more than 100 mm to attain required load bearing capacity and it is necessary to strengthen other members of the bridge to carry increased self-weight of overlaid bridge decks.

Following the above-mentioned situation, the development of a new bridge deck overlay method has begun for the expressway renewal project were strengthening of bridge decks is carried out. The new method uses Ultra-High Performance fiber Reinforced cementitious Composites (UHPFRC) as overlay material. UHPFRC has very low permeability against liquid/gas and high resistance to freezing and thawing action. In addition, UHPFRC is known for its high strength and high elastic modulus. By using UHPFRC instead of SFRC, the overlay thickness can be made thinner and increase of the concrete bridge deck thickness can be made even unnecessary by replacing a certain depth of the top surface concrete with UHPFRC [1] (Fig. 1). In addition, UHPFRC functions as a protective layer and application of waterproofing material is maybe unnecessary. Thus, application of UHPFRC to concrete bridge decks for upgrading is an efficient and effective method as increase of the load bearing capacity and enhancement of the durability are achieved simultaneously. The authors have conducted various detailed laboratory studies after confirming the feasibility of the overlay method and identifying issues of the construction method through small-scale construction experiment. As a final step, Full-scale construction experiment was conducted with the aim of developing a bridge upgrade method using cast-in-place UHPFRC. The UHPFRC is cast using construction machinery and is intended to be applied under two-lane regulations. In this paper was reported the feasibility of the developed method by confirming its workability and quality.

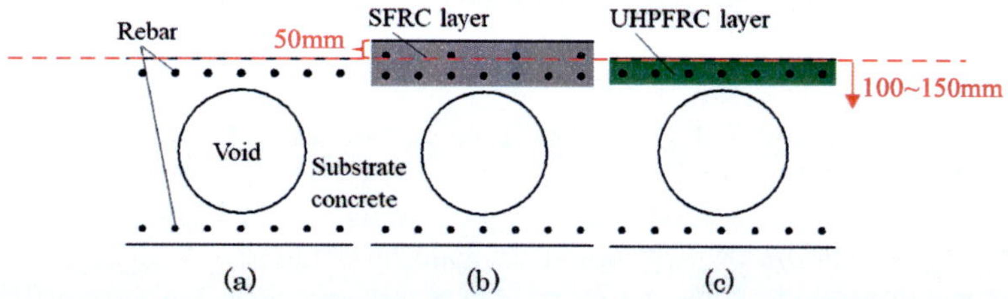

Fig. 1. Cross sections of (a) RC void slab and RC void slab overlay strengthened with (b) SFRC and (c) UHPFRC

1.2 AFt-UHPFRC

The mix proportion of UHPFRC in this experiment which is called AFt-UHPFRC [2] is characterized by a matrix densified by ettringite (AFt) formation. Microstructure of the AFt-UHPFRC matrix is basically formed by decreasing water/binder ratio using spherical pozzolan particles and superplasticizer and packing ultrafine particles optimally. In addition, numerous needle-shaped ettringite crystals of 1–2 μmlength fill micropores of hydration structure together with inert and reactive fine fillers. The porosity of AFt-UHPFRC is 4.0 to 6.0%, which constitutes a very dense cured product. Table 1 shows the mix proportion of AFt-UHPFRC.

Table 1. Mix proportion of AFt-UHPFRC

Component	Mass(kg/m^3)	Remarks
Portland cement	927	
Premixed materials	360	Pozzolanic material, ettringite formation additives
Sand	905	Crushed sand, d_{max} <2.5 mm
Steel fiber[b]	235.5	3.0 vol.%, diameter = 0.2 mm, length = 15 mm
Superplasticiser	36	
Shrinkage reducing admixture	12.9	
Defoaming agent	6.4	
Water[a]	195	W/B = 0.152

[a]Including water in superplasticizer
[b]Not including in the unit volume

2 Full-Scale Construction Experiment

2.1 Outline

Following our previous investigation, it was understood that the AFt-UHPFRC, manufactured with a 100 L mixerand cured in air, has almost the same mechanical properties and durability as the steam curing [3, 4]. However, when AFt-UHPFRC is applied to multi-span or large bridges which comprise a majority of highway bridges, it is planned to use larger mixers and construction machines for manufacturing and casting AFt-UHPFRC for enhancement of work efficiency. In order to investigate the behavior of the AFt-UHPFRC manufactured with a large mixer and spread and compacted with a construction machine, a Full-scale construction experiment with a construction area of 390 m^3 (width 5 m, length 78 m) was conducted. Full-scale structure modeling RC void slab bridge decks and steel decks were built as test specimens in a testing field and the AFt-UHPFRC was cast on top of them.

2.2 Test Conditions

Figure 2 shows the work performed during the Full-scale construction experiment. In the experiment, a twin-shaft forced action mixer with a capacity of 3.0 m^3 was used for manufacturing AFt-UHPFRC. A bating plant was built at the experiment field, in which 1.5 m^3 AFt-UHPFRC was manufactured for every batch, taking 30 min. The consistency is controlled by adjusting the amount of mineral-based inorganic powder used according to the transverse slope. AFt-UHPFRC was transported and placed by a wheel loader and was spread and compacted by a machine that has been used for concrete paving. The moving speed of the machine was kept constant to be 0.5 m/min. In order to prevent plastic shrinkage cracking after casting, sheet membrane curing was applied on top of AFt-UHPFRC immediately after the surface finish, and sheet membrane was removed the next day. The width of the specimens was 5 m, and the length of RC void slab bridge deck specimens and steel deck specimens was 62 m and 16 m, respectively. Since AFt-UHPFRC layer thickness varies depending on the condition of concrete bridge decks, two different thickness of AFt-UHPFRC layer (thickness of 100 mm and 150 mm) was planned on top of the RC void slab bridge deck specimen. Top surface of the concrete was roughened with water jets for interfacial bonding between AFt-UHPFRC and concrete. Parameters and points of interest in the construction experiment are summarized in Fig. 3.

Fig. 2. Full-scale construction experiment

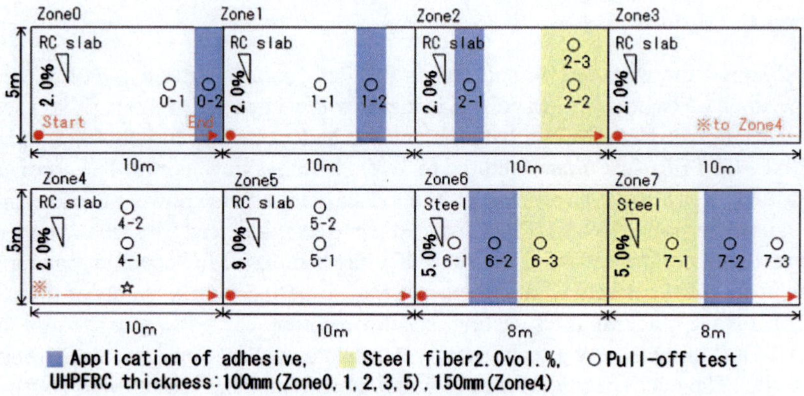

Zone0-2: Filling ability below steel rebars, Interfacial bonding
Zone3: Assume loss of flowability (limit of fluidity / fillability)
Zone4: Effect of increase in thickness (energy propagation distance)
Zone5: Anti-drafting and workability against slope
Zone7: Filling ability around shear studs and splice-plates(drip / inclusion into UHPFRC), Bonding on the interface

Fig. 3. Schematic of full-scale construction experiment

3 Evaluation and Material Test

3.1 Outline

The Full-scale construction experiment was evaluated during and after the experiment by visual observation, measurements of specimen behavior and tests of material properties. In addition, shrinkage deformation of AFt-UHPFRC was measured with mold strain gauges, which was started immediately after casting of AFt-UHPFRC. Properties of fresh AFt-UHPFRC were also tested. After the experiment, mechanical and durability properties of AFt-UHPFRC were investigated by performing laboratory tests using specimens produced from AFt-UHPFRC fabricated by the large mixer. Interfacial bond strength between AFt-UHPFRC and concrete/steel was tested on site. Table 2 shows material testing mainly performed after the experiment. In this paper, (1) fresh state properties, (2) strength, (3) length change, (4) bond strength, (5) porosity, and (6) chloride penetration resistance are presented.

Table 2. Strength test results

Property			Material age			
			7 days	28 days	91 days	257 days
Compressive strength (MPa)		Ave.	131	164	184	189
		Max.	137	177	196	198
		Min.	120	144	166	181
Elastic modulus (GPa)		Ave.	41	45	47	46
Cracking strength (MPa)	Splitting tensile test (cylinder)	Ave.	7.6	8.7	9.3	11.3
		Max.	9.3	9.9	10.1	11.8
		Min.	5.8	6.4	8.5	10.7
	Static tensile tests [3]	Ave.	–	–	–	12.1
		Max.	–	–	–	12.6
		Min.	–	–	–	11.6
Tensile strength (MPa)	Static tensile tests [3]	Ave.	–	–	–	12.8
		Max.	–	–	–	14.6
		Min.	–	–	–	11.7

3.2 Fresh Properties of AFt-UHPFRC

The fresh properties of AFt-UHPFRC were adjusted according to the slope of the slab and the compaction energy of the paving machine, and a flow value suitable for thixotropic AFt-UHPFRC was determined. Figure 4 shows the results of the mortar flow test performed in compliance with JIS R 5201. It was found that when the flow value of AFt-UHPFRC is in the range of 200 ± 25 mm, it has good properties for slope.

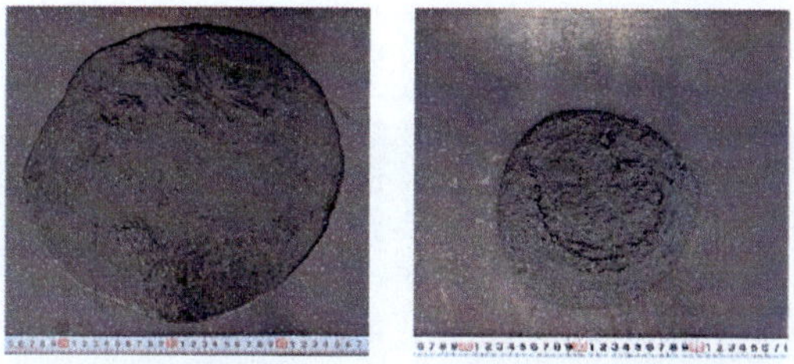

Fig. 4. Thixotropic formulation (flow)

3.3 Mechanical Properties of AFt-UHPFRC

Table 2 shows the strength test results of specimens cured at an outdoor environment as with the floor slab. At the age of 28 days, the compressive strength reached over 150 MPa, and the cracking strength reached over 8 MPa. It is notable that the compressive strength reached over 120 MPa even at seven days. When repair and strengthening are performed on bridges of busy highways, there is a possibility that sufficient time is not secured for curing of AFt-UHPFRC because minimizing traffic disruption is required. Rapid hardening property is conferred to the AFt-UHPFRC by controlled ettringite formation, and the property is considered favorable for construction with time constraints.

In order to investigate the tensile property of the AFt-UHPFRC, direct tension tests were performed. Figure 5 shows specimen geometry and the test set-up configuration. The method of direct tension test was developed in the previous study [3]. Stress deformation relationships obtained from the static tensile tests are shown in Fig. 5.

The averages of elastic limit strength and ultimate tensile strength were 12.1 MPa and 12.8 MPa, respectively. In the average stress-deformation curve of AFt-UHPFRC, strain-hardening was hardly observed.

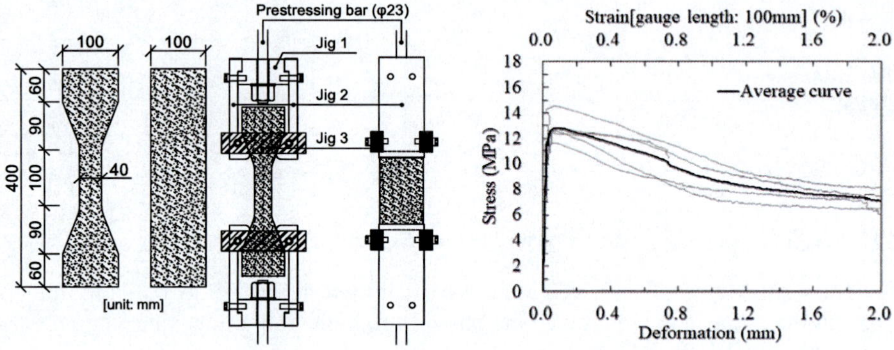

Fig. 5. Static tensile tests (fiber 3.0 vol%)

3.4 Length Change of AFt-UHPFRC

Figure 6 shows the results of AFt-UHPFRC strain measurement using an embedded strain gauge (number of samples = 1/Zone). Strain measurement was commenced immediately after casting, and it could be seen that the AFt-UHPFRC contracted rapidly in one day. This shrinkage was thought to be caused by the autogenous shrinkage and drying shrinkage of AFt-UHPFRC. Comparing the strain values of AFt-UHPFRC cast on top of the RC void slab bridge deck with the ones of AFt-UHPFRC cast on top of the steel deck, the former is smaller than the latter. In addition, the result of the specimen was similar to those cast on the steel deck. From these facts, it is thought that contraction of the AFt-UHPFRC was restrained more significantly by concrete substrates and steel rebars (higher degree of restraint) and mitigated by creep of the AFt-UHPFRC. Strain values and evolution were almost the same between 100 and 150 mm thick AFt-UHPFRC layer on top of the RC void slab bridge deck specimen and between AFt-UHPFRC layer on top of the steel deck specimen with and without shear studs on it.

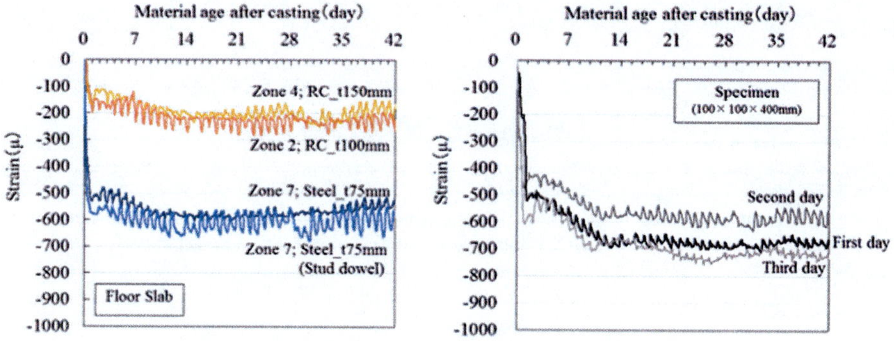

Fig. 6. Strain measurement results (shrinkage strain)

3.5 Bond Strength

The interfacial bond strength between AFt-UHPFRC and concrete and steel was assessed on site using the pull-off test method specified in ASTM C 1583 (Fig. 7). The thickness of the AFt-UHPFRC layer was 100 mm and 150 mm for the RC void slab and 40 mm and 75 m for steel decks. A partial core was drilled through the bonded material and into the substrate material; furthermore, a steel pull-off disc that is approximately 80 mm in diameter was bonded at the desired test location.

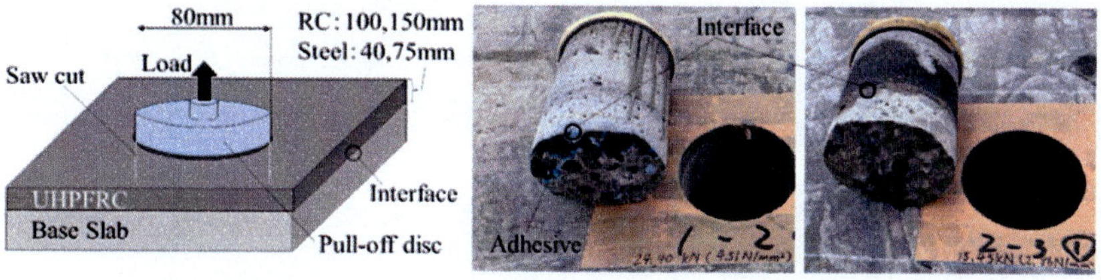

Fig. 7. Pull-off test of AFt-UHPFRC cast on top of concrete slab

Table 3 shows the bond test results. In case of the RC void slab bridge deck specimen, failure occurred mostly at the substrate concrete irrespective of application of adhesive on the AFt-UHPFRC-concrete interface.

In addition, there was concern that the compaction energy at the edges of the slab was lower than at the center, and that the slab would warp upwards by shrinkage at the edges. However, the adhesive strength obtained in all tests was almost the same (about 2.8 MPa) regardless of the test position and the thickness of AFt-UHPFRC. In case of the steel deck specimen, fracture occurred at the interface in all tests. At the location where interfacial bonding adhesive was used, the bond strength of the interface with smaller amount of adhesive was lower than that with larger amount of adhesive. The bond strength at the interface without adhesive was about the same as with adhesive, but with greater variability. In particular, it is probable that the screed vibration was not properly transmitted to the ends of the test piece and sufficient compaction was not performed because the adhesive strength at both ends of the steel deck test piece was low.

Table 3. Results of bond tests

Zone	Conditions							Test results		
	Slab	UHPFRC thickness (mm)	Slope (%)	Fiber (vol.%)	Compaction Energy (rpm)	Test position[a]	Adhesive (kg/m^2)	Bond strength (MPa)[b]	Porosity (%)	
								Ave	Surface	Interface
0	RC	100	2	3	2000 ~ 3000	C	–	2.64	–	-
							1.2	3.06		
1					2000	C	–	2.61	5.87	3.26
							1.2	3.52		
2				2	3000	C	1.2	2.73	4.21	1.91
							–	2.83		
						E	–	3.09		
4		150		3		C	–	2.74	6.22	4.50
						E	–	2.99		
5		100	9			C	–	2.67	5.28	3.96
						E	–	2.91		
6	Steel	40		3	2000	C	–	1.42	–	3.18
							1.2	4.95		
				2			–	2.44		
7		75				C	–	3.61	–	2.46
				3			0.4	3.12		
							–	0.97		

[a]Test Position C:center, E:edge
[b]Fracture □: Substrate concrete, ■: Interface

3.6 Porosity of the Specimen

The core was collected from the deck and the porosity was measured by the mercury intrusion method (number of samples = 1/Zone). Table 3 shows measurement results of porosity and pore size distribution of the surface and inside of the AFt-UHPFRC layer. The specimens were made from cores taken from the Full-scale structure 91 days after AFt-UHPFRC casting and cured at 20 °C–60% RH for about 270 days. Porosity of all specimens was lower than approximately 6% due to the effect of ettringite crystal growth over time.

In all zones, the porosity of the inside of the AFt-UHPFRC layer was smaller than that of the surface of the AFt-UHPFRC layer. This is considered that this is because the surface was dried for a long time due to the early removal of the sheet curing, and more hydration reactions occurred inside than the surface of the UHPFRC layer.

3.7 Chloride Penetration Resistance

The crack resistance when AFt-UHPFRC is completely cured is very high. However, AFt-UHPFRC shrinks significantly in the early stages as explained in Sect. 3.3. So, if moisturizing is not performed in the early stage of material age, it will increase the probability of crack during construction, similar to general concrete.

In this Full-scale construction experiment, cracks were confirmed relatively early on the surface at slab when the curing sheet was removed 1 day after casting. Specifically, cracks of 0.1 to 0.15 mm were confirmed. In addition, at some construction joints, it was confirmed that the joints had an opening of 0.2 mm or more.

In this study, a primer containing a fluorescent paint was applied to the cracks for the purpose of closing the cracks and checking the depth. Several types of primers were prepared and verified. As a result of irradiating the cut surface with ultraviolet rays, it was confirmed that cracks had reached the reinforcing bars and the underlying concrete (Table 4). Afterwards, an accelerated test was conducted in which dry and wet was repeated in warm seawater, and EPMA analysis was performed. By confirming the permeation state of chloride ions, it was evaluated whether the application of the primer was effective from the viewpoint of salt shielding property.

Table 4. Chloride ions penetration status after accelerated test

<table>
<tr><td></td><td colspan="2">Crack Part</td><td>Joints Part</td></tr>
<tr><td></td><td>Irradiating with Ultraviolet Rays</td><td>EPMA Analysis</td><td>EPMA Analysis</td></tr>
<tr><td>Primer A</td><td>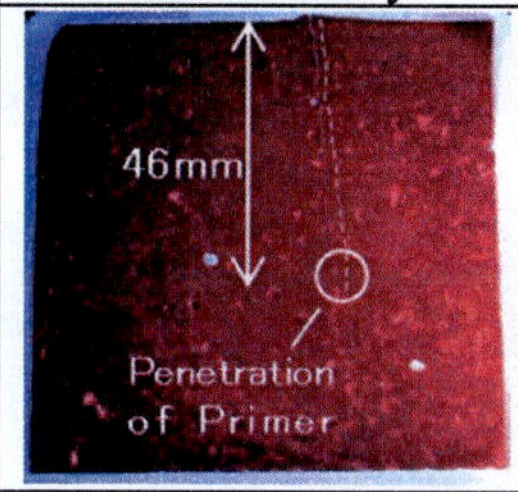
</td><td>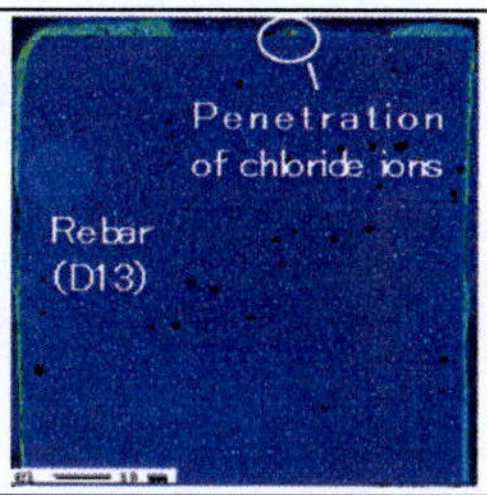
</td><td>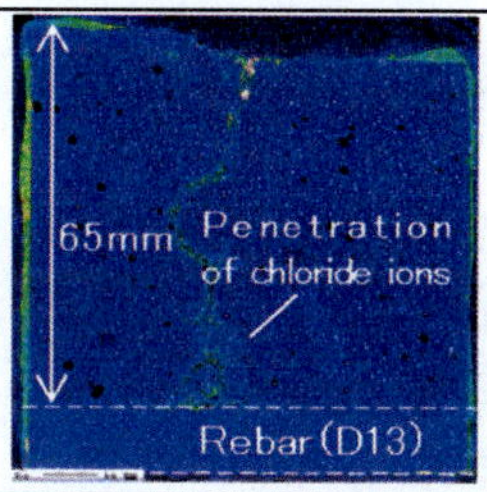
</td></tr>
</table>

Table 4 shows the primer penetration position after the accelerated test and some of the results of EPMA analysis. In the case of cracks of 0.1 to 0.15 mm, chloride ions stayed within 5 mm from the surface, and penetration could be suppressed. However, in the case of construction joints had an opening of 0.2 mm or more, none of the primers could suppress the penetration of chloride ions.

Therefore, it is possible to suppress the penetration of chloride ions for cracks smaller than 0.15 mm by applying a primer. However, depending on the type of primer, there are some that have good penetration but insufficient salt tolerance. So, it is necessary to select an appropriate primer in advance. In addition, it is considered that the penetration of the primer only is not sufficient to ensure waterproofing of the construction joints where the gaps are large. In order to ensure durability of the construction joints, it is important to prevent opening by devising the shape and using adhesive, after recommending the combined use of waterproofing.

4 Conclusion

In this paper, a Full-scale construction experiment using AFt-UHPFRC was performed. The results obtained by this experiment are summarized below.

- The fresh properties of AFt-UHPFRC were adjusted according to the slope of the Full-scale structure and the compaction energy of the construction equipment and a flow value (200 ± 25 mm) suitable for UHPFRC was determined.
- At the age of 28 days, the compressive strength reached over 150 MPa, and the cracking strength determined by splitting tensile tests reached over 8 MPa. Averages of the cracking strength and ultimate tensile strength were determined to be 12.1 MPa and 12.8 MPa, respectively, from the direct tension test.
- The shrinkage strain of AFt-UHPFRC cast on top of the Full-scale structure was smaller than that of the specimen. This is thought to be because contraction of AFt-UHPFRC was restrained more significantly by concrete substrates and steel rebar (higher degree of restraint) and mitigated by creep of AFt-UHPFRC.
- Failure of specimens of AFt-UHPFRC-concrete interfacial bonding tests occurred mostly in the substrate concrete irrespective of application of adhesive on the interface. All bond strengths were almost the same (about 2.8 MPa) regardless of the AFt-UHPFRC thickness and test position (center or edge).
- The porosity of the inside of the AFt-UHPFRC layer was smaller than that of the surface of the AFt-UHPFRC layer. It is considered that this is because the surface was dried for a long time due to the early removal of the sheet curing.
- For cracks smaller than 0.15 mm, the penetration of chloride ions could be sup pressed by using an appropriate primer. However, if there is an opening of 0.2 mm or more, such as at the joints, it is considered necessary to also use measures to prevent the opening.

References

1. Brühwiler, E.: “Structural UHPFRC”: welcome to the post-concrete era!’. In: Proceedings of the First International Interactive Symposium on Ultra-High Performance Concrete, Des Moines, Iowa, July 18–20 (2016)
2. Watanabe, Y., Ichinomiya, T., Yanai, S., Iriuchijima, K., Suhara, K.: Development of cast-in-place method of ultra high strength fiber reinforced concrete. In: Proceedings of the Fifth International Conference on Construction Materials (CONMAT15), Whistle, British Columbia, August 19–21 (2015)
3. Makita, T., Watanabe, Y., Yanai, S., Ichinomiya, T.: Upgrading of existing bridge decks using UHPFRC densified by Ettringite Formation (AFt-UHPFRC): Preliminary Investigation. In: AFGC-ACI-fib-RILEM International Symposium on Ultra-High Performance Fiber-Reinforced Concrete, UHPFRC 2017, Montpellier, October 2–4 (2017)
4. Watanabe, Y., Fujiwara, H., Makita, T., Yanai, S.: An experimental study on the mechanical and durability properties of UHPFRC for cast-in-place method. In: Second International Interactive Symposium on Ultra-High Performance Concrete, Albany, New York, June 2–5 (2019)

Characteristic of UHPFRC and New Applications

Satoru Kobayashi(✉), Tomoko Takagi, Manato Nakamura, Takuya Iwamoto, Naoki Sogabe, and Shinichi Yamanobe

Kajima Corporation, Tokyo, Japan
kobasato@kajima.com

Abstract. Public awareness of the seismic performance of civil engineering structures has increased against earthquakes. The authors are developing a seismic retrofit method for RC piers that utilizes ultra high performance fiber reinforced concrete (hereafter, UHPFRC). The placement of UHPFRC in the cover part imparts resistance against the large bending and compressive stresses that occur during earthquakes and prevents crushing of the cover, and we have devised a method to retrofit existing RC piers by using two types of UHPFRC that can be applied in situ. it is possible to plaster UHPFRC onto vertical surfaces by adjusting the mix proportion. The ductility of the pier was evaluated by cyclic loading test on mock-up specimens of a pier base retrofitted by this method. It was confirmed that reinforcing the pier base with UHPFRC that can be applied in situ increased the ductility by approximately 1.4 times and improved the seismic performance. A sprayable UHPFRC was developed and required to allow its application to vertical surfaces by spraying. UHPFRC can be applied in situ to existing piers by plastering or spraying for retrofitting purposes, and it was found that the seismic performance of the pier can be easily improved.

Keywords: UHPFRC · RC piers · Seismic performance · Plastering · Spraying

1 Introduction

In earthquake-prone Japan, ensuring the seismic performance of civil engineering structures, which form the social infrastructure, is a major imperative for building a resilient nation. Further, public awareness of the seismic performance of civil engineering structures has increased against the backdrop of the Great East Japan Earthquake that occurred in 2011 and the Nankai Trough Earthquake whose occurrence in the future is of concern. On the other hand, a lot of social infrastructure built during the period of rapid economic growth is lacking in seismic performance and urgently needs seismic retrofitting along with measures against aging deterioration, which is becoming increasingly apparent.

Against this backdrop, the authors are developing a seismic retrofit method for RC piers that utilizes UHPFRC. UHPFRC is classified as a ductile fiber reinforced cementitious composite (DFRCC) and is considered to be a material with high strength and the tendency to strain harden after cracking. This paper presents an overview of a seismic retrofit method using UHPFRC and its construction method.

M. Kunieda et al. (Eds.): SHCC 2022, RILEM Bookseries 39, pp. 321–329, 2023.
https://doi.org/10.1007/978-3-031-15805-6_33

2 Overview of High Seismic Performance RC Piers Using UHPFRC

When seismic forces act on an RC pier, large bending moment is generated at the base of the RC pier, and the longitudinal reinforcement in that portion of the pier yields to form a plastic hinge. As bending deformation progresses further, the concrete in the base of the pier fails in compression and the longitudinal reinforcement buckles and ruptures, resulting in a reduction in flexural capacity and ultimate failure. In other words, the seismic performance of the entire RC pier largely depends on the deformation performance of the plastic hinge. Therefore, as a new technology to improve the seismic performance of RC piers, we have developed high seismic performance RC piers using UHPFRC (Fig. 1) [1, 2].

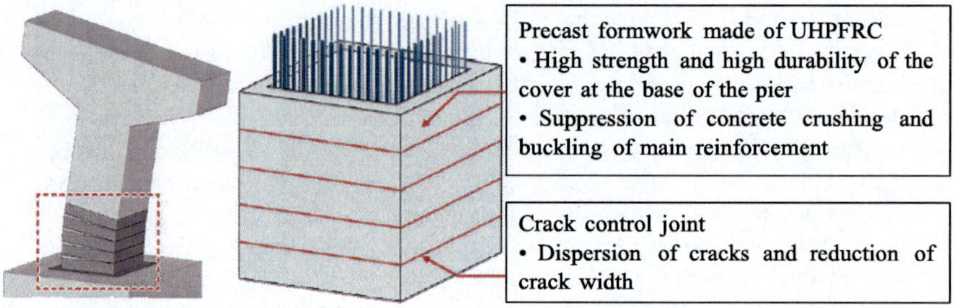

Fig. 1. Overview of high seismic performance RC piers using UHPFRC.

This structure dramatically improves the ductility of RC piers by applying precast formwork made of UHPFRC with compressive strength of 150 N/mm^2 or higher as buried formwork at the base of the pier, which is subject to significant damage during earthquakes as described above. The placement of UHPFRC in the cover part imparts resistance against the large bending and compressive stresses that occur during earthquakes and prevents crushing of the cover. Moreover, this material, which is mixed with steel fiber, restrains the longitudinal reinforcement and thus delays its buckling. Furthermore, by dividing the precast formwork into upper and lower tiers and having the inter-tier boundaries function as horizontal joints, flexural cracks in the part where the precast formwork is applied are distributed, with the effect that a stable plastic hinge is formed in that part. This technology can improve ductility without changing the cross-section dimensions or weight of the pier, thus increasing seismic resistance without increasing the load on the foundation or the impediment ratio of river flow.

3 Seismic Retrofit Method by UHPFRC for Existing Piers

The technology described above was developed for the construction of new RC piers. Here, we have devised a method to retrofit existing RC piers by using UHPFRC that can be applied in situ [3]. Figure 2 shows a schematic diagram of the seismic retrofit of an existing pier using applied-in-situ UHPFRC. The basic procedure for seismic retrofitting of existing piers is as follows: (1) Determine the extent of repair required through preliminary structural analysis. (2) Remove the cover concrete from the base of the pier. (3) Install crack control joints. (4) Perform cross-section restoration with UHPFRC. Since RC piers requiring seismic retrofit are of various sizes, as the method of application in situ, both the plastering method and the spraying method were considered to accommodate various conditions.

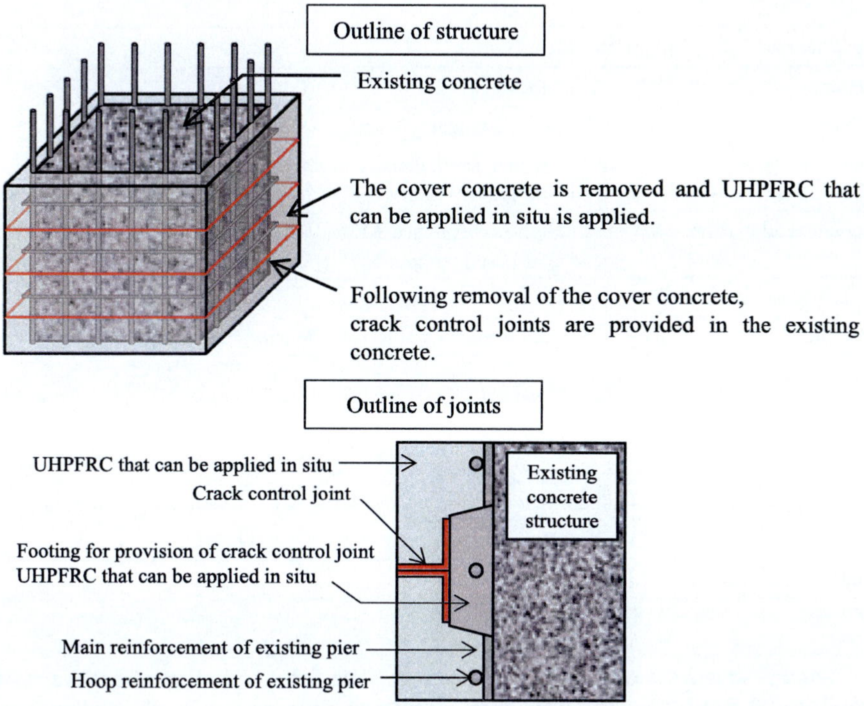

Fig. 2. Overview of a schematic diagram of the seismic retrofit of an existing pier using applied-in-situ UHPFRC.

4 Plastering Method

4.1 Mix Proportion of UHPFRC for Plastering Method

Use of UHPFRC with the plastering method requires its application to vertical surfaces. As existing UHPFRC is highly fluid with a mortar flow (without tapping) of approximately 250 mm [4], it was decided to add an inorganic viscosity agent to the UHPFRC mixture to confer viscosity. Table 1 shows the materials used and Table 2 shows the mix proportion. Various investigations have experimentally confirmed that it is possible to plaster UHPFRC onto vertical surfaces by adjusting the amount of inorganic viscosity agent added and setting the control value of mortar flow (15 taps) to 110 ± 0 mm [5].

Table 1. Constituent materials.

Material name	Symbol	Application
Water	W	Tap water
Binder	B	Special binder, density: 2.98 g/cm^3
Fine aggregate	S	Crushed sand, density in saturated surface-dry condition: 2.62 g/cm^3
Chemical admixture	SP	High-performance AE water reducing agent (polycarboxylic acid-based compound)
Viscosity agent	VA	Inorganic viscosity agent
Steel fiber	SF	φ0.2 mm, length: 10 mm, density: 7.85 g/cm^3

Table 2. Mix proportion.

W/B (%)	Air (%)	Unit content (kg/m^3)			SP (B × %)	VA (kg)	SF (vol%)
		W	B	S			
15.2	2.0	195	1287	898	2.5	21	2.5

Figure 3 shows the outline of the specimen. The formwork dimensions were 500 × 500 × 100 mm (thickness). The base concrete was placed to a depth of 50 mm and its surface was roughened to secure a gap of 10 mm, and reinforcement (D16) was arranged horizontally at intervals of 150 mm. Figure 4 shows the coating process. It was experimentally confirmed that one coat of this material can be applied to a thickness of up to 35 mm, and that the next layer can be applied after approximately one hour [5]. Thus, this material can be applied in a working time of one day even with a coating thickness of 100 mm, and the construction period can be significantly shortened compared to the conventional method.

Figure 5 shows a cut cross-section of a specimen prepared by the plastering method. The UHPFRC was densely compacted and integrated with the base concrete with no overlap lines or dripping at the bottom of the reinforcement, even though 100 mm thickness was applied in one day's work time. The specimens were cut from the UHPFRC portion of the specimen and tested, and the compressive strength and flexural strength were confirmed to be greater than 150 N/mm^2 and 25 N/mm^2, respectively.

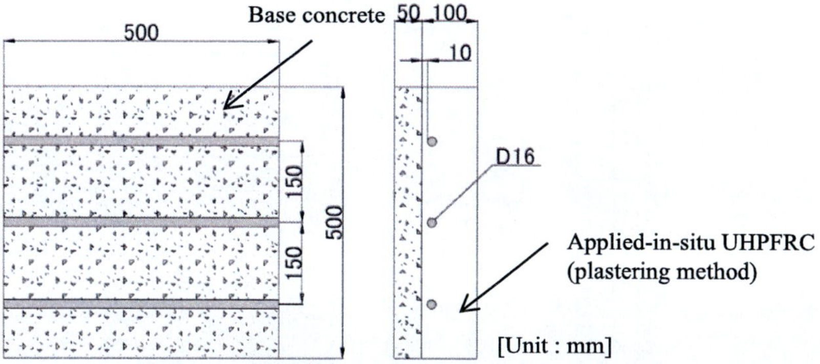

Fig. 3. Outline of specimen.

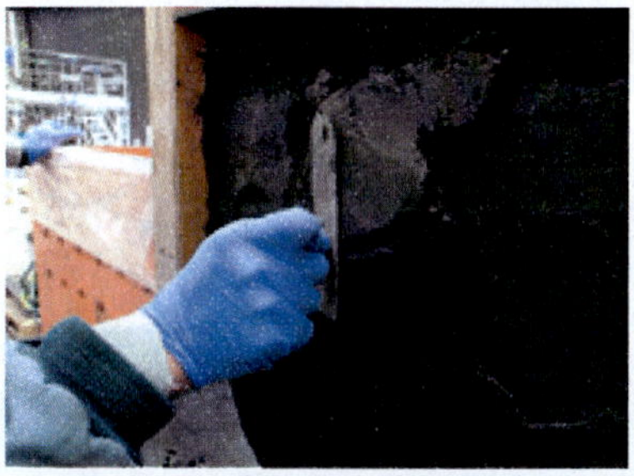

Fig. 4. Application process.

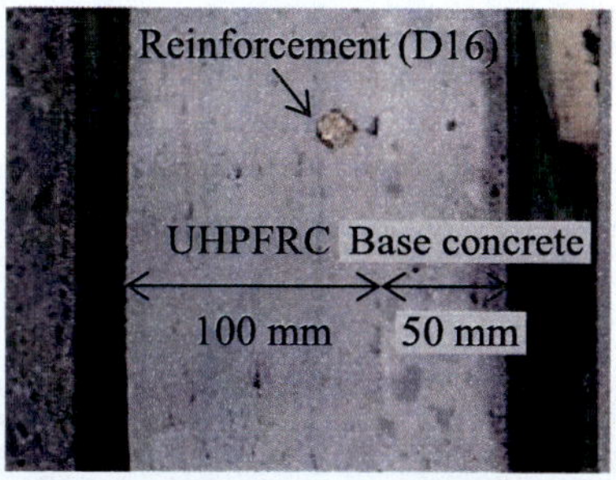

Fig. 5. Cut cross-section of specimen.

4.2 Cyclic Loading Test

The ductility of the pier was evaluated by cyclic loading test on mock-up specimens of a pier base retrofitted by this method. Figure 6 shows an outline of the pier mock-up specimen and Fig. 7 shows the retrofit process with UHPFRC that can be applied in situ. Figure 8 shows the loading apparatus. Figure 9 shows the horizontal load-loading point displacement relationship in the cyclic loading test. It was confirmed that reinforcing the pier base with UHPFRC that can be applied in situ increased the ductility by approximately 1.4 times and improved the seismic performance.

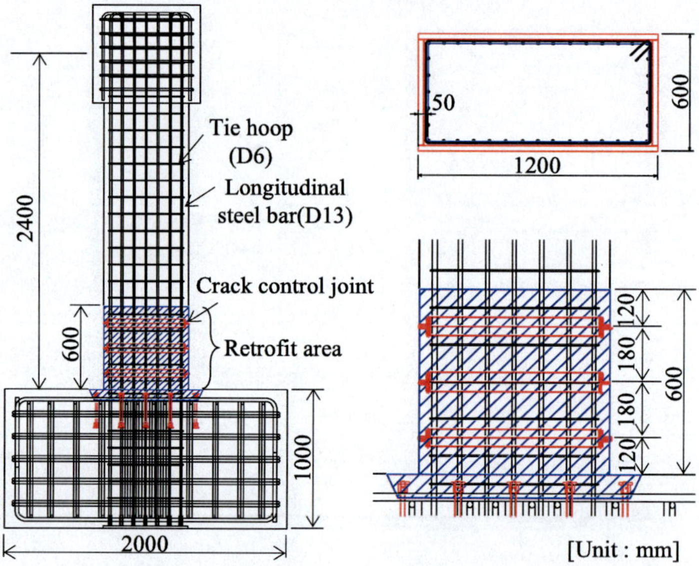

Fig. 6. Outline of the pier mock-up specimen.

Fig. 7. Retrofit process using UHPFRC that can be applied in situ.

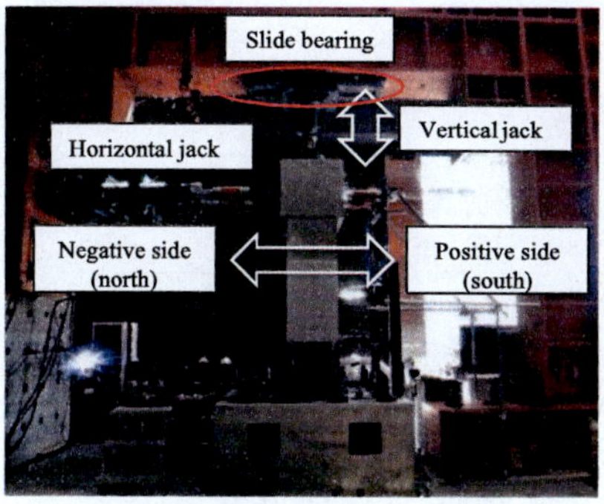

Fig. 8. Loading apparatus.

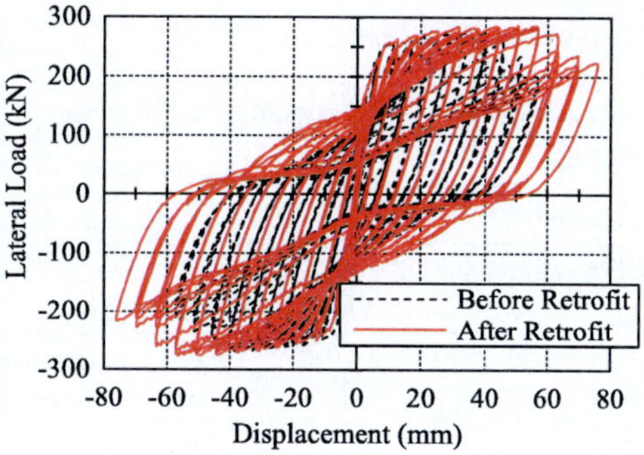

Fig. 9. Hysteresis curves.

5 Spraying Method

A sprayable UHPFRC was developed for the purpose of reinforcing areas with a larger construction area than can be accommodated by the plastering method [6]. Reducing the flowability of UHPFRC is required to allow its application to vertical surfaces by spraying, In the case of existing UHPFRC, a large amount of high-performance water reducing agent is added to give good flowability. Here, however, the amount of high-performance water reducing agent was reduced to 1.0%, which is the lower dosage limit for mixing, to achieve a hard fresh property. Table 3 shows the materials used and Table 4 shows the mix proportion. In the case of the spraying method, the material needs to be pumped out with a pump, so a surfactant-type pumpability improving agent used for transporting sand through conveying pipes, is applied to the binder at the dosage of 0.04% to ensure good pumpability.

To check the post-spray hardening properties of the selected UHPFRC, the UHPFRC was sprayed onto a 500 × 500 mm flat plate to a thickness of 100 mm. An ordinary squeeze type pump (200 V, maximum discharge rate 100ℓ/min, maximum discharge pressure 2 MPa) was used for spraying. The mortar hose had an inner diameter of 40 mm and length of 9 m, and the discharge rate was adjusted to keep the discharge pressure at 2.5 MPa or less. Figure 10 shows the spraying process and Fig. 11 shows the finished state after spraying. Trowel finishing is possible as soon as one hour after spraying. After spraying, the flat plate was subjected to sealed curing indoors at 20 °C, and at 7 days of age, φ50 × 100 mm specimens for the compression test and 40 × 40 × 160 mm specimens for the bending test were cut out from the flat plate, and the specimens were subjected to the planned tests at 28 days of age. Table 5 shows the measured hardening properties. Compressive strength and flexural strength of more than 150 N/mm^2 and 15 N/mm^2, respectively, were achieved. Going forward, as with the plastering method, we plan to evaluate the improvement in seismic performance yielded by the spraying method using pier base mock-up specimens.

Table 3. Constituent materials.

Material name	Symbol	Application
Water	W	Tap water
Binder	B	Special binder, density: 2.98 g/cm^3
Fine aggregate	S	Crushed sand, density in saturated surface-dry condition: 2.62 g/cm^3
Chemical admixture	SP	High-performance AE water reducing agent (polycarboxylic acid-based compound)
	P	Pumpability improving agent (surfactant-type)
Steel fiber	SF	φ0.2 mm, length: 10 mm, density: 7.85 g/cm^3

Table 4. Mix proportion.

W/B (%)	Air (%)	Unit content (kg/m^3)			SP (B × %)	SP (B × %)	SF (vol%)
		W	B	S			
18.0	2.0	208	1156	1021	1.0	0.04	2.5

Fig. 10. Spraying process.

Fig. 11. Finished state after spraying.

Table 5. Measured physical properties.

Material age (days)	Compressive strength (N/mm^2)	Young's modulus (kN/mm^2)	Flexural strength (N/mm^2)
14	152.6	34.1	16.2
28	168.7	36.7	19.2
91	186.8	40.5	20.2

6 Conclusion

This paper has introduced the application of ultra high strength fiber reinforced concrete for seismic retrofitting. Besides its use to improve the seismic performance of new piers, UHPFRC can be applied in situ to existing piers by plastering or spraying for retrofitting purposes. We intend to promote seismic retrofitting of aging piers in Japan using this technology.

References

1. Yamanobe, S., Sogabe, N., Iemura, H., Takahashi, Y.: Development of high-seismic-performance reinforcement concrete bridge pier with high-performance plastic hinge. Proceedings of JSCE **64**(2), 317–332 (2008). (in Japanese)
2. Yamanobe, S., Kono, T., Saito, K., Masumoto, K., Moro, T., Tateoka, M.: Application of high seismic performance piers using ultra high strength fiber reinforced concrete formwork—Hanshin expressway Yamatogawa line Sambo Junction. Bridge Found. Eng. **46**(5), 19–24 (2012). (in Japanese)
3. Iwamoto, T., Kobayashi, S., Sogabe, N., Yamanobe, S.: Cyclic alternating positive-negative loading test for Retrofitted RC Pier mock-up specimen using in-situ applicable UFC. Proc. Jpn. Concrete Inst. **42**(2), 811–816 (2020). (in Japanese)
4. SUQCEM Design and Construction Manual (2014). (in Japanese)
5. Kobayashi, S., et al.: Experimental study on the application of ultra high strength fiber reinforced concrete to the plastering method. In: 2018 JSCE Annual Meeting, vol. 104 (2018). (in Japanese)
6. Yoshida, Y., Kobayashi, S., Nakamura, M., Matsuda, Y., Gunawan, D., Sogabe, N.: Study of fiber reinforced mortar for spraying to form outer shell of structures. In: 2021 JSCE Annual Meeting, vol. 127 (2021). (in Japanese)

Author Index

M. Kunieda et al. (Eds.): SHCC 2022, RILEM Bookseries 39, pp. 331–332, 2023.
https://doi.org/10.1007/978-3-031-15805-6